AF368534

Sustainable Agricultural Practices—Impact on Soil Quality and Plant Health

Sustainable Agricultural Practices—Impact on Soil Quality and Plant Health

Editors

Nikolaos Monokrousos
Efimia M. Papatheodorou

MDPI • Basel • Beijing • Wuhan • Barcelona • Belgrade • Manchester • Tokyo • Cluj • Tianjin

Editors
Nikolaos Monokrousos
International Hellenic
University
Greece

Efimia M. Papatheodorou
Aristotle University of
Thessaloniki
Greece

Editorial Office
MDPI
St. Alban-Anlage 66
4052 Basel, Switzerland

This is a reprint of articles from the Special Issue published online in the open access journal *Agronomy* (ISSN 2073-4395) (available at: https://www.mdpi.com/journal/agronomy/special_issues/sustainable_soil_plant).

For citation purposes, cite each article independently as indicated on the article page online and as indicated below:

LastName, A.A.; LastName, B.B.; LastName, C.C. Article Title. *Journal Name* **Year**, *Volume Number*, Page Range.

ISBN 978-3-0365-3216-5 (Hbk)
ISBN 978-3-0365-3217-2 (PDF)

Contents

About the Editors

Nikolaos Monokrousos is Assistant Professor of Soil Ecology at the University Center of International Programs of Studies of the International Hellenic University. Before joining IHU, he was a researcher at the Institute of Soil and Water Resources, Hellenic Agricultural Organization – Demeter. He has been active in the field of Soil Ecology for over 15 years, and his research interests are focused on the study of the soil microbial community structure and functionality, soil enzyme activity, and their role in soil organic matter decomposition and nutrient cycling. Part of his research is also focused on the study of soil free-living nematode communities, aiming to study in depth the relations and interactions of organisms that belong to different levels of the soil food web. His goal is to better understand how agricultural management practices affect organisms of different soil food web levels (indicators of soil health) and to improve soil functions optimizing agricultural productivity and sustainability. He has published more than 50 research papers in international peer-reviewed journals and has participated in several national and international projects.

Efimia M. Papatheodorou is Professor of Soil Ecology at the Ecology Department of Biology School, Aristotle University of Thessaloniki. Her research focuses on the response of soil microbes to disturbances, microbial interactions in soil and their effects on crop productivity, the relation between the structure and function of soil microbial communities, soil enzymes, and restoration of soils. She has published 56 peer-reviewed articles, eight book chapters and two textbooks. She has participated in 15 national and international research projects, of which she served as PI in three. In 2017, she was recognized for her outstanding contribution in reviewing for the journal *Applied Soil Ecology*. She was Managing Editor of the journal *Web Ecology*, supported by the European Ecological Federation, for the period 1999–2005, Head of the Ecology Department for two years, member of the Editorial Board of the Hellenic Ecological Society from 2014 to 2018, Head of the Master program "Sustainable Agriculture and Business" in International Hellenic University from 2017 to 2020, and has served as Reviewer Board Member for the *Applied Sciences* journal (MDPI) since 2019.

 *agronomy*

Editorial

Crop Yield and Soil Quality Are Partners in a Sustainable Agricultural System

Efimia M. Papatheodorou [1,*] **and Nikolaos Monokrousos** [2]

[1] Department of Ecology, School of Biology, Aristotle University, 54124 Thessaloniki, Greece
[2] University Center of International Programmes of Studies, International Hellenic University, 57001 Thessaloniki, Greece; nmonokrousos@ihu.gr
* Correspondence: papatheo@bio.auth.gr

Citation: Papatheodorou, E.M.; Monokrousos, N. Crop Yield and Soil Quality Are Partners in a Sustainable Agricultural System. *Agronomy* **2022**, *12*, 140. https://doi.org/10.3390/agronomy12010140

Received: 21 December 2021
Accepted: 5 January 2022
Published: 7 January 2022

Publisher's Note: MDPI stays neutral with regard to jurisdictional claims in published maps and institutional affiliations.

1. Introduction

Agricultural practices involving the excessive use of chemical fertilizers and pesticides pose major risks to the environment and human health. Over the last two decades, great attention has been paid to the development of sustainable eco-friendly agricultural management practices, aiming to improve soil quality and increase crop yields while maintaining environmental sustainability. For this purpose, management practices such as intercropping, crop rotation, precision fertilization, no-till, and chisel plowing are employed, as well as the addition of amendments and the use of plant-growth-promoting rhizobacteria (PGPR), arbuscular mycorrhizal fungi (AMF), endophytes, and biostimulants. This Special Issue contains original contributions detailing cutting-edge research and review articles that help to assess the effect of sustainable agricultural practices on soil quality and plant health. The latter is related to the alleviation of abiotic or biotic stresses that plants undergo.

2. Inoculation with Beneficial Soil Microbes

Most studies dealing with the application of ecofriendly management practices are focused on plant growth and development [1,2], but studies of their effects on soil communities are sparse. Exceptions are the soil bacterial communities that are under extensive investigation due to the development of metagenomics tools. The effects of different inoculants on plant biomass and the indigenous rhizosphere microbial community of *Lactuca sativa* are presented in two studies [3,4]. According to Nikolaidou et al. [3], the inoculation of *L. sativa* roots with *Bacillus subtilis* (PGPR) inhibited the inoculation by *Rhizophagus irregularis* (AMF) at a percentage of 50%, while both inoculants acted competitively to free-living fungi. Compared to relevant studies, the authors mentioned that the negative interactions were due to limitations in soil nutrients (P and K) recorded in this specific soil. However, both inoculants acted synergistically, inducing a positive effect on plant growth. In the same line, Angelina et al. [4] inoculated the roots of *L. sativa* plants with two PGPR species (*B. subtilis*, Gram-positive and *P. fluorescens*, Gram-negative) and developed them in soils of similar soil texture but of different management history: an organic and a conventional cultivated soil. The differences in plant growth were related only to soil management history, while the effect of the inoculants on the communities of the free-living microbes depended on soil history. Distinct rhizosphere microbial communities induced by inoculants were identified only in organically cultivated soil.

3. Intercropping and Organic Farming Effect on Soil Bacteria and Nutrient Stocks

Intercropping increases the productivity per unit of land through better utilization of resources, control of disease, and reduction of soil erosion [5]. Yang et al. [6] examined whether an intercropping system of maize and mushroom had an influence on the soil bacterial diversity, richness, and community structure compared to maize monoculture. Additionally, the effect of intercropping on soil nutrients was considered. Intercropping

Agronomy **2022**, *12*, 140. https://doi.org/10.3390/agronomy12010140

increased soil organic matter and the richness and diversity of the bacterial community but had no effect on the amount of soil-available nitrogen compared to monocultures. Due to these positive effects, the authors suggest that this is a sustainable agricultural practice for the black soil zone in Northeast China.

Relevant to soil quality but mainly concerning the soil nutrient stocks was the study by Brock et al. [7]. Although the use of organic farming is considered an eco-friendly approach if stocking rates in mixed-crop livestock farms are too low (which is very common in farms in Germany), sustainability may be threatened since a declined in C content is recorded [8]. The use of an additive source of carbon such as compost produced by plant residues in the interior of the farms and the fertilization with potassium sulfate (in the case of a potassium-fixing soil) seemed promising. The study showed that the application of compost and fertilizer in farms with low stocking rate compensated the soil organic matter loss and the nutrient exports, increased the productivity of legume plants by mitigating the stress imposed by P deficiency, and facilitated the overall crop production, although it has still remained at moderate levels.

4. Irrigation, Fertilization, and Environmental Conditions Affecting Crop Production

Field production of melon largely depends on the thermal conditions and precipitation during the growing season due to the production of aboveground biomass with a high coefficient of transpiration [9]. However, water soil availability is limited in light soils with low water-holding capacity and nutrient retention. In order to identify the best water and fertility conditions for melon production, Rolbiecki et al. [10] examined the effect of drip irrigation plus nitrogen fertigation on the yield and the nutritional characteristics of fruits of two species of Cucurbitaceae compared to the effects of drip irrigation plus broadcast nitrogen fertilization. The method of drip irrigation–nitrogen fertigation proved to be the most suitable practice for the cultivation of Cucurbitaceae in very light soil in Poland. It increased the total marketable fruit yield by almost 13% and positively influenced the melons' nutritive value by affecting the concentration of total sugars, monosaccharides, ascorbic acid, total carotenoids, and polyphenols.

For the improvement of crop yield under specific environmental conditions, the necessity for studying the interactions between plant genotypes and multienvironmental traits (GEN) is globally recognized. In this line, Katsenios et al. [11] studied the quality and the quantity of grain production in four maize hybrids under different soil environmental conditions. The authors aimed to identify the conditions that are most suitable for maize productivity and whether these were differentiated in relation to maize genotype and/or to the characteristic under consideration (e.g., yield or protein content). The authors found idiosyncratic responses, showing that general rules concerning maize productivity could not be applied. The GEN effect explained a low percentage of the variation and could not lead to the selection of a stable genotype for all environments. However, the authors suggested the division of the region into mega-environments and introduced the most suitable genotype for every environment.

5. Plant Based Nematicides and Zeolites' Amendments

Biostimulants are substances that could be used as biological control agents against plants' pests and diseases. The essential oils (EOs) of aromatic plants are used in many fields because of their antimicrobial, antifungal, antioxidant, and antibacterial activities [12]. The biological activity of EOs is related to their chemical composition, which is influenced by the specific climatic, seasonal, and geographic conditions affecting the aromatic species from which EOs derive [13]. In this line, Ntalli et al. [14] examined in vitro the nematicidal activity of seven Greek Lamiaceae species used as hydrosols or as water extracts against the phytoparasitic nematodes *Meloidogyne javanica* and *M. incognita*. *Meloidogyne incognita*, *M. javanica*, and *M. arenaria* infect *Solanaceae* and *Malvaceae*, have broad host ranges, and are in the list of the most economically damaging root-knot nematodes [15]. The *Origanum vulgare* plant was the one with stronger effects on nematodes' suppression; nevertheless, applications at high concentrations in pots became phytotoxic for tomato plants. *O. vulgare* seems to be a sustainable solution against nematodes' pathogenicity, and it could be a sustainable solution for essential oil leftovers use. Further, the authors suggested that the preparation and application of this nematicide is quite easy for non-experts (e.g., farmers).

Apart from biostimulants, other amendments could also provide positive impacts on many functions such as plants' metabolism, germination, photosynthesis, and nutrient absorption. Zeolites are naturally occurring, alkaline-hydrated aluminosilicates with a wide range of applications on soil physicochemical variables (water-holding capacity, infiltration rate, cation exchange capacity, hydraulic conductivity) and plants' nutrient use efficiency. Nutrients that are tied to zeolites because of zeolites' structural complexity are released slowly into soil, enabling the synchronization between nutrient release and plant absorption, thus reducing leaching. The improvement of the wide range of agronomic and horticultural crops with respect to growth, yield, and quality traits with the application of zeolites has been well reported by various researchers [16,17]. A literature review of the structure and the applications of zeolite in agriculture is presented by Mondal et al. [18], who recorded a "boost" in relevant studies since 2018.

6. Sustainability in CH$_4$ Emissions

A sustainable agricultural system must rely on low emissions of gases such as CO_2, NO_2, N_2O, and CH_4, as they are involved in climate change. More specifically, CH_4 has a relatively high global warming potential [19]. The conditions that determined CH_4 emission in boreal and temperate peatlands are not similar to those occurring in tropical peatlands because of differences in environmental factors, peat soil properties, peat temperature, peatland-use practices, vegetation composition and structure, and microbial diversity and population. Since tropical peatlands are commonly used in agriculture, inappropriate drainage may lead to increased CH_4 emissions in these areas. The examination of the changes that the fluctuations in water table imposed on CH_4 emission [20], in different periods of a year, showed that CH_4 emission is a quite complicated process depending on many agents: the microbial CH_4 production and consumption and its transportation by molecular diffusion and by non-microbial processes taking place in peat soils. Water table fluctuations affected emissions, but the emissions in an open field area were significantly lower than in field lysimeters. Due to this inconsistency and the complexity of CH_4 production, the authors proposed more long-term studies to be designed in the future.

7. Land Suitability for Specific Crop Production

Said et al. [21] approached the concept of agricultural sustainability in Egypt by emphasizing land capability. The assessment of land capability depends on the evaluation of the soil quality and expresses the capacity of a soil ecosystem to function in order to sustain crop production in parallel with reduced soil degradation and increased ability to perform a number of soil functions [22,23]. Such an approach integrates soil characteristics, topography, vegetation cover, land use, and climate conditions [24]. The mapping of soils on the Northwestern coast of Egypt in terms of soil fertility, their discrimination in classes by using multivariate analysis, and the fitting between these classes and the needs of specific crops gives valuable information to decision-makers that are involved in the development of the agricultural sector in Egypt.

8. Conclusions

The effects of different management practices even when these are considered "sustainable", they appear quite idiosyncratic and depend on environmental conditions, plant identity, soil fertility, and previous management practices. Although in most cases positive effects on crop productivity were recorded, the effects on microbial soil communities varied. Additionally, these studies revealed the lack of knowledge concerning the effects of biostimulants and biofertilizers on indigenous soil communities' structure and function. However, to consider an agricultural management system as a sustainable one, both partners (plant and soil) must exhibit an enhanced ability to recover after disturbances such as fire, grazing, climate change, and pesticide application. Advanced knowledge must be developed in this direction.

Author Contributions: E.M.P., writing—original draft; N.M., writing—review and editing. All authors have read and agreed to the published version of the manuscript.

Funding: This research received no external funding.

Institutional Review Board Statement: Not applicable.

Informed Consent Statement: Not applicable.

Data Availability Statement: Not applicable.

Conflicts of Interest: The authors declare no conflict of interest.

References

1. Gagné-Bourque, F.; Mayer, B.F.; Charron, J.B.; Vali, H.; Bertrand, A.; Jabaji, S. Accelerated growth rate and increased drought stress resilience of the model grass brachypodium distachyon colonized by bacillus subtilis B26. *PLoS ONE* **2015**, *10*, e0130456. [CrossRef]
2. Hashem, A.; Tabassum, B.; Fathi Abd_Allah, E. Bacillus subtilis: A plant-growth promoting rhizobacterium that also impacts biotic stress. *Saudi J. Biol. Sci.* **2019**, *26*, 1291–1297. [CrossRef]
3. Nikolaidou, C.; Monokrousos, N.; Kapagianni, P.D.; Orfanoudakis, M.; Dermitzoglou, T.; Papatheodorou, E.M. The Effect of *Rhizophagus irregularis*, *Bacillus subtilis* and Water Regime on the Plant–Microbial Soil System: The Case of *Lactuca sativa*. *Agronomy* **2021**, *11*, 2183. [CrossRef]
4. Angelina, E.; Papatheodorou, E.M.; Demirtzoglou, T.; Monokrousos, N. Effects of *Bacillus subtilis* and *Pseudomonas fluorescens* Inoculation on Attributes of the Lettuce (*Lactuca sativa* L.) Soil Rhizosphere Microbial Community: The Role of the Management System. *Agronomy* **2020**, *10*, 1428. [CrossRef]
5. Fu, Z.; Zhou, L.; Chen, P.; Du, Q.; Pang, T.; Song, C.; Wang, X.; Liu, W.; Yang, W.; Yong, T. Effects of maize-soybean relay intercropping on crop nutrient uptake and soil bacterial community. *J. Integr. Agric.* **2019**, *18*, 2006–2018. [CrossRef]
6. Yang, X.; Wang, Y.; Sun, L.; Qi, X.; Song, F.; Zhu, X. Impact of Maize–Mushroom Intercropping on the Soil Bacterial Community Composition in Northeast China. *Agronomy* **2020**, *10*, 1526. [CrossRef]
7. Brock, C.; Oltmanns, M.; Matthes, C.; Schmehe, B.; Schaaf, H.; Burghardt, D.; Horst, H.; Spieß, H. Compost as an Option for Sustainable Crop Production at Low Stocking Rates in Organic Farming. *Agronomy* **2021**, *11*, 1078. [CrossRef]
8. Brock, C.; Oberholzer, H.-R.; Schwarz, J.; Fließbach, A.; Hülsbergen, K.-J.; Koch, W.; Pallutt, B.; Reinicke, F.; Leithold, G.; Fliessbach, A. Soil organic matter balances in organic versus conventional farming—modelling in field experiments and regional upscaling for cropland in Germany. *Org. Agric.* **2012**, *2*, 185–195. [CrossRef]
9. Kaniszewski, S. *Nawadnianie Warzyw Polowych (Irrigation of Field Vegetables)*; Plantpress: Kraków, Poland, 2005; pp. 1–85.
10. Rolbiecki, R.; Rolbiecki, S.; Figas, A.; Jagosz, B.; Wichrowska, D.; Ptach, W.; Prus, P.; Sadan, H.A.; Ferenc, P.-F.; Stachowski, P.; et al. Effect of Drip Fertigation with Nitrogen on Yield and Nutritive Value of Melon Cultivated on a Very Light Soil. *Agronomy* **2021**, *11*, 934. [CrossRef]
11. Katsenios, N.; Sparangis, P.; Chanioti, S.; Giannoglou, M.; Leonidakis, D.; Christopoulos, M.V.; Katsaros, G.; Efthimiadou, A. Genotype × Environment Interaction of Yield and Grain Quality Traits of Maize Hybrids in Greece. *Agronomy* **2021**, *11*, 357. [CrossRef]
12. Karamanoli, K.; Menkissoglu-Spiroudi, U.; Bosabalidis, A.M.; Vokou, D.; Constantinidou, H.I.A. Bacterial colonization of the Phyllosphere of nineteen plant species and antimicrobial activity of their leaf secondary metabolites against leaf associate bacteria. *Chemoecology* **2005**, *15*, 59–67. [CrossRef]
13. Figueiredo, A.C.; Barroso, G.J.; Pedro, L.G.; Scheffer, J.J.C. Factors affecting secondary metabolite production in plants: Volatile components and essential oils. *Flavour Fragr. J.* **2008**, *23*, 213–226. [CrossRef]
14. Ntalli, N.G.; Ozalexandridou, E.X.; Kasiotis, K.M.; Samara, M.; Golfinopoulos, S.K. Nematicidal Activity and Phytochemistry of Greek Lamiaceae Species. *Agronomy* **2020**, *10*, 1119. [CrossRef]
15. Jones, J.T.; Haegeman, A.; Danchin, E.G.J.; Gaur, H.S.; Helder, J.; Jones, M.G.K.; Kikuchi, T.; Manzanilla-López, R.; Palomares-Rius, J.E.; Wesemael, W.M.L.; et al. Top 10 plant-parasitic nematodes in molecular plant pathology. *Mol. Plant Pathol.* **2013**, *14*, 946–961. [CrossRef]
16. Shahsavari, N.; Jais, H.M.; Shirani Rad, A.H. Responses of canola oil quality characteristics and fatty acid composition to zeolite and zinc fertilization under drought stress. *Int. J. Agric. Sci.* **2014**, *4*, 49–59.
17. Chen, T.; Guimin, X.; Qi, W.; Zheng, J.; Jin, Y.; Sun, D.; Wang, S.; Chi, D. The Influence of Zeolite Amendment on Yield Performance, Quality Characteristics, and Nitrogen Use Efficiency of Paddy Rice. *Crop. Sci.* **2017**, *57*, 2777–2787. [CrossRef]

18. Mondal, M.; Biswas, B.; Garai, S.; Sarkar, S.; Banerjee, H.; Brahmachari, K.; Bandyopadhyay, P.K.; Maitra, S.; Brestic, M.; Skalicky, M.; et al. Zeolites Enhance Soil Health, Crop Productivity and Environmental Safety. *Agronomy* **2021**, *11*, 448. [CrossRef]
19. Collins, W.J.; Fry, M.M.; Yu, H.; Fuglestvedt, J.S.; Shindell, D.T.; West, J.J. Global and regional temperature-change potentials for near-term climate forcers. *Atmos. Chem. Phys.* **2013**, *13*, 2471–2485. [CrossRef]
20. Luta, W.; Ahmed, O.H.; Omar, L.; Heng, R.K.J.; Choo, L.N.L.K.; Jalloh, M.B.; Musah, A.A.; Abdu, A. Water Table Fluctuation and Methane Emission in Pineapples (*Ananas comosus* (L.) Merr.) Cultivated on a Tropical Peatland. *Agronomy* **2021**, *11*, 1448. [CrossRef]
21. Said, M.E.S.; Ali, A.M.; Borin, M.; Abd-Elmabod, S.K.; Aldosari, A.A.; Khalil, M.M.N.; Abdel-Fattah, M.K. On the Use of Multivariate Analysis and Land Evaluation for Potential Agricultural Development of the Northwestern Coast of Egypt. *Agronomy* **2020**, *10*, 1318. [CrossRef]
22. AbdelRahman, M.A.; Shalaby, A.; Mohamed, E.S. Comparison of two soil quality indices using two methods based on geographic information system. *Egypt. J. Remote Sens. Space Sci.* **2019**, *22*, 127–136. [CrossRef]
23. Mohamed, E.S.; Schütt, B.; Belal, A. Assessment of environmental hazards in the north western coast-Egypt using RS and GIS. *Egypt. J. Remote Sens. Space Sci.* **2013**, *16*, 219–229. [CrossRef]
24. De la Rosa, D.; Moreno, J.A.; García, L.V.; Almorza, J. MicroLEIS: A microcomputer-based Mediterranean land evaluation information system. *Soil Use Manag.* **1992**, *8*, 89–96. [CrossRef]

 *agronomy*

Article

The Effect of *Rhizophagus irregularis*, *Bacillus subtilis* and Water Regime on the Plant–Microbial Soil System: The Case of *Lactuca sativa*

Charitini Nikolaidou [1], Nikolaos Monokrousos [2,*], Pantelitsa D. Kapagianni [1], Michael Orfanoudakis [3], Triantafyllia Dermitzoglou [4] and Efimia M. Papatheodorou [1,*]

[1] Department of Ecology, School of Biology, Aristotle University of Thessaloniki, 54124 Thessaloniki, Greece; char.nikolaidou@gmail.com (C.N.); kapagianni@gmail.com (P.D.K.)
[2] University Center of International Programmes of Studies, International Hellenic University, 57001 Thessaloniki, Greece
[3] Department of Forestry and Management of the Environment and Natural Resources, Democritus University of Thrace, 68200 Orestiada, Greece; morfan@fmenr.duth.gr
[4] R&D Department of HUMOFERT S.A., 14452 Athens, Greece; fdemi@humofert.gr
* Correspondence: nmonokrousos@ihu.gr (N.M.); papatheo@bio.auth.gr (E.M.P.); Tel.: +30-2310-807572 (N.M.); +30-2310-998313 (E.M.P.)

Abstract: Inoculation with beneficial microbes represents a promising solution for sustainable agricultural production; however, knowledge on the effects of inoculants on the indigenous microbial communities remains limited. Here, we evaluated the impact of the arbuscular mycorrhizal fungus *Rhizophagus irregularis* and the promoting rhizobacterium *Bacillus subtilis* on the growth of *Lactuca sativa*. The biomass, the composition, and the enzyme activity (urease, acid phosphatase, and β-glycosidase) of the rhizosphere microbial community at two soil moisture levels (5 and 10% soil water content) were evaluated. Fungal colonization was lower in co-inoculated plants than those only inoculated with *R. irregularis*. Plant growth was enhanced in co-inoculated and *B. subtilis* inoculated soils. Bacterial biomass and the composition of the microbial communities responded to the joint effect of inoculant type × water regime while the biomass of the other microbial groups (fungi, actinomycetes, microeukaryotes) was only affected by inoculant type. Co-inoculation enhanced the activity of acid phosphatase, indicating a synergistic effect of the two inoculants. Co-inoculation positively impacted the index reflecting plant–microbial soil functions under both water regimes. We concluded that the interactions between the two inocula as well as between them and the resident rhizosphere microbial community were mainly negative. However, the negative interactions between *R. irregularis* and *B. subtilis* were not reflected in plant biomass. The knowledge of the plant and rhizosphere microbial responses to single and co-inoculation and their dependency on abiotic conditions is valuable for the construction of synthetic microbial communities that could be used as efficient inocula.

Keywords: AMF; enzyme activity; microbial communities; PGPR; plant growth; PLFAs

Citation: Nikolaidou, C.; Monokrousos, N.; Kapagianni, P.D.; Orfanoudakis, M.; Dermitzoglou, T.; Papatheodorou, E.M. The Effect of *Rhizophagus irregularis, Bacillus subtilis* and Water Regime on the Plant–Microbial Soil System: The Case of *Lactuca sativa. Agronomy* **2021**, *11*, 2183. https://doi.org/10.3390/agronomy11112183

Academic Editor: Nicolas Chemidlin Prevost-Boure

Received: 17 August 2021
Accepted: 25 October 2021
Published: 29 October 2021

Publisher's Note: MDPI stays neutral with regard to jurisdictional claims in published maps and institutional affiliations.

1. Introduction

In recent years, climate change has affected global agricultural production [1], especially in the Mediterranean region, through the increased frequency of seasonal drought events [2]. Drought is one of the greatest natural disasters that affect agricultural productivity [3,4]. Dry conditions affect plant growth directly and indirectly; indirectly through the effect of dryness on the activity and the composition of the rhizosphere microbial communities [5]. Plant health is closely linked to the activity of these microbes and in turn, plants play a key role in determining their community composition [6]. As a result of this close interconnection, changes in the abiotic environment affecting plants are expected to affect the rhizosphere microbiota and vice versa [7]. To increase crop production, the main practices applied include the extensive use of agrochemicals and pesticides [8], which

Agronomy **2021**, *11*, 2183. https://doi.org/10.3390/agronomy11112183   https://www.mdpi.com/journal/agronomy

can boost plant growth, and suppress pathogenic organisms. However, these products are expensive and toxic to humans and soil health [9], while they cause deleterious side effects on soil beneficial microorganisms that are involved in nutrient cycling processes and biotransformation [10]. There are also environmental concerns about their accumulation in soil [11]. Nevertheless, in recent years, alternative practices that involve the use of microbial inocula consisting of plant growth-promoting rhizobacteria (PGPR) [12], arbuscular mycorrhizal fungi (AMF) [13], or their combination, are gaining ground [14,15]. These are environmentally friendly, enhance plant growth, as well as enable the plant's ability to overcome diseases and mitigate the adverse effects caused by drought.

PGPR plays an important role in the resistance of plants to drought, through the production of phytohormones, osmolytes, exopolysaccharides, the modification of plants' antioxidant defense systems, and the alteration of the host plants' root morphology [16,17]. Many *Bacillus* species that belong to PGPR are widely used for inducing tolerance in water-stressed environments [18,19]. In addition, AMF symbiosis enhances plant tolerance through various morphological and physiological mechanisms [13,20,21]. The AM fungus species *Rhizophagus irregularis* is used to mitigate the deleterious effects of drought [22]; it regulates the signaling pathways and the proteins' production involved in plants' responses to drought [23]. However, competitive interactions may occur among inoculants, due to the increased competition for nutrients and/or the production of secondary metabolites that can harm either PGPR or AMF populations [24,25]. These interactions vary between species [26] and under different abiotic conditions. Thus, the best approach to improve plant growth, especially under unfavorable conditions, is to apply the most effective combination of plant species, bacteria, and fungi [27].

Apart from their interactions with plants and between themselves, the introduced microorganisms also exhibit synergistic and/or competitive interactions with the indigenous microbial communities. It was shown that AMF, affected indirectly the growth of rhizosphere microbes, by increasing [28] or reducing roots exudates [29], while PGPR impact on the existing microbial community [30–32] through the production of antibiotics [18]. These complex interactions between plants, microbial inoculants, and indigenous microbial communities are poorly understood and need further clarification, as they could be vital to sustainable agriculture and the mitigation of abiotic stresses.

Over the last years, most studies focused on the beneficial role of PGPR, AMF, and their co-inoculation on plant growth [18,30,33,34]. Much of the research in this field was conducted using sterilized soil and therefore only a few studies have analyzed the interactions of these microorganisms with the indigenous soil microbial communities and especially the communities of indigenous bacteria [35–38]. This study aimed to investigate the effect of *Bacillus subtilis* (PGRB), *Rhizophagus irregularis* (AM fungus), and their co-inoculation on the composition, the structure, and the functions of lettuce (*Lactuca sativa* L.) rhizosphere microbial community at two soil humidity levels indicative of dry and optimum soil water conditions. Kohler et al. [30] studied the effects of these inoculants separately and jointly on the same plant species measuring plant and soil variables without considering the rhizosphere microbial community; nevertheless, in the present study, we aimed to extend the existing knowledge considering the effects of these inoculants on the indigenous microbial community under different soil water conditions. We hypothesized that the co-inoculation of lettuce plants with these beneficial microorganisms will promote plant growth and improve the plant–microbial soil functions mainly under dry conditions compared to optimum ones since in dry soils the limitations for plant and microbial growth are expected more pronounced.

2. Materials and Methods

2.1. Experimental Design

The soil that was used for the cultivation of lettuce seeds was collected from an abandoned agricultural area between Lake Koroneia and Volvi, Northern Greece (40.4021° N, 23.1533° E). The climate of the area is hot-summer Mediterranean (Csa according to Köppen

Climate Classification) with an average annual precipitation of 554.38 mm and a mean annual temperature of 15.24 °C. After sampling, the soil was homogenized and passed through a sieve of 6 mm in order to remove residues of roots or leaves that could alter the soil microbial load. The soil texture was characterized as sandy loam soil (sand: 74%, clay: 6%, silt: 20%), and the physicochemical characteristics are presented in Table 1.

Table 1. Physicochemical characteristics of the soil used in our study.

Soil Physicochemical Properties	
pH	5.72
Electrical conductivity (mS/cm)	0.35
Organic C (%)	1.26
NO_3-N (mg/kg)	15.68
P (ppm)	20
K (ppm)	125
Ca (ppm)	577
Mg (ppm)	100

Lactuca sativa seeds were placed in a seed container that was sterilized with ethanol and then filled with sieved soil, and they grew for two months in room temperature conditions. When seedlings reached the stage of five leaves, they were transplanted into 2.5 L pots that had been previously sterilized with ethanol. During the transplantation, 2 kg of sieved soil and four seedlings were added to each pot. The seedlings were kept for one week in a soil water content of 10% (considered sufficient for plant growth [39]) in order to recover from the transplantation stress. After this period, only one seedling per plot remained.

Afterward, plants were inoculated with the rhizobacterium *Bacillus subtilis* and the AM fungus *Rhizophagus irregularis* separately or jointly or remained non-inoculated (control). *Bacillus subtilis* was isolated from the rhizosphere soil of *Solanum lycopersicum*. The preliminary identification was confirmed by molecular classification with polymerase chain reaction (PCR), using ERIC1f /ERIC2 (ERIC1 5′-ATGTAAGCTCCTGGGGATTCAC-3′ ERIC2: 5′-AAGTAAGTGACTGGGGTGAGCG-3′) oligonucleotide primers, and based on the sequence of 16S rDNA gene. The preparation of the inoculum involved the growth of bacteria in nutrient broth (100 mL) on the rotary shaker for 24–48 h at 30 °C. Once reached the stationary phase, the cell density of each strain was determined spectrophotometrically (Hitachi U-2800A, Hitachi High-Technologies Corporation, Tokyo, Japan) by comparing the obtained 600 nm optical densities with the growth calibration curves. After incubation, the bacterial cultures are diluted with 0.85% NaCl water solution at the final cell density of about 10^{14-15} CFU cm^{-3}. The *Bacillus subtilis* inoculant contained 5 mL of bacterial solution (10^{14-15} CFU mL^{-1}) diluted in 45 mL of deionized and sterilized water. The AM fungal inoculum consisted of spores and hyphal fragments of *R. irregularis* (BEG 141, 1000 propagules/g-TERI (Energy and Resources Institute, Delhi, India). The viability of the inoculant was tested before the application following the procedure described at Monokrousos et al. [36]. AMF inoculant contained 0.2 g of powder (10^6 propagules g^{-1}) diluted in 50 mL of water. The inoculants were injected in the area near the plant roots and left for 4 days without watering to prevent the washout of inoculants. Pots with dual inoculation received 25 mL of bacterial plus 25 mL of AM fungal inocula while control pots received 50 mL of deionized and sterilized water.

In each treatment half of the pots were kept at 5% *w/w* soil water content (WC) (dry conditions) throughout the experiment [40], and the rest were kept at 10% *w/w* WC (optimum conditions) that impose no limitation to plant growth [39]. Pots were weighed prior to each watering and the amount of added water required was estimated. The experiment lasted for two months (mid-December to mid-February) and took place in a glasshouse at the Aristotle University of Thessaloniki, under natural light conditions with an average temperature of 18 °C. The experimental design included the cultivation of *Lactuca sativa* plants in two water regimes (dry and optimum water conditions) $\times$ 4 types of inoculants [*Bacillus subtilis* inoculant (B), *Rhizophagus irregularis* inoculant (M), co-inoculation (MB) and without inoculant; control (C)], with four replicates per treatment, giving a total of 32 pots.

After 60 days, a destructive harvest was conducted. Soil samples were collected from the rhizosphere of the plants by shaking the soil that was attached to the roots after the plant's extraction. The rhizosphere soil samples were passed through a 2 mm sieve and stored at 4 °C until the end of the analyses (within two weeks after harvest) [41]. Shoots and roots were dried for 48 h at 70 °C and weighed. From each pot, at least 50 fresh root fragments (1 cm length) were stored at −4 °C to be used for the determination of the AMF colonization percentage.

2.2. AMF Colonization

Root samples were cleaned and stained according to Koske and Gemma [42], modified by Orfanoudakis et al. [43]. The stained samples were examined under a compound microscope (Axio Lab A1, Zeiss, Germany) and AMF root colonization was evaluated according to Trouvelot et al. [44]. The percentage of mycorrhizal colonization was calculated with the MycoCalc software [45].

2.3. Phospholipid Fatty Acid Analysis

Microbial communities in soil samples were determined by the phospholipid fatty acids (PLFAs) method as described by Papadopoulou et al. [46]. After the isolation of PLFAs, gas chromatography was performed to determine the chemical compounds present in the samples. A Trace GC Ultra gas chromatograph (ThermoFinnigan, SanJose, CA, USA) was used in conjunction with a Trace ISQ mass spectrometry detector, and a separator-injector automatic sampler. The chromatographic results were then processed on the XcaliberMS platform to identify the major fatty acids found in the samples. Each fatty acid was measured (in nmol g^{-1}) by a one-point calibration against the GC response of the internal standard (19:0 methyl ester).

A total of 24 phospholipid fatty acids methyl esters were identified in all 32 samples, which are characteristic of specific microbial groups. Given the standard nomenclature that describes phospholipid fatty acids methyl esters, the ones found in our samples are presented in Table 2 [36,47–49]. Biomasses of the microbial groups Gram-positive, Gram-negative, actinomycetes, and microeukaryotes resulted from the sum of the concentration of the fatty acid corresponding to each group of microorganisms, while for fungi biomass all fungal PLFAS were summed except for 16:1ω5. Total bacterial biomass resulted from the sum of the fatty acids which are representative of Gram-positive, Gram-negative, and actinomycetes. In addition, ratios such as Gram$^+$/Gram$^-$ and bacteria/fungi which reflect the structure of the soil community were estimated.

Table 2. The phospholipid fatty acids methyl ester biomarkers found in our study.

PLFA Biomarkers	
i15:0, a15:0, 15:0, i16:0, i17:0, 17:0	Gram-positive bacteria
16:1ω9cis, 16:1ω9trans, cy17:0	Gram-negative bacteria
10Me16:0, 10Me17:0, 10Me18:0	Actinomycetes
20:0, 22:0, 24:0	Microeukaryotes
18:2ω9,12	Fungi
18:1ω9t, 18:1ω9c	Gram-negative bacteria and fungi
16:0	Bacteria and fungi
11:0, 13:0, 14:0, 18:0	Microbial origin
16:1ω5	AM fungal mycelium

2.4. Enzyme Activity Analysis

We estimated the enzyme activity of urease, β-glucosidase (BG), and acid phosphatase (AP) involved in N, C and P-cycles respectively. The activities of AP and BG were determined according to the methodology of Allison and Jastrow [50], modified for 96 well microplates. We used 5 mM p-nitrophenyl-phosphate and 5 mM p-nitrophenyl-β-glucopyranoside solutions for AP and BG, respectively. The p-nitrophenol (pNP) reaction product for AP and BG was measured at 405 nm. The microplate method of Sinsabaugh et al. [51] was used to assess urease activity. The concentration of urea in the assay 96 well microplate was 20 mM. The plates were incubated at 20 °C for 18 h. Ammonium released by the reaction was quantified using colorimetric salicylate and cyanurate reagent packages from Hach (absorbance was measured at 610 nm by using LT-4500 96-well microplate reader, Labtech, UK).

2.5. Statistical Analysis

To determine the effect of water regime, type of microbial inoculant and their interaction on rhizosphere microbial biomasses and their activity, we applied a two-way ANOVA. Prior to analyses, the data were transformed, where necessary, to meet the assumptions of the ANOVA (normal distribution, independence among means and variances, homogeneity of variances). Wherever statistically significant effects were recorded, post hoc comparisons (Fisher LSD) were further applied. To detect changes in microbial community composition between treatments, we applied a one-way analysis of similarities (ANOSIM) based on the Bray–Curtis similarity index (PAST 3 software). Community composition was described by the relevant abundances of individual PLFAs.

Finally, an index that gives an overall picture of the plant–microbial soil functions was estimated separately for each treatment. The higher the value of the index, the better the functionality of the system under consideration [52]. To quantify the index, the z-score of nine variables recorded in each treatment was calculated. The variables were the abundance of Gram-positive, Gram-negative, actinomycetes, fungi, microeukaryotes, the plant dry biomass, and the enzyme activity of AP, BG, and urease. We used the microbial biomasses for the construction of the index because that microbial groups are involved in the biotransformation of recalcitrant (actinomycetes, fungi) and labile substrates (bacteria) and their presence reflects a well-functioning soil. Z-score was calculated for each sample and afterward, the average and standard error of the index for the samples of each treatment were calculated. The effects of water regime, type of microbial inoculant, and their interaction on the index values were determined by a two-way analysis of Variance. All analyses were performed using the SPSS software (version 25).

3. Results

3.1. AMF Colonization

In relation to humidity level, plants are grown in 5% WC presented a significantly lower percent of colonization ($p < 0.000$) compared to those growing in 10% WC. In addition, plants inoculated with fungus and *B. subtilis* showed a lower colonization percentage than those inoculated only with AM fungus (Table 3).

Table 3. Mean % colonization ($\pm$SE) of *L. sativa* roots by AMF after inoculation with *R. irregularis* or co-inoculation with *R. irregularis* and *B. subtilis* in dry (5% WC) and wet (10% WC) pots. Different superscripts letters denote the differences in % colonization in relation to the joint effect of water regime x inoculant type.

	% AMF Colonization Rate	
Inoculant	5% WC	10% WC
R. irregularis	31.67 $\pm$ 1.67 [b]	58.80 $\pm$ 1.39 [a]
R. irregularis + *B. subtilis*	16.40 $\pm$ 0.93 [d]	22.00 $\pm$ 1.38 [c]

3.2. Plant Growth

The single effects of water regime and inoculant type on plant growth were significant ($p < 0.001$ for both effects) while a non-significant joint effect was recorded. Plants grown at higher WC exhibited significantly higher values of dry biomass than the ones grown at lower WC (Table S1, Figure 1). Moreover, co-inoculated plants (MB) presented significantly higher biomass than the control (C) while *B. subtilis* and *R. irregularis* single inoculants had no significant effect on plant growth compared to the control (Table S2).

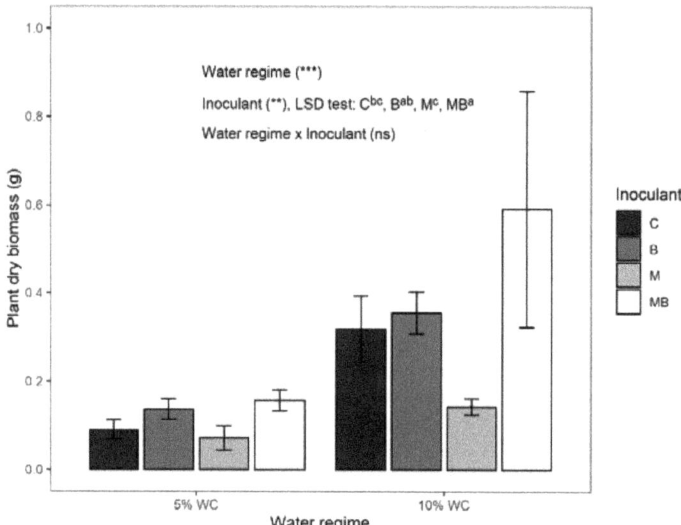

Figure 1. Mean values (±SE) of dry biomass (g) of *Lactuca sativa* plants grown under different water regimes. In the case of significant effects (** $p < 0.01$, *** $p < 0.001$), the results of the post hoc comparisons are presented; different superscript letters denote significant differences (C = Control, B = *B. subtilis*, M = AMF, MB = AMF and *B. subtilis*).

3.3. Biomass and Activity of Soil Microbial Community

The water regime did not affect microbial biomasses, while inoculant type significantly affected the biomass of all microbial groups (Table 4). In addition, the interaction of water regime × inoculant type significantly affected the biomass of bacteria, Gram$^+$ ($p < 0.05$) and Gram$^-$ bacteria ($p < 0.001$) and the bacteria to fungi ratio ($p < 0.05$).

Table 4. Effects of water regime, inoculant type and their interaction on the biomass of individual microbial groups, and the bacteria/fungi and Gram$^+$/Gram$^-$ ratios as revealed by two-way ANOVA (n = 4, df: degree of freedom).

Microbial Groups	Water Regime (1 df)		Inoculant Type (3 df)		Water Regime × Inoculant Type (3 df)	
	F Value	*p*-Value	F Value	*p* Value	F Value	*p* Value
Bacteria	6.22	0.018	8.18	0.000	3.79	0.020
Gram$^+$	2.09	0.158	10.64	0.000	7.90	0.000
Gram$^-$	46.33	0.000	3.14	0.039	3.69	0.022
Gram$^+$/Gram$^-$	66.56	0.000	1.63	0.202	1.72	0.183
Bacteria/Fungi	6.35	0.017	9.65	0.000	3.48	0.027
Total microbial biomass	2.60	0.116	8.03	0.000	2.03	0.129
Fungi	0.00	0.958	12.03	0.000	1.21	0.322
Actinomycetes	0.74	0.395	7.19	0.000	2.14	0.115
Microeukaryotes	2.05	0.162	4.21	0.013	2.46	0.081

Specifically, the total microbial biomass and the biomass of fungi, actinomycetes, and microeukaryotes exhibited an almost similar pattern of response to inoculant type (Figure 2a,e–g). Higher biomasses were recorded in controls and co-inoculated samples compared to biomasses recorded in rhizosphere soils with single inoculants, while significant differences between the soils inoculated with AM fungus and those with *B. subtilis* were recorded only in the case of fungi.

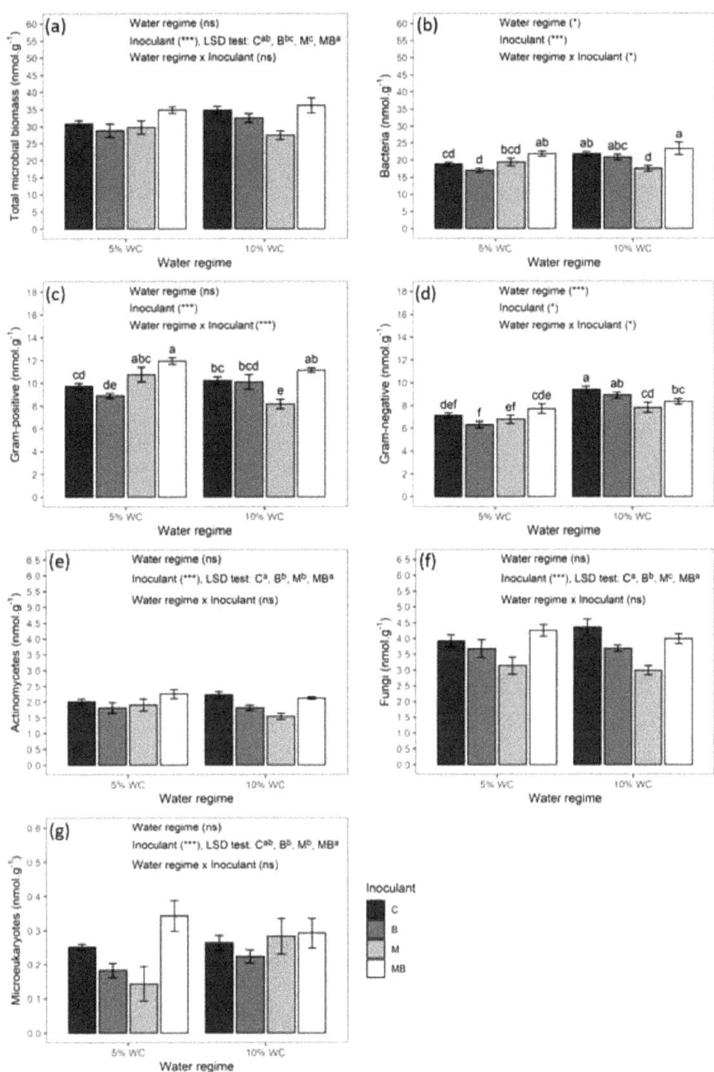

Figure 2. Mean (±SE) biomass of total microbes (**a**), bacteria (**b**), Gram-positive (**c**), Gram-negative (**d**), actinomycetes (**e**), fungi (**f**) and microeukaryotes (**g**) (nmol g^{-1}) in the rhizosphere soil of *L. sativa* under different water regimes. The significant effect of "inoculant type × water regime" is presented by the letters above the bars; bars topped by the same letter do not differ significantly. In the case of a single effect of the inoculum type, different superscript letters produced by LSD, denote significant differences (a: corresponds to the highest value, C = Control, B = *B. subtilis*, M = AMF, MB = AMF and *B. subtilis*). (* $p < 0.05$, *** $p < 0.001$, ns: non-significant).

Bacterial biomasses (total biomass as well as that of Gram$^+$ and Gram$^-$) showed an idiosyncratic response to the joint effect of the inoculant x water regime. Bacterial biomass was high in co-inoculated soils (no matter the water status), and in controls and *B. subtilis* soils under optimum water conditions (Figure 2b). The biomass of Gram-positive bacteria was highest in co-inoculated soils in both water regimes and in AMF inoculated dry soils, (Figure 2c). The controls and *B. subtilis*-inoculated soils under optimum water conditions exhibited the highest biomass of Gram-negative bacteria (Figure 2d).

Changes in the ratios Gram$^+$/Gram$^-$ and bacteria/fungi denote changes in the structure of the microbial community. The Gram$^+$/Gram$^-$ ratio was only significantly influenced by the water regime ($p < 0.001$, Table 4). The high Gram$^+$/Gram$^-$ ratio in 5% WC (Figure 3a) was mainly caused by a decrease in Gram-negative bacteria, rather than an increase in Gram-positive bacteria, because the biomass of the latter remained constant regardless of the water regime (Table 2). Although fungi were not affected by the water regime, the bacteria/fungi ratio was jointly affected by the inoculant type $\times$ water regime (Table 4). High ratios were reported in AMF-inoculated pots, as well as in co-inoculated and *B.subtilis* inoculated pots under optimum water conditions (Figure 3b).

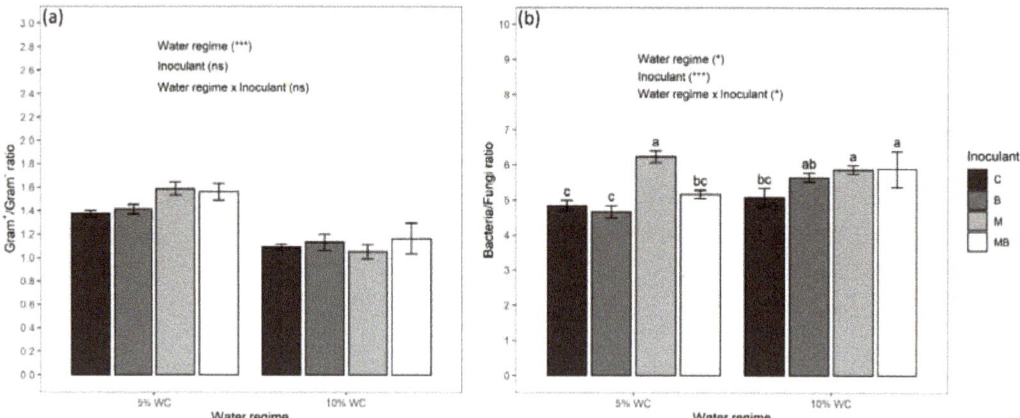

Figure 3. Mean ($\pm$SE) Gram$^+$/Gram$^-$ (**a**) and bacteria/fungi (**b**) ratios in soil samples of *Lactuca sativa* under different water regimes. In the case of significant effects (* $p < 0.05$, *** $p < 0.001$), the results of the post hoc comparisons are presented. The significant effect of "inoculant type $\times$ water regime" is presented by the letters above the bars; bars topped by the same letter do not differ significantly (C = Control, B = *B. subtilis*, M = AMF, MB = AMF and *B. subtilis*).

Water regime did not significantly affect the activity of any of the three enzymes (Table S3, Figure S1) whereas inoculant type only affected the activity of acid phosphatase. Co-inoculated soils exhibited the highest acid phosphatase activity (Figure 4).

Dissimilarities between treatments as concern the composition of the rhizosphere microbial communities are presented in Table S4. For most cases, no significant differences in composition were revealed. Significant dissimilarities in the composition of the microbial communities were recorded for controls (C: $p = 0.047$), and between the control (C5) and the jointly inoculated soils (MB5) developed at 5% WC.

3.4. Index of Plant–Microbial Soil Functions

The index was significantly affected by the single effect of water regime and inoculant type (Table S5; Figure 5). The z-score values were higher in 10% WC compared to 5% while the highest index values were recorded in co-inoculated soils and the lowest in AMF inoculated ones (Figure 5).

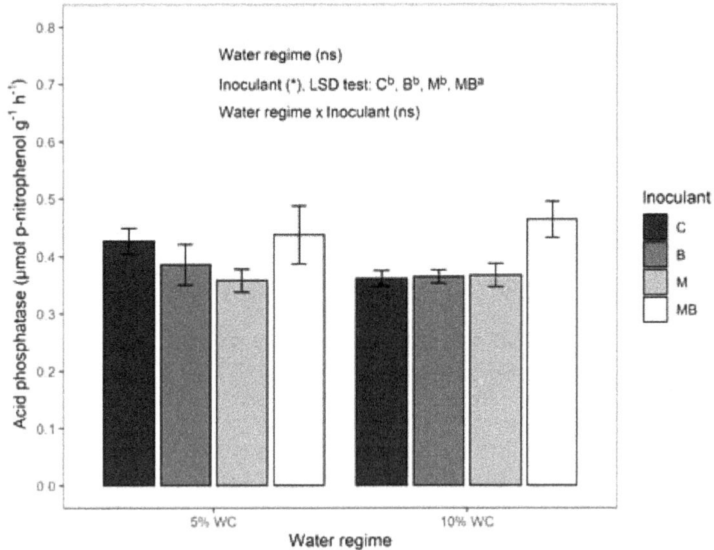

Figure 4. Mean values (±SE) of acid phosphatase activity in rhizosphere soil samples of *L. sativa* under different water regimes. In the case of significant effects (* $p < 0.05$), the results of the post hoc comparisons are presented; different superscript letters denote significant differences (C = Control, B = *B. subtilis*, M = AMF, MB = AMF and *B. subtilis*).

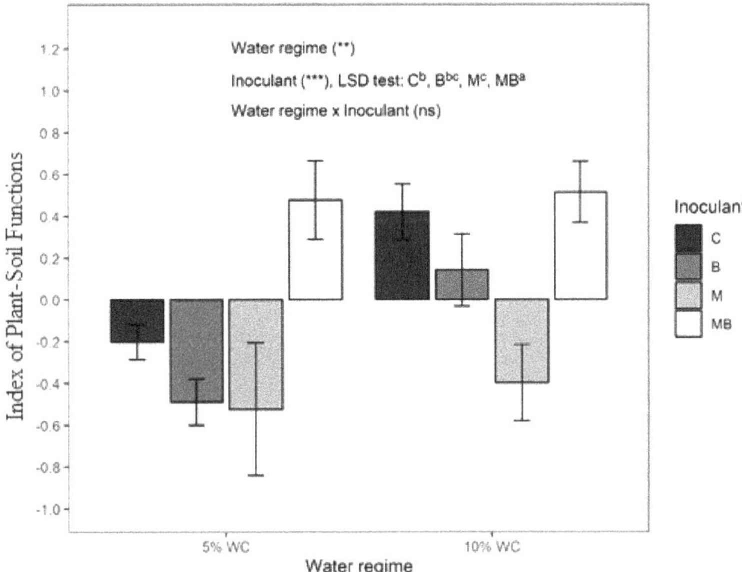

Figure 5. Mean values (±SE) of the index of plant-soil interactions obtained from the z-scores of nine variables in soil rhizosphere samples of *L. sativa* under different water regimes. In the case of significant effects (** $p < 0.01$, *** $p < 0.001$), the results of the post hoc comparisons are presented; different superscript letters denote significant differences (C = Control, B = *B. subtilis*, M = AMF, MB = AMF and *B. subtilis*).

4. Discussion

The effect of co-inoculation on plant, soil, and rhizosphere microbial traits depends on the inoculants involved and the prevailing abiotic conditions [38,53]. In this study, the effects of water regime and type of microbial inoculants on *L. sativa* traits and its rhizosphere microbial community were examined. Although there are relevant studies, most of them examined the effect of inoculants on the composition of the rhizosphere bacterial communities at the OTUs level. Herein, an independent effect of water regime and inoculant type for most of the variables were recorded. Dry conditions inhibited the growth of *L. sativa* and its roots' colonization by AMF. The findings of Saia et al. [54] partially diverse from ours; the reduction in water availability from well- to moderate watered conditions did not affect the lettuce yield but reduced both AMF and *Trichoderma* presence in the roots and soil. The low soil humidity could impact directly the survival and germination of fungal spores limiting the effectiveness of fungal colonization [23], or indirectly by inhibiting root development [15].

The biomasses of the various microbial groups, the total microbial biomass, and the soil enzyme activity were not affected by the water regime, a fact that implies the tight correlation between biomass and activity. Usually, at low water levels, overall soil microbial biomass decreases [55]. However, microbial biomass could remain stable [56], because of the balance between microbial growth and death [57]. Alternatively, the lack of difference in microbial biomass in dry soils might be explained by the fact that microbial communities originating from dry ecosystems, such as the present system, which was a Mediterranean one, have been adapted to be resilient to long periods of drought [58] since drought is one of the main characteristics of the Mediterranean climate. However, the lack of water effect on total microbial biomass was followed by changes in microbial community structure as indicated by changes in the Gram$^+$/Gram$^-$ ratios. The increased Gram$^+$/Gram$^-$ ratio under dry conditions was likely due to the reduction in drought-vulnerable Gram-negative bacteria, rather than the increase in Gram-positive ones. Moreover, Fanin et al. [59] mentioned that the Gram$^+$/Gram$^-$ ratio could be used as a coarse indicator of the relative C availability for bacterial communities; it increases with decreasing C availability [60,61]. We could assume that the prevailing dry conditions-imposed limitations in decomposition or inhibited C transfer to bacteria due to water limitation.

Contrary to the response of most microbial biomasses, bacteria are affected significantly by the joint effect of water regime and inoculant type. High values of total bacterial biomass were recorded in optimum watered soils, except those inoculated with *R. irregularis*. This outcome showed that high populations of AM fungus (percentage of colonization 59%) could negatively affect the biomass of indigenous bacteria. In both water regimes, the presence of *B. subtilis* inhibited the colonization by *R. irregularis*, a result that is similar to that reported by Xiao et al. [28]. It is attributed to the ability of *B. subtilis* strains to synthesize antibiotics with antifungal properties, such as iturin [62].

Similarly to bacterial biomass, the bacteria/fungi ratio of the resident microbial community responded to the joint effect of the two independent variables. In dry soils, the ratio decreased in co-inoculated soils and those inoculated with *B. subtilis*, indicating a shift towards fungal dominance in the presence of *B. subtilis*. This shift in dominance could be indirectly induced by *B. subtilis*; the latter inhibited the growth of AM fungus which acted competitively to the free-living fungi.

Inocula' Effects

Plants inoculated with *R. irregularis*, exhibited above-ground biomass similar to the control, but lower than that of the co-inoculated or *B. subtilis* treatments. These differences were similar in both water regimes. The limited ability of *R. irregularis* to enhance plant growth in *L. sativa* plants was also reported by Kohler et al. [30]. AM fungi do not always form mutualistic relationships with their host plants. Johnson et al. [63] suggested that a commensalistic or, even, parasitic relationship might occur between the two symbionts. The type of relationship is defined by the benefit/cost ratio, which, in turn, is affected by developmental, environmental, and genotypic factors [64]. In our case, the cost of maintaining AMF symbiosis seems to be too high for the plant, a fact that imposed limitations on its growth. Kohler et al. [30] also reported that the dual inoculation of *L. sativa* with *Glomus intraradices* and *B. subtilis* promoted plant growth. However, in our study, the plant biomass in co-inoculated soils was similar to that recorded in the *B. subtilis* inoculated ones. Based on these findings, we assumed that plant growth was enhanced by the presence of *B. subtilis* per se, rather than the presence of both inoculants. The study of Kohler et al. [30] was conducted in more fertile soil than the one used in the current study (available P: 70 vs 15.68 $\mu g\, g^{-1}$; available K: 440 vs 125 $\mu g\, g^{-1}$). In soils with low P and K availability, *B. subtilis* which is phosphate- and potassium-solubilizing rhizobacterium, might enhance mineral uptake by plants by solubilizing insoluble P and releasing K from silicate [65], resulting in increased plant growth.

Low rhizosphere microbial biomass was observed in soils inoculated with *R. irregularis*, or with *B. subtilis* implying the development of negative interactions among the inoculants and the members of the resident microbial community. *B. subtilis* exerted a negative effect on the biomass of the various microbial groups; fungi and actinomycetes were decreased no matter the water regime while $Gram^+$ and $Gram^-$ bacteria were decreased in dry soils. In fact, PGPR is characterized by exhibiting highly competitive ability over indigenous microbial communities [10]. Similar to the bacterium, *R. irregularis* reduced the total microbial biomass and the biomass of actinomycetes, microeukaryotes, and fungi (the latter to levels lower than *B. subtilis*). Competitive interactions between free soil fungi and *R. irregularis* were documented by Tian et al. [66]. The authors attributed the reduction in fungal abundance to the death of some pathogenic fungi by the development of mycorrhizal mycelium. Alternatively, competition between mycorrhizae and free fungi for nitrogen might exist [67]. Fungal hyphae absorb NH_4 and/or NO_3 to meet their high nitrogen requirements [68] in order to absorb soil phosphorus. It is also possible that *R. irregularis* by affecting the metabolism of plants induces changes to the profile of root exudates. The inoculation of olive trees with the AM fungi *Glomus intraradices* followed by the increase in glucose and trechalose and the decrease in fructose, galactose, sucrose, raffinose, and mannitol and by changes in the rhizosphere microbial community; increase in actinomycetes and decrease in bacteria [69].

Co-inoculation resulted in higher than the control plant biomass, but to almost similar to the control, microbial biomasses. The latter could be due to competitive interactions between microbial inoculants during colonization. *B. subtilis* decreased the ability of the fungus to colonize roots at a percentage of 50%. However, co-inoculation increased the activity of acid phosphatase, a fact that was in accordance with the results showing that certain bacteria and/or AM fungi secrete phosphatases [70]. According to the microbial resource allocation theory, microbes expend energy to produce enzymes when nutrients are short in supply [71], such as the P-deficient soil environment that existed in this study. Further, the changes in the available soil P could adjust the establishment of the P-solubilizing strains [38].

Changes in the composition of the resident bacterial communities induced by inocula are usually reported [37,38,72]. For instance, the introduction of *Bacillus amyloliquefaciens* into soil resulted in the enrichment of two bacterial taxa and the inhibition of other 18 taxa [38] in the cucumber rhizosphere bacterial community. In this study, the analysis of similarities between the microbial communities revealed minor changes in composition because of treatments. Since the water regime had a significant effect only on control pots, it seems that the changes in composition induced by differences in soil water were mitigated by inoculation (single or dual). Also only at 5% WC, the communities in jointly inoculated pots differed from those in controls, partly supporting our initial hypothesis that differences because of inoculation are expected more pronounced under water-stressed conditions. Similarly, Chao et al. [72] found that the effects of dark septate endophyte inoculation (DSE) on the microbial community composition of licorice rhizosphere soil were dependent on the DSE species and the water regime as well.

5. Conclusions

The interactions between the two inocula as well as between them and the resident rhizosphere microbial community were mainly negative, a fact that could be attributed to the initial low amount of soil nutrients. However, the negative interactions between *R. irregularis* and *B. subtilis* were not reflected in plant biomass. Co-inoculation improved considerably the functionality of the plant–soil system due to its positive effect on plant growth while the inoculation with the AM fungus deteriorated the index. *R. irregularis* adversely affected plant growth, microbial biomasses, and the enzyme activity in the plant rhizosphere, especially when it occupied a large proportion of the roots. Remarkably, the positive impact of co-inoculation on the index of plant–soil microbial functions was independent of the soil water status, a finding that contradicts our initial hypothesis. The knowledge of the plant and rhizosphere microbial responses to single and co-inoculation and their dependency on abiotic conditions is valuable for the construction of synthetic microbial communities that could be used as efficient inocula.

Supplementary Materials: The following are available online at https://www.mdpi.com/article/10.3390/agronomy11112183/s1, Table S1: Effects of water regime, inoculant type and their interaction on dry biomass per plant. Table S2: Effects of water regime, inoculant type and their interaction on soil enzyme activity. Table S3: *p*-values of ANOSIM analysis based on Bray-Curtis index applied on values of individual PLFAs, Figure S1: Mean values ($\pm$SE) of urease and β-glucosidase activity in soil samples of *L. sativa* under different water regimes (5 and 10% soil water content).

Author Contributions: Conceptualization, E.M.P. and N.M.; formal analysis, C.N. and N.M.; investigation, C.N. and T.D.; methodology, P.D.K., M.O. and T.D.; resources, P.D.K., M.O. and T.D.; supervision, E.M.P.; writing—original draft, C.N.; writing—review and editing, E.M.P., N.M. and P.D.K. All authors have read and agreed to the published version of the manuscript.

Funding: This research received no external funding.

Institutional Review Board Statement: Not applicable.

Informed Consent Statement: Not applicable.

Data Availability Statement: The data presented in this study are available on request from the corresponding authors.

Conflicts of Interest: The authors declare no conflict of interest.

References

1. Lobell, D.B.; Gourdji, S.M. The influence of climate change on global crop productivity. *Plant Physiol.* **2012**, *160*, 1686–1697. [CrossRef]
2. Alamanos, A.; Loukas, A.; Mylopoulos, N.; Xenarios, S.; Vasiliades, L.; Latinopoulos, D. Climate change effects on agriculture in southeast Mediterranean: The case of Karla Watershed in Central Greece. *Geophys. Res. Abstr.* **2019**, *21*, 1.
3. Gornall, J.; Betts, R.; Burke, E.; Clark, R.; Camp, J.; Willett, K.; Wiltshire, A. Implications of climate change for agricultural productivity in the early twenty-first century. *Philos. Trans. R. Soc. B Biol. Sci.* **2010**, *365*, 2973–2989. [CrossRef]
4. Lesk, C.; Rowhani, P.; Ramankutty, N. Influence of extreme weather disasters on global crop production. *Nature* **2016**, *529*, 84–87. [CrossRef]
5. Naylor, D.; Coleman-Derr, D. Drought stress and root-associated bacterial communities. *Front. Plant Sci.* **2018**, *8*, 2223. [CrossRef] [PubMed]
6. Berendsen, R.L.; Pieterse, C.M.J.; Bakker, P.A.H.M. The rhizosphere microbiome and plant health. *Trends Plant Sci.* **2012**, *17*, 478–486. [CrossRef] [PubMed]
7. Wardle, D.A.; Bardgett, R.D.; Klironomos, J.N.; Setälä, H.; Van Der Putten, W.H.; Wall, D.H. Ecological linkages between aboveground and belowground biota. *Science* **2004**, *304*, 1629–1633. [CrossRef]
8. Goswami, D.; Thakker, J.N.; Dhandhukia, P.C. Portraying mechanics of plant growth promoting rhizobacteria (PGPR): A review. *Cogent Food Agric.* **2016**, *2*, 1127500. [CrossRef]
9. Dodd, I.C.; Belimov, A.A.; Sobeih, W.Y.; Safronova, V.I.; Grierson, D.; Davies, W.J. Will modifying plant ethylene status improve plant productivity in water-limited environments. In Proceedings of the 4th International Crop Science Congress, Brisbane, Australia, 26 September–1 October 2004; p. 134.
10. Mandal, A.; Sarkar, B.; Mandal, S.; Vithanage, M.; Patra, A.K.; Manna, M.C. Impact of Agrochemicals on Soil Health. In *Agrichemicals Detection, Treatment and Remediation*; Butterworth-Heinemann: Oxford, UK, 2020; pp. 161–187. ISBN 9780081030172.

11. Silva, V.; Mol, H.G.J.; Zomer, P.; Tienstra, M.; Ritsema, C.J.; Geissen, V. Pesticide residues in European agricultural soils—A hidden reality unfolded. *Sci. Total Environ.* **2019**, *653*, 1532–1545. [CrossRef]

12. Glick, B.R. Bacteria with ACC deaminase can promote plant growth and help to feed the world. *Microbiol. Res.* **2014**, *169*, 30–39. [CrossRef] [PubMed]

13. Lenoir, I.; Fontaine, J.; Lounès-Hadj Sahraoui, A. Arbuscular mycorrhizal fungal responses to abiotic stresses: A review. *Phytochemistry* **2016**, *123*, 4–15. [CrossRef]

14. Nadeem, S.M.; Ahmad, M.; Zahir, Z.A.; Javaid, A.; Ashraf, M. The role of mycorrhizae and plant growth promoting rhizobacteria (PGPR) in improving crop productivity under stressful environments. *Biotechnol. Adv.* **2014**, *32*, 429–448. [CrossRef]

15. Karthikeyan, B.; Abitha, B.; Henry, A.J.; Sa, T.; Joe, M.M. Interaction of rhizobacteria with arbuscularmycorrhizal fungi (AMF) and their role in stress abetment in agriculture. In *Recent Advances on Mycorrhizal Fungi*; Springer: Cham, Switzerland, 2016; pp. 117–142. ISBN 978-3-319-24353-5.

16. Ngumbi, E.; Kloepper, J. Bacterial-mediated drought tolerance: Current and future prospects. *Appl. Soil Ecol.* **2016**, *105*, 109–125. [CrossRef]

17. Vurukonda, S.S.K.P.; Vardharajula, S.; Shrivastava, M.; SkZ, A. Enhancement of drought stress tolerance in crops by plant growth promoting rhizobacteria. *Microbiol. Res.* **2016**, *184*, 13–24. [CrossRef] [PubMed]

18. Hashem, A.; Tabassum, B.; Fathi Abd_Allah, E. Bacillus subtilis: A plant-growth promoting rhizobacterium that also impacts biotic stress. *Saudi J. Biol. Sci.* **2019**, *26*, 1291–1297. [CrossRef]

19. Gagné-Bourque, F.; Mayer, B.F.; Charron, J.B.; Vali, H.; Bertrand, A.; Jabaji, S. Accelerated growth rate and increased drought stress resilience of the model grass brachypodium distachyon colonized by bacillus subtilis B26. *PLoS ONE* **2015**, *10*, e0130456. [CrossRef]

20. Wu, Q.S.; Zou, Y.N. Arbuscular Mycorrhizal Fungi and Tolerance of Drought Stress in Plants. In *Arbuscular Mycorrhizas and Stress Tolerance of Plants*; Springer: Singapore, 2017; pp. 25–41. ISBN 9789811041150.

21. Wu, Q.S.; Srivastava, A.K.; Zou, Y.N. AMF-induced tolerance to drought stress in citrus: A review. *Sci. Hortic.* **2013**, *164*, 77–87. [CrossRef]

22. Li, T.; Lin, G.; Zhang, X.; Chen, Y.; Zhang, S.; Chen, B. Relative importance of an arbuscular mycorrhizal fungus (Rhizophagus intraradices) and root hairs in plant drought tolerance. *Mycorrhiza* **2014**, *24*, 595–602. [CrossRef] [PubMed]

23. Porcel, R.; Aroca, R.; Cano, C.; Bago, A.; Ruiz-Lozano, J.M. Identification of a gene from the arbuscular mycorrhizal fungus Glomus intraradices encoding for a 14-3-3 protein that is up-regulated by drought stress during the AM symbiosis. *Microb. Ecol.* **2006**, *52*, 575–582. [CrossRef]

24. Xiao, X.; Chen, H.; Chen, H.; Wang, J.; Ren, C.; Wu, L. Impact of Bacillus subtilis JA, a biocontrol strain of fungal plant pathogens, on arbuscular mycorrhiza formation in Zea mays. *World J. Microbiol. Biotechnol.* **2008**, *24*, 1133–1137. [CrossRef]

25. Trivedi, P.; Pandey, A.; Palni, L.M.S. Bacterial inoculants for field applications under mountain ecosystem: Present initiatives and future prospects. In *Bacteria in Agrobiology: Plant Probiotics*; Springer: Berlin/Heidelberg, Germany, 2012; pp. 15–44. ISBN 9783642275159.

26. Jäderlund, L.; Arthurson, V.; Granhall, U.; Jansson, J.K. Specific interactions between arbuscular mycorrhizal fungi and plant growth-promoting bacteria: As revealed by different combinations. *FEMS Microbiol. Lett.* **2008**, *287*, 174–180. [CrossRef] [PubMed]

27. Larsen, J.; Cornejo, P.; Barea, J.M. Interactions between the arbuscular mycorrhizal fungus Glomus intraradices and the plant growth promoting rhizobacteria Paenibacillus polymyxa and P. macerans in the mycorrhizosphere of Cucumis sativus. *Soil Biol. Biochem.* **2009**, *41*, 286–292. [CrossRef]

28. Azaizeh, H.A.; Marschner, H.; Römheld, V.; Wittenmayer, L. Effects of a vesicular-arbuscular mycorrhizal fungus and other soil microorganisms on growth, mineral nutrient acquisition and root exudation of soil-grown maize plants. *Mycorrhiza* **1995**, *5*, 321–327. [CrossRef]

29. Gupta Sood, S. Chemotactic response of plant-growth-promoting bacteria towards roots of vesicular-arbuscular mycorrhizal tomato plants. *FEMS Microbiol. Ecol.* **2003**, *45*, 219–227. [CrossRef]

30. Kohler, J.; Caravaca, F.; Carrasco, L.; Roldán, A. Interactions between a plant growth-promoting rhizobacterium, an AM fungus and a phosphate-solubilising fungus in the rhizosphere of Lactuca sativa. *Appl. Soil Ecol.* **2007**, *35*, 480–487. [CrossRef]

31. Ramos, B.; García, J.A.L.; Probanza, A.; Barrientos, M.L.; Gutierrez Mañero, F.J. Alterations in the rhizobacterial community associated with European alder growth when inoculated with PGPR strain Bacillus licheniformis. *Environ. Exp. Bot.* **2003**, *49*, 61–68. [CrossRef]

32. Jha, C.K.; Saraf, M. Plant growth promoting Rhizobacteria (PGPR): A review. *J. Agric. Res. Dev.* **2015**, *5*, 108–119. [CrossRef]

33. Adesemoye, A.O.; Torbert, H.A.; Kloepper, J.W. Enhanced plant nutrient use efficiency with PGPR and AMF in an integrated nutrient management system. *Can. J. Microbiol.* **2008**, *54*, 876–886. [CrossRef]

34. Subramanian, K.S.; Santhanakrishnan, P.; Balasubramanian, P. Responses of field grown tomato plants to arbuscular mycorrhizal fungal colonization under varying intensities of drought stress. *Sci. Hortic.* **2006**, *107*, 245–253. [CrossRef]

35. Malviya, M.K.; Sharma, A.; Pandey, A.; Rinu, K.; Sati, P.; Palni, L.M.S. Bacillus subtilis NRRL B-30408: A potential inoculant for crops grown under rainfed conditions in the mountains. *J. Soil Sci. Plant Nutr.* **2012**, *12*, 811–824. [CrossRef]

36. Monokrousos, N.; Papatheodorou, E.M.; Orfanoudakis, M.; Jones, D.G.; Scullion, J.; Stamou, G.P. The effects of plant type, AMF inoculation and water regime on rhizosphere microbial communities. *Eur. J. Soil Sci.* **2020**, *71*, 265–278. [CrossRef]

37. Sakineh, A.; Ayme, S.; Akram, S.; Safaie, N. *Streptomyces strains* modulate dynamics of soil bacterial communities and their efficacy in disease suppression caused by Phytophthora capsici. *Sci. Rep.* **2021**, *11*, 9317.

38. Wang, J.; Xu, S.; Yang, R.; Zhao, W.; Zhu, D.; Zhang, X.; Huang, Z. *Bacillus amyloliquefaciens* FH-1 significantly affects cucumber seedlings and the rhizosphere bacterial community but not soil. *Sci. Rep.* **2021**, *11*, 12055. [CrossRef] [PubMed]

39. Troelstra, S.R.; Wagenaar, R.; Smant, W.; Peters, B.A.M. Interpretation of bioassays in the study of interactions between soil organisms and plants: Involvement of nutrient factors. *New Phytol.* **2001**, *150*, 697–706. [CrossRef]

40. Ratnayaka, D.D.; Brandt, M.J.; Johnson, K.M. Hydrology and Surface Supplies. In *Water Supply*; Butterworth-Heinemann: Oxford, UK, 2009; pp. 63–107. ISBN 978-0-7506-6843-9.

41. Carter, M.R.; Gerard, G. *Soil Sampling and Methods of Analysis*; CRC Press: Boca Raton, FL, USA, 2007.

42. Koske, R.E.; Gemma, J.N. A modified procedure for staining roots to detect VA mycorrhizas. *Mycol. Res.* **1989**, *92*, 486–488. [CrossRef]

43. Orfanoudakis, M.; Wheeler, C.T.; Hooker, J.E. Both the arbuscular mycorrhizal fungus Gigaspora rosea and Frankia increase root system branching and reduce root hair frequency in Alnus glutinosa. *Mycorrhiza* **2010**, *20*, 117–126. [CrossRef] [PubMed]

44. Trouvelot, A.; Kough, J.L.; Gianinazzi-pearson, V. Mesure du taux de mycorhization VA d'un systeme radiculaire. Recherche de methodes d'estimation ayant une signification fonctionnelle. In *Physiological and Genetical Aspects of Mycorrhizae: Proceedings of the 1st European Symposium on Mycorrhizae, Dijon, 1–5 July 1985*; Gianinazzi, S., Gianinazzi-Pearson, V., Eds.; Institut National de la Recherche Agronomique, INRA: Paris, French, 1986; pp. 217–221.

45. Available online: http://www2.dijon.inra.fr/mychintec/Mycocalc-prg/download.html (accessed on 20 March 2019).

46. Papadopoulou, E.S.; Karpouzas, D.G.; Menkissoglu-Spiroudi, U. Extraction parameters significantly influence and the quantity and the profile of PLFAs extracted from soil. *Microb. Ecol.* **2011**, *62*, 704–714. [CrossRef]

47. Ntalli, N.; Monokrousos, N.; Rumbos, C.; Kontea, D.; Zioga, D.; Argyropoulou, M.D.; Menkissoglu-Spiroudi, U.; Tsiropoulos, N.G. Greenhouse biofumigation with *Melia azedarach* controls *Meloidogyne* spp. and enhances soil biological activity. *J. Pest Sci.* **2018**, *91*, 29–40. [CrossRef]

48. Stamou, G.; Konstadinou, S.; Monokrousos, N.; Mastrogianni, A.; Orfanoudakis, M.; Hassiotis, C.; Menkissoglu-Spiroudi, U.; Vokou, D.; Papatheodorou, E.M. The effects of arbuscular mycorrhizal fungi and essential oil on soil microbial community and N-related enzymes during the fungal early colonization phase. *AIMS Microbiol.* **2017**, *3*, 938–959. [CrossRef]

49. Olsson, P.A. Signature fatty acids provide tools for determination of the distribution and interactions of mycorrhizal fungi in soil. *FEMS Microb. Ecol.* **2006**, *29*, 303–310. [CrossRef]

50. Allison, S.D.; Jastrow, J.D. Activities of extracellular enzymes in physically isolated fractions of restored grassland soils. *Soil Biol. Biochem.* **2006**, *38*, 3245–3256. [CrossRef]

51. Sinsabaugh, R.L.; Reynolds, H.; Long, T.M. Rapid assay for amidohydrolase (urease) activity in environmental samples. *Soil Biol. Biochem.* **2000**, *32*, 2095–2097. [CrossRef]

52. Maestre, F.T.; Quero, J.L.; Gotelli, N.J.; Escudero, A.; Ochoa, V.; Delgado-Baquerizo, M.; García-Gómez, M.; Bowker, M.A.; Soliveres, S.; Escolar, C.; et al. Plant species richness and ecosystem multifunctionality in global drylands. *Science* **2012**, *335*, 214–218. [CrossRef]

53. Ambrosini, A.; de Souza, R.; Passaglia, L.M.P. Ecological role of bacterial inoculants and their potential impact on soil microbial diversity. *Plant Soil* **2016**, *400*, 193–207. [CrossRef]

54. Saia, S.; Colla, G.; Raimondi, G.; Di Stasio, E.; Cardarelli, M.; Bonini, P.; Vitaglione, P.; De Pascale, S.; Rouphael, Y. An endophytic fungi-based biostimulant modulated lettuce yield, physiological and functional quality responses to both moderate and severe water limitation. *Sci. Hortic.* **2019**, *256*, 108595. [CrossRef]

55. Hueso, S.; García, C.; Hernández, T. Severe drought conditions modify the microbial community structure, size and activity in amended and unamended soils. *Soil Biol. Biochem.* **2012**, *50*, 167–173. [CrossRef]

56. Hartmann, M.; Brunner, I.; Hagedorn, F.; Bardgett, R.D.; Stierli, B.; Herzog, C.; Chen, X.; Zingg, A.; Graf-Pannatier, E.; Rigling, A.; et al. A decade of irrigation transforms the soil microbiome of a semi-arid pine forest. *Mol. Ecol.* **2017**, *26*, 1190–1206. [CrossRef]

57. Schimel, J.P. Life in Dry Soils: Effects of Drought on Soil Microbial Communities and Processes. *Annu. Rev. Ecol. Evol. Syst.* **2018**, *49*, 409–432. [CrossRef]

58. Landesman, W.J.; Dighton, J. Response of soil microbial communities and the production of plant-available nitrogen to a two-year rainfall manipulation in the New Jersey Pinelands. *Soil Biol. Biochem.* **2010**, *42*, 1751–1758. [CrossRef]

59. Fanin, N.; Kardol, P.; Farrell, M.; Nilsson, M.-C.; Gundale, M.J.; Wardle, D.A. The ratio of Gram-positive to Gram-negative bacterial PLFA markers as an indicator of carbon availability in organic soils. *Soil Biol. Biochem.* **2019**, *128*, 111–114. [CrossRef]

60. Fierer, N.; Schimel, J.P.; Holden, P.A. Variations in microbial community composition through two soil depth profiles. *Soil Biol. Biochem.* **2003**, *35*, 167–176. [CrossRef]

61. Breulmann, M.; Masyutenko, N.P.; Kogut, B.M.; Schroll, R.; Dorfler, U.; Buscot, F.; Schulz, E. Short-term bioavailability of carbon in soil organic matter fractions of different particle sizes and densities in grassland ecosystems. *Sci. Total Environ.* **2014**, *497–498*. [CrossRef]

62. Citernesi, A.S.; Filippi, C.; Bagnoli, G.; Giovannetti, M. Effects of the antimycotic molecule Iturin A2, secreted by *Bacillus subtilis* strain M51, on arbuscular mycorrhizal fungi. *Microbiol. Res.* **1994**, *149*, 241–246. [CrossRef]

63. Johnson, N.C.; Graham, J.H.; Smith, F.A. Functioning of mycorrhizal associations along the mutualism-parasitism continuum. *New Phytol.* **1997**, *135*, 575–585. [CrossRef]
64. Janos, D.P. Plant responsiveness to mycorrhizas differs from dependence upon mycorrhizas. *Mycorrhiza.* **2007**, *442*, 75–91. [CrossRef] [PubMed]
65. Toro, M.; Azcón, R.; Barea, J.M. Improvement of arbuscular mycorrhiza development by inoculation of soil with phosphate-solubilizing rhizobacteria to improve rock phosphate bioavailability (32P) and nutrient cycling. *Appl. Environ. Microbiol.* **1997**, *63*, 4408–4412. [CrossRef] [PubMed]
66. Tian, L.; Shi, S.; Ma, L.; Zhou, X.; Luo, S.; Zhang, J.; Lu, B.; Tian, C. The effect of *Glomus intraradices* on the physiological properties of *Panax ginseng* and on rhizospheric microbial diversity. *J. Ginseng Res.* **2019**, *43*, 77–85. [CrossRef]
67. Konstantinou, S.; Monokrousos, N.; Kapagianni, P.; Menkissoglu-Spiroudi, U.; Gwynn-Jones, D.; Stamou, G.P.; Papatheodorou, E.M. Instantaneous responses of microbial communities to stress in soils pretreated with *Mentha spicata* essential oil and/or inoculated with arbuscular mycorrhizal fungus. *Ecol. Res.* **2019**, *34*, 701–710. [CrossRef]
68. Hodge, A.; Fitter, A.H. Substantial nitrogen acquisition by arbuscular mycorrhizal fungi from organic material has implications for N cycling. *Proc. Natl. Acad. Sci. USA* **2010**, *107*, 13754–13759. [CrossRef]
69. Mechri, B.; Manga, A.G.B.; Tekaya, M.; Attia, F.; Cheheb, H.; Meriem, F.B.; Braham, M.; Boujnah, D.; Hammami, M. Changes in microbial communities and carbohydrate profiles induced by the mycorrhizal fungus (*Glomus intraradices*) in rhizosphere of olive trees (*Olea europaea* L.). *Appl. Soil Ecol.* **2014**, *75*, 124–133. [CrossRef]
70. Kim, Y.; Jordan, D.; McDonald, G.A. Effect of phosphate-solubilizing bacteria and vesicular-arbuscular mycorrhizae on tomato growth and soil microbial activity. *Biol. Fertil. Soils* **1997**, *26*, 79–87. [CrossRef]
71. Stone, M.M.; Plante, A.F.; Casper, B.B. Plant and nutrient controls on microbial functional characteristics in a tropical Oxisol. *Plant Soil* **2013**, *373*, 893–905. [CrossRef]
72. He, C.; Wang, W.; Hou, J. Plant growth and soil microbial impacts of enhancing licorice with inoculating dark septate endophytes under drought stress. *Front. Microbiol.* **2019**, *10*, 2277. [CrossRef] [PubMed]

Article

Effects of *Bacillus subtilis* and *Pseudomonas fluorescens* Inoculation on Attributes of the Lettuce (*Lactuca sativa* L.) Soil Rhizosphere Microbial Community: The Role of the Management System

Eirini Angelina [1], Efimia M. Papatheodorou [1,2,*], Triantafyllia Demirtzoglou [3] and Nikolaos Monokrousos [1,*]

[1]   Lab of Molecular Ecology, University Center of International Programmes of Studies, International Hellenic University, 57001 Thessaloniki, Greece; e.angelina@ihu.edu.gr
[2]   Department of Ecology, School of Biology, Aristotle University, 54124 Thessaloniki, Greece
[3]   Department of Research & Development, of HUMOFERT Co., 14452 Athens, Greece; fdemi@humofert.gr
*   Correspondence: papatheo@bio.auth.gr (E.M.P.); nmonokrousos@ihu.gr (N.M.);
    Tel.: +30-231-0998-313 (E.M.P.); +30-231-0807-572 (N.M.)

Received: 25 August 2020; Accepted: 17 September 2020; Published: 19 September 2020

Abstract: Inoculation with beneficial microbes has been proposed as an effective practice for the improvement of plant growth and soil health. Since soil acts as a physicochemical background for soil microbial communities, we hypothesized that its management will mediate the effects of microbial inoculants on the indigenous soil microbes. We examined the effects of bacterial inoculants [*Bacillus subtilis* (Ba), *Pseudomonas fluorescens* (Ps), and both (BaPs)] on the growth of *Lactuca sativa* cultivated in soils that originated from an organic maize (OS) and a conventional barley (CS) management system. Moreover, the biomass and the community structure of the rhizosphere microbial communities and the soil enzyme activities were recorded. The root weight was higher in CS than OS, while the foliage length was greater in OS than CS treatments. Only in OS pots, inoculants resulted in higher biomasses of bacteria, fungi, and actinomycetes compared to the control with the highest values being recorded in Ps and BaPs treated soils. Furthermore, different inoculants resulted in different communities in terms of structure mainly in OS soils. For soil enzymes, the effect of the management system was more important due to the high organic matter existing in OS soils. We suggest that for microbial inoculation to be effective it should be considered together with the management history of the soil.

Keywords: microbial inoculants; soil enzyme activities; soil microbes

---

1. Introduction

Several conventional farming management practices, such as extensive use of inorganic fertilizers and pesticides, have a significant impact on the environment [1,2]; increasing the greenhouse effect [3], reducing biodiversity [4], and enhancing toxicity in the food chain [5]. Such practices could have serious impacts on the soil environment and more specifically on soil microbial communities. Insecticides, such as pyrethroids altered the composition of the soil microbial community [6], fungicides lowered the abundance of *Bacillus* species by 63% [7] or affected negatively the populations of soil fungi and bacteria [8]. Moreover, pesticide residues could remain in the soil for a long period (even for eight years) after their application [9]. On the contrary, organic farming practices, such as manure application and lack of tillage could improve soil fertility, increase microbial biomass [10] and activity [11], and elevate soil microbial diversity [12], thus affecting the soil microbial community structure [13].

The need for more environmentally friendly viable alternatives to traditional fertilizers for enhancing plant productivity and improving soil quality is growing [14]. The use of bacteria and fungi as inoculants for enhancing crop production is a sustainable approach that has gained ground over the last years, as inoculation with beneficial microorganisms can reduce the requirements for chemical fertilizers and pesticides [15]. Inoculation of soil with plant growth-promoting rhizobacteria (PGPR) proved to have beneficial effects on plant growth, as PGPR have a wide range of activities such as the enhancement of nutrient availability, the biocontrol of soil-borne pathogens, the release of plant hormones, and the alleviation of various types of abiotic stress [16,17]. PGPR are either exogenous bacteria introduced into agricultural ecosystems or autochthonous bacteria that are being enhanced. Both types of bacteria act positively upon plant development in annual and perennial crops and under different abiotic conditions [18,19]. The genera of PGPR that are most often used in sustainable agriculture are *Azospirillum, Azotobacter, Bacillus, Enterobacter, Pseudomonas,* and *Rhizobium* [20,21].

Among the PGPR genera, several *Pseudomonas* and *Bacillus* species are the most widely known ones and those that have been frequently commercialized due to their survival within a diverse range of biotic and abiotic environments [22]. *Pseudomonas* is one of the most abundant Gram-negative soil bacterial genus in rhizosphere soil [23]. *Pseudomonas* species can either suppress the growth of soil pathogens by inducing the plant's systemic resistance [24] and by producing antibiotics and siderophores [23] or play a vital role in plant growth by modifying plant phytohormone concentrations [25]. The *Bacillus*-based inoculants have demonstrated significant biocontrol properties by inhibiting the growth of plant pathogens (e.g., *Rhizoctonia solani* and *Botrytis cinerea*) due to the secretion of antifungal compounds [26], or toxins [27], or by inducing the plant's defense systems through the synthesis of plant growth hormones [28]. *Bacillus* species can also increase the concentration of soil essential nutrients, such as P and N, by converting complex nutrient compounds to more simple available forms easily accessible by the plant roots [29].

Over the last 10 years, many studies have been conducted to investigate the effect of PGPR inoculants on various lettuce varieties aiming at increasing productivity and nutritional quality, improving plant abiotic stress tolerance, facilitating nutrient uptake, and controlling pests [30,31]. *Bacillus* and *Pseudomonas*-based inoculants have presented promising results for lettuce crop yield [32,33]. Nevertheless, only a few studies have focused on the effect of PGPR inoculants' application on the indigenous soil microbial communities [33,34]. The incorporation of PGPR inoculants into the soil could have a significant impact on indigenous soil microbial community because these exogenous microbes are involved in a wide range of interactions with the autochthonous microbes such as competition, synergy, or prey-predator interactions. For that reason, PGPR may increase, decrease, or not affect the indigenous microorganisms [35].

The present study aimed to assess the effect of *Pseudomonas fluorescens* and *Bacillus subtilis* inoculation on the biomass, community composition, and functionality of the lettuce rhizosphere microbial community in plants cultivated in soils that originated from two different management systems (conventional barley and organic maize cultivations). Since soil acts as a physicochemical template for microbial communities due to nutrient supply (availability and movement of nutrients to microbes) and by providing refugia for them, we expect that inoculants applied in the conventional system will impose suppressing effects on microbial biomass, community, and enzyme activity resulting in a narrower range of responses of microbial attributes compared to relevant effects recorded in the organic management system. We hypothesize that in the conventional system, exogenous microbes would be involved mainly in competitive interactions due to limitations in resource availability. In contrast, the lack of nutrient limitation and tillage effect in the organic system would enable inoculants to express their influence, a fact that would be identified by a wide range of inoculant-specific responses of soil microbial community structure, biomass, and functionality in this system.

2. Materials and Methods

2.1. Experimental Design and Sampling

The experimental design consisted of two management systems [organically (OS) and conventionally (CS) managed soils] × four inoculation treatments [*Bacillus subtilis* (Ba), *Pseudomonas fluorescens* (Ps), *Bacillus* and *Pseudomonas* consortia (BaPs), and non-inoculated (control, C)], with four replicates per treatment, giving a total of 32 pots arranged in a randomized block design. The soils used in the experiment were collected from two different fields in Lithia, Kastoria, Western Macedonia, Greece (40.5208° N, 21.4097° E). The OS originated from an organically managed maize crop cultivation, involving the application of sheep manure (1 t ha^{-1}) every 2 years, no-tillage, and no use of pesticides for more than 10 years, while plant residues were left on the soil surface after harvest. The CS originated from a conventionally managed barley crop cultivation that involved the use of inorganic fertilizers (NPK 20-10-0; 20–25 kg ha^{-1}), extensive tillage, and extensive use of pesticides (Phenylpyrazolines), insecticides (Pyrethroids), and fungicides (Triazoles and Pyrazoles) for more than 10 years. We selected these specific fields as they presented similar soil texture properties [CS: SL soil texture (S: 44%, C: 6%, Si: 50%), pH: 6.60 and EC: 0.93 mS cm^{-1} and OS: SL soil texture (S: 24%, C: 14%, Si: 62%), pH 7.81 and EC 1.19 mS cm^{-1}]. The soil collected from the top 15 cm from each management system was passed through a 2 mm sieve and 1500 g of either OS or CS were put into plastic pots (15 cm diameter and 11 cm depth) that have been previously disinfected with ethanol. Before the experiment, lettuce seeds (*Lactuca sativa* var. *longifolia*) were surface sterilized with sodium hypochlorite 10% (*v/v*) for 20 min and then washed repeatedly with deionized water; then the sterilized seeds were sowed in seedbeds containing soil from either CS or OS to grow for 60 days, before being transplanted in the experimental pots. The last watering was 24 h before their transplantation. Five seeds of lettuce were planted in each seedbed. After their germination, plants were thinned, leaving only one lettuce plant per seedbed and these were transplanted to pots. The experiment was conducted in outdoor conditions under natural light and temperature conditions. The experimental period was November to December with mean temperatures 12.1 and 6.3 °C, respectively (Florina Meteorological Station). Both seedling trays and the plastic pots (after the transplantation) were watered every 3 days with 50 mL of deionized/distilled water to keep the soil water content around 10% *w/w* that imposes no limitation to plant growth [36]. A destructive sampling was conducted 60 days after the transplantation.

We manually separated the roots from the soil, while the rhizosphere soil attached to the roots was collected on a sterilized surface. Fresh rhizosphere soil samples were sieved to remove small roots and then were stored at 4 °C until further analysis. We analyzed the rhizosphere soil samples for microbial community structure, abundance, and extracellular enzyme activity as well. We also determined lettuce root weight, foliage weight, and foliage length.

2.2. Inoculum Preparation

Bacillus subtilis and *Pseudomonas fluorescens* were isolated from different soil samples collected from the rhizosphere of the *Solanum lycopersicum (tomato)* and the *Gossypium hirsutum* (cotton), respectively. The rhizobacteria were isolated on their respective media; *B. subtilis* on nutrient agar and *P. fluorescence* on King's B agar. The pure isolates were further cultured on new plates for colony morphology. The colony morphological characteristics including margins, shape, raised, and pigmentation were observed. Biochemical studies such as Gram reaction, endospore staining catalase production, oxidase, indole production, citrate utilization, methyl red test were carried out for the confirmation of *Pseudomonas fluorescens* and *Bacillus subtilis* [37]. The preliminary identification was confirmed by molecular classification with polymerase chain reaction (PCR), using ERIC1f/ERIC2 (ERIC 1: 5′-ATGTAAGCTCCTGGGGATTCAC-3′, ERIC2: 5′-AAGTAAGTGACT GGGGTGAGCG-3′) oligonucleotide primers [38], and based on the sequence of the 16S rDNA gene. The preparation of the inocula involved the growth of the bacteria in nutrient broth (100 mL) for 48 h on a rotary shaker at 28–30 °C. After incubation, the inoculum reached the final cell density of about 1010 cfu cm^{-3} by diluting the bacterial cultures with a 0.85% NaCl water solution. Bacterial inoculation was applied once. Bacteria were applied by drenching of the soil in a dose of 10 cm^3 of bacterial suspension per 1 cm^3 of soil. The suspensions of *B. subtilis*, *P. fluorescens*, or their mixture (1:1 *v/v*) were used for inoculation. The concentration of the suspensions of B and P was 5 mL of the inoculant in 45 mL of water, while the consortia suspension of Ba/Ps was constituted of 2 × 2.5 mL and 45 mL of water. The control was treated with 50 mL of water.

2.3. Analyses of Soil Chemical and Biochemical Variables

Soil pH was determined by the 1:2 soil/water suspension method. Soil texture was estimated by the Bouyoucos method. Soil organic C was determined by a wet oxidation titration procedure using an acid dichromate system. NO3–N concentration was determined by distillation and titration. Mg and K were estimated in soil extracts by the atomic absorption spectrophotometer (Perkin–Elmer 2380). For inorganic P the Olsen method was used. All soil chemical variables were determined as described in Allen [39].

2.4. Phospholipid Fatty Acid Analysis

Soil samples were analyzed for phospholipid (PLFAs) bioindicators according to the method presented in Ntalli et al. [40]. A Trace GC Ultra gas chromatograph (Thermo Finnigan, San Jose, CA, USA) coupled with a Trace ISQ mass spectrometry detector, an autosampler with a split–splitless injector, and an Xcalibur MS platform was used for the chromatographic separation and identification of the main components. We quantified each fatty acid (in nmol/g) by one-point calibrating against the GC response of the internal standard (19:0 methyl ester). Overall, in all the samples, we consistently found 24 fatty acid methyl esters which were included in all further analyses. These fatty acids were assigned to functional groups as follows [40,41]: i-15:0, a-15:0, 15:0, i-16:0, i-17:0, 17:0 (Gram-positive bacteria); 16:1ω9c, 16:1ω9t, cy17:0 (Gram-negative bacteria); 10Me16:0, 10Me17:0, 10Me18:0 (actinomycetes); 18:2ω9,12 (fungi); 20:0, 22:0, and 24:0 (microeukaryotes, e.g., algae, nematodes). The remaining PLFAs may derive from several sources and were considered only for the estimation of total microbial biomass. For example, 18:1ω9t, 18:1ω9c may derive from both Gram-negative bacteria and fungi, 16:0 from bacteria and fungi, while 11:0, 13:0, 14:0, 18:0, 18:2ω6 are mainly of microbial origin. We also estimated the bacteria/fungi (B/F) and Gram$^+$/Gram$^-$ ratios.

2.5. Enzymatic Activity Analysis

Urease, β-glucosidase, and acid phosphatase play key roles in the N, C, and P cycles, respectively [42]. We determined the acid phosphatase (AP) and β-glucosidase (BG) activity according to the procedures of Allison and Jastrow [43], modified for 96-well microplates. We used 5 mM p-nitrophenyl-phosphate and 5 mM p-nitrophenyl-β-glucopyranoside substrate solutions for AP and BG, respectively. The p-nitrophenol (PNP) reaction product from the AP and BG assays was measured at 405 nm. The method of Sinsabaugh et al. [44] was used for the estimation of the urease activity. The urea concentration in the wells was 20 mM. The ammonium released by the reaction was measured spectrophotometrically at 610 nm.

2.6. Data Analysis

We applied a two-way analysis of variance (ANOVA) to determine the effect of the management system, inoculum type, and their interaction, on plant variables, microbial biomasses, and enzyme activity data. In the case of significant effects, a post hoc test was performed at $p < 0.05$. An independent t-test was used to compare the mean soil physicochemical variables between the two management systems (CS and OS). Before analyses, we logarithmically transformed the data when considered necessary, to meet the assumptions of t-test and ANOVA.

To further explore whether the management system or inoculum type exerted the greatest influence on microbial community structure, we applied a principal component analysis (PCA). Moreover, we applied the analysis of similarities (ANOSIM), based on the similarity index of Bray-Curtis, to the data of individual PLFAs to detected similarities in community structure between the treatments.

3. Results

A t-test revealed that the OS pots presented significantly higher values of pH, organic matter, nitric nitrogen, P, K, and Mg compared to the CS. On the contrary, the concentration of Zn, Mn, and B did not differ significantly (Table 1).

Table 1. Mean values (±SE) of the soil physicochemical variables in conventionally and organically managed systems. The independent *t*-test was used for the comparison of the mean values. (**: $p < 0.01$; ***: $p < 0.001$; ns: Non-significant).

	Conventionally Managed System	Organically Managed System	*p*-Value
pH	6.60 ± 0.054	7.81 ± 0.035	***
EC (mS cm^{-1})	0.93 ± 0.107	1.19 ± 0.049	ns
Organic matter (%)	2.29 ± 0.159	3.24 ± 0.351	**
Nitric Nitrogen (mg kg^{-1})	14.13 ± 6.955	32.36 ± 2.377	**
Pext (mg kg^{-1})	26.33 ± 0.609	133.75 ± 3.609	***
K (mg kg^{-1})	153.66 ± 9.769	347.33 ± 10.170	***
Mg (mg kg^{-1})	188.33 ± 7.264	352.33 ± 7.838	***
Zn (mg kg^{-1})	5.64 ± 0.065	5.61 ± 0.196	ns
Mn (mg kg^{-1})	28.04 ± 2.256	26.40 ± 1.307	ns
B (mg kg^{-1})	0.84 ± 0.070	0.81 ± 0.051	ns

The effects of the management system, inoculum type, and their interactions on plant growth are shown in Supplementary Materials Table S1 and Figure 1. The root weight and the foliage length were affected only by the management type. The root weight was higher in CS than in OS, while the foliage length was greater in OS than in CS.

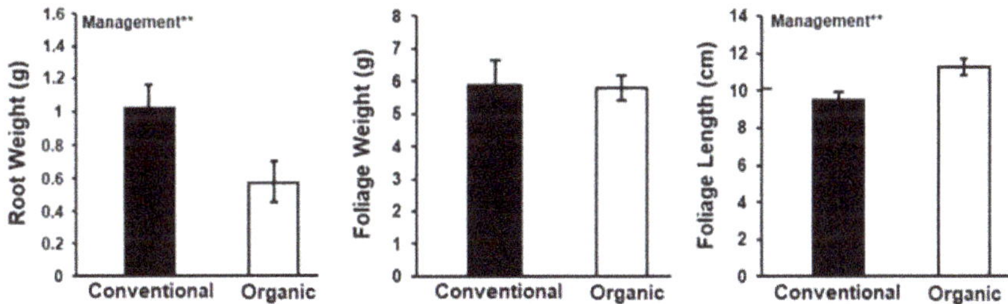

Figure 1. Mean values (±SE) of root weight, shoot weight, and shoot length in conventionally and organically managed systems. "Management" on the top of the graphs, indicates a significant effect of the management type, as revealed by two-way analysis of variance (ANOVA). (**: $p < 0.01$)

For the abundance of most microbial groups (Gram-positive and Gram-negative bacteria, actinomycetes, fungi), the interaction of management system × inoculum type was significant (Table S2). The same was held also for the total microbial biomass. The effect of inoculum type was much more pronounced in OS pots, while in the CS pots no differences in biomasses were recorded between the different inoculants (Figure 2). Ba or Ps inoculants resulted in significantly higher biomasses of Gram-positive bacteria in the OS pots compared to CS ones, while the BaPs inoculant induced similar effects in both management systems. The biomass of Gram-negative bacteria was higher in OS pots inoculated with Ps or with both inoculants (Figure 2). For actinomycetes, the highest biomass was recorded in Ps inoculated organic pots. The highest value of fungal biomass was in Ps OS pots, while the biomass in BaPs pots was similar to Ba and Ps ones. Finally, the total microbial biomass showed non-significant differences in organic pots inoculated with Ba, Ps, or BaPs. The abundance of microeukaryotes was higher in the CS samples, regardless of the inoculum type. The bacteria/fungi ratio was affected only by the management system; the organic pots presented significantly higher values than the conventional ones (Figure 3). On the contrary, inoculants resulted in increased values of Gram$^+$/Gram$^-$ ratios compared to the controls in both CS and OS pots. The ratios recorded in the Ps and BaPs conventional pots were the highest, whereas in organic pots the Ba inoculant exhibited the highest ratios.

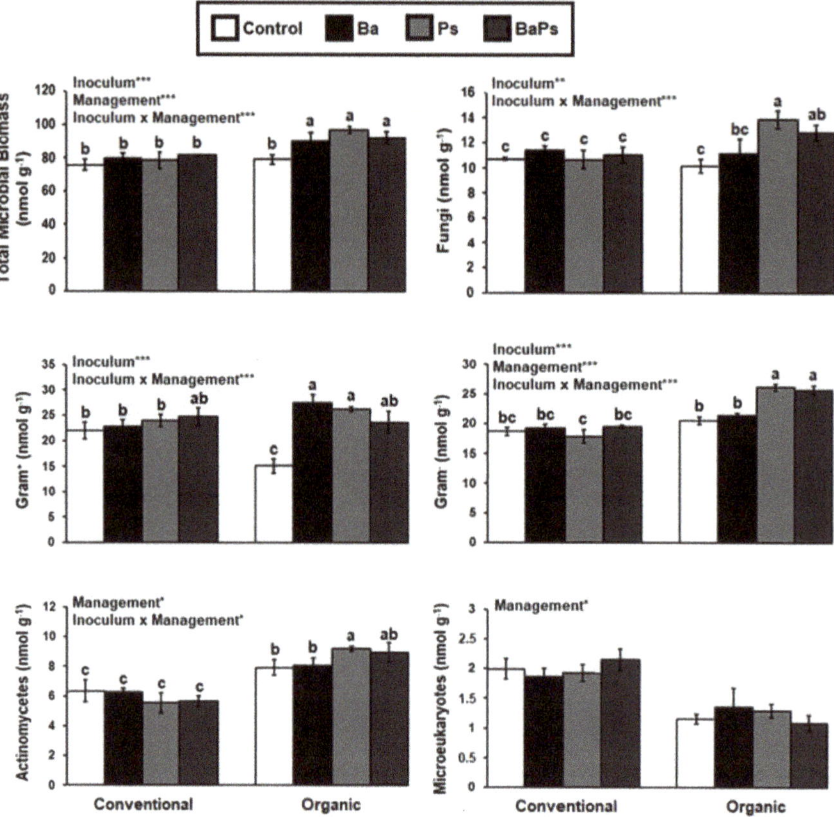

Figure 2. Mean biomass (±SE) of total microbes, fungi, Gram$^+$ and Gram$^-$ bacteria, actinomycetes, and microeukaryotes in the conventionally and organically managed systems, at the four treatments. "Inoculum", "Management", and "Inoculum × Management" at the top of the graphs, indicate a significant effect of these factors and their interaction, respectively, as revealed by two-way ANOVA (*: $p < 0.05$, **: $p < 0.01$, ***: $p < 0.001$, for all cases $n = 4$). The different letters above bars represent statistically significant differences between treatments described by the inoculum type × management system as emerged from Fisher's test (Fisher post hoc; a: Corresponds to the highest value).

For acid phosphatase and β-glucosidase, the independent effect of the management system and inoculum type was significant (Table S3), while for urease only the effect of management system was significant. All enzymes presented higher activity values in the OS compared to the CS pots (Figure 4). While for phosphatase and urease the co-inoculated pots exhibited significantly higher activity compared to the controls in both management systems.

PCA depicting the ordination of samples and individuals PLFAs is presented in Figure 5. The first axis explained 36.7% of the data variability and the second one explained 16.9% (53.6% in total). The OS control samples were ordinated at the right end of the first axis, while all the other samples from the OS treatments (Ba, Ps, BaPs) were ordinated at the left side of the first axis showing a positive correlation to all PLFA biomarkers, that correspond to Gram-positive, Gram-negative bacteria, actinomycetes, and fungi. Along the second axis, the samples of CS are ordinated in the upper part showing a positive correlation to PLFA biomarkers describing microeukaryotes (20:0, 22:0, 24:0) and were clearly separated from the OS samples that are ordinated in the lower part. The classification of all the CS samples close to each other indicated that the addition of any type of inoculum did not affect significantly the structure of the soil microbial community. On the contrary, the distance between the control and the inoculated OS samples showed a great impact of the inoculants on the microbial community structure.

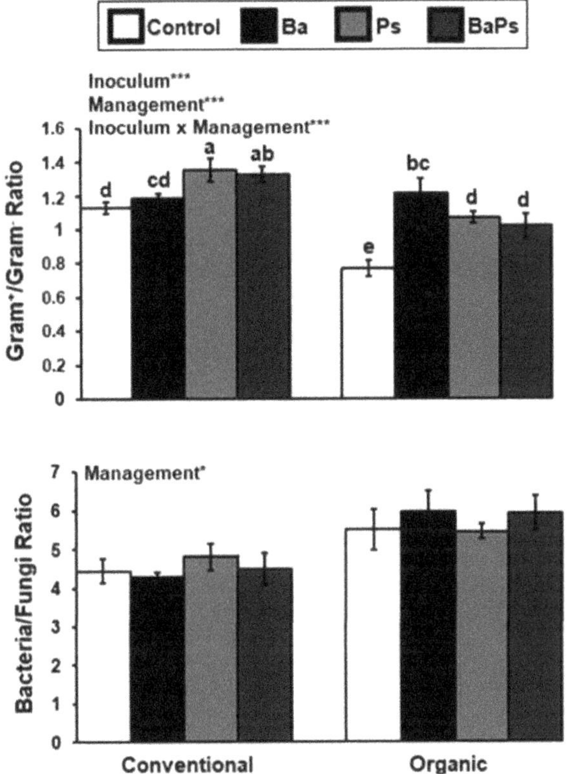

Figure 3. Mean values (±SE) of microbial ratios in the conventionally and organically managed systems, at the four treatments. "Inoculum", "Management", and "Inoculum × Management" at the top of the graphs, indicate a significant effect of these factors and their interaction, respectively, as revealed by two-way ANOVA (*: $p < 0.05$, ***: $p < 0.001$, for all cases $n = 4$). The different letters above bars represent statistically significant differences between treatments described by the inoculum type × management system as emerged from Fisher's test (Fisher post hoc; a: Corresponds to the highest value).

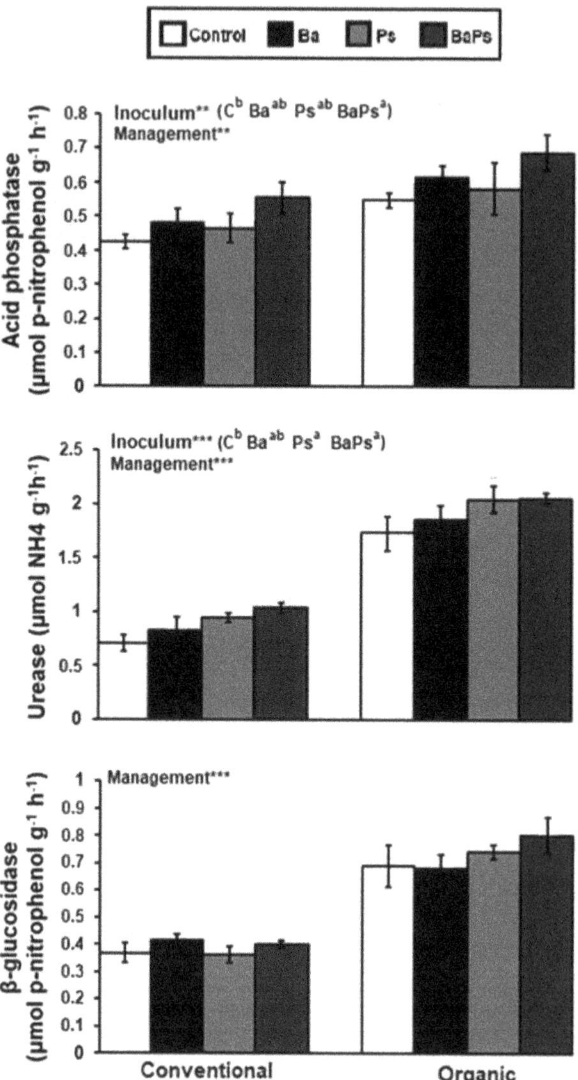

Figure 4. Mean values (±SE) of acid phosphatase, urease, and β-glucosidase enzymes in the conventionally and organically managed systems, at the four treatments. "Inoculum", "Management" at the top of the graphs, indicate a significant effect of these factors, as revealed by two-way ANOVA (**: $p < 0.01$, ***: $p < 0.001$, for all cases $n = 4$). Superscript letters in parentheses following the term "Inoculum" denote the significant differences between inoculum types (C: Control, Ba: *B. subtilis*, Ps: *P. fluorescens* and BaPs: *B. subtilis* and *P. fluorescens*) as emerged from Fisher's test (Fisher post hoc; a: Corresponds to the highest value).

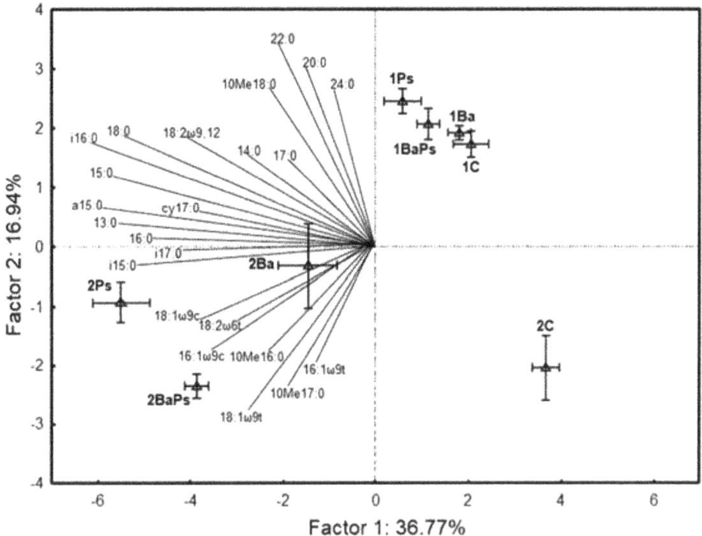

Figure 5. Ordination of the soil samples and the phospholipid (PLFA) biomarkers on a principal component analysis (PCA) biplot. Each point corresponds to the mean value of the loadings of the four samples belonging to the same treatment at the first and second axis (the first symbol corresponds to the management system (1: Conventional system (CS), 2: Organic system (OS)) and the second symbol corresponds to the inoculum treatment; Ba: *B. subtilis*; Ps: *P. fluorescens*; BaPs: Both inoculants; C: Control). Error bars indicate standard errors at both axes ($n = 4$).

To get a deeper insight into the differences in microbial community structure between treatments, we applied an ANOSIM analysis on individual PLFAs. As shown in Table 2, in organic pots, the Ps community structure differed significantly from the Ba community and both differed from the control. Moreover, the community in co-inoculated organic pots differed from the control. On the contrary, in conventional pots, Ba and Ps communities' structure was similar to the control community. In this management system, when both inoculants were added together (BaPs), a significantly different microbial community structure emerged compared to the control.

Table 2. Results of ANOSIM analysis (p-values) based on the similarity index of Bray-Curtis applied to the data of individual PLFAs. The numbers in bold ($p < 0.05$) indicate statistically significant differences (the first symbol is the acronym for the management system (CS: Conventional system; OS: Organic system) and the second symbol corresponds to the inoculum type; Ba: *B. subtilis*, Ps: *P. fluorescens*, BaPs: Both inoculants; C: Control).

CS-C		0.174	0.116	**0.024**	**0.030**	**0.027**	**0.026**	0.061
CS-Ba	0.174		0.318	0.284	**0.029**	0.061	**0.030**	**0.027**
CS-Ps	0.116	0.318		0.055	**0.030**	**0.028**	**0.030**	**0.027**
CS-BaPs	**0.024**	0.284	0.055		**0.029**	**0.028**	**0.029**	**0.028**
OS-C	**0.030**	**0.029**	**0.030**	**0.029**		**0.028**	**0.031**	**0.033**
OS-Ba	**0.027**	0.061	**0.028**	**0.028**	**0.028**		**0.026**	0.057
OS-Ps	**0.026**	**0.030**	**0.030**	**0.029**	**0.031**	**0.026**		0.222
OS-BaPs	0.061	**0.027**	**0.027**	**0.028**	**0.033**	0.057	0.222	

4. Discussion

4.1. Plant Growth

The effects of different microbial inoculants on lettuce growth and attributes of the indigenous microbial community were examined in two soils of similar soil texture but of different management histories (organic maize and conventional barley).

The root biomass of the lettuce plants grown in the conventional soil was increased significantly in relation to the organic one while lettuce's growth was unaffected by inoculation. As the conventional soil had lower soil nutrient concentrations (e.g., organic matter, nitric N, P, K, and Mg), plants may have allocated more photosynthetic products to their roots to increase their exploitative capacity [45]. Previous studies showed that the lettuce plant's inoculation with *Bacillus* and *Pseudomonas* strains, either separately or as co-inoculants resulted in increased plant growth parameters compared to the control [33,46]. In those studies, different results were recorded, as the temperature of the experimental areas was higher than 20 °C while the pots had been fertilized before the plant transplantation either with chemical fertilizers or compost and green manure. In our experiment, the poor plant growth performance in the inoculated treatments could be related to the low winter air and soil temperatures (6 to 12 °C). Plant inoculation during winter has frequently resulted in restricted PGPR colonization [47,48]. Specifically, *Bacillus* species are favored by temperatures above 20 °C, enhancing plant growth through hormone production and nutrient solubilization [48]. Along this line, Nguyen et al. (2019) [47] reported that the root growth of the inoculated plants with a *B. velezensis* strain was lower in October when the temperature was low and the natural daytime was short, in comparison to the root growth in May. Further, due to the short daytime in this experiment, plants grown in organically managed system pots with no limitation in nutrient availability probably increased significantly their foliage length to increase their photosynthetic surface. Contrary to *B. subtilis*, Lynch [49] detected functional growing cells of the psychrotrophic *P. fluorescens* at a temperature from 3 to 35 °C. Based on these studies, although there is a difference in the response of the two microbial species to the temperature spectrum, this had no consequence for the lettuce plant performance.

4.2. Microbial Biomass, Composition, and Activity

The inoculants in the conventionally managed pots had a limited effect on the biomass of all of the microbial groups; in most cases, the abundances did not differ significantly compared to the control. This may be attributed to the pesticide residues of the conventional management practice before this experiment. More specifically, the conventional management of barley crops included the extensive use of triazole fungicides (propiconazole), which reduced significantly the bacterial and fungal populations, even after long incubation periods [50]. On the contrary, in the organically managed pots, the application of the Ps and BaPs inoculants increased the abundance of most rhizosphere microbial groups, while the abundances in *B. subtilis*-inoculated pots were similar to the control.

The co-inoculation with *B. subtilis* and *P. fluorescens* enhanced the biomass of Gram-positive and Gram-negative bacteria in the organic system. This suggests a lack of competitive interactions between the inoculants and the indigenous microbes, which could be due to the enhanced availability of resources. Similarly, *P. fluorescens per se* increased the size of the populations of both Gram-positive and Gram-negative bacteria in the organic system. *P. fluorescens* is a Gram-negative opportunistic bacterium that adapts rapidly to the environment and becomes the dominant species [51]. Moreover, according to Ke et al. [52], *P. fluorescens* could affect positively the indigenous diazotroph populations, the majority of which are Gram-negative bacteria, while Kozdroj et al. [53] found that pots inoculated with *Pseudomonas* sp. presented higher populations of the slow-growing Gram-positive bacterial classes compared to the untreated controls. Apart from *P. fluorescens*, *B. subtilis per se* increased the biomass of Gram-positive bacteria in the organic system. This increase could be attributed to the increased *B. subtilis* biomass. However, the low temperatures prevailing during the experiment, compared to the optimum temperature for this species (30–35 °C, makes this explanation less possible. Alternatively,

B. subtilis could act synergistically with the other Gram–positive bacteria supporting their growth. The lack of *B. subtilis* effect on Gram-negative bacteria contradicts the results of Han et al. [54] who mentioned the antibacterial activity of *B. subtilis* against specific Gram-negative phytopathogens.

The abundance of microeukaryotes was affected only by the management type and exhibited higher values in conventional pots. This finding contradicted the idea that the increased biomass of microeukaryotes is strongly associated with an increased number of bacteria, due to the prey-predator relationship [55]. Nevertheless, our results may be explained by the increased plant's root biomass in the conventional pots, which may have led to an enhanced number of plant-feeding nematodes (parasitic and non-parasitic), which are associated with plant roots.

The addition of any of the three inoculants in the organic pots increased the fungal abundance, while the Ps treatment presented the highest values; similar results were recorded by Viollet et al. [56]. Inoculation of the organic soil with Ba resulted in increased fungal biomass, but not to the levels of the Ps treatment. This may be explained by the negative relationship between the *Bacillus* species and some strains of fungi, either by synthesizing chitinases, which are hydrolytic enzymes that break down glycosidic bonds in chitin, a component of the fungal cell walls or by producing antifungal lipopeptides [57]. The fact that the co-inoculated pots presented intermediate fungal biomass between the other inoculated treatments further supports our suggestion.

The Gram$^+$/Gram$^-$ ratio showed an idiosyncratic response to the management type × inoculum type treatment. The Ps and BaPs treatments in the pots with conventionally managed soil and the Ba ones of the organically managed pots had the highest values. According to Moeskops et al. [58], Gram-negative bacteria are negatively affected by chemical fertilizers. Another possible reason could be the capacity of the Gram-positive for spore formation under extreme conditions. Further in Ps conventional pots, *Pseudomonas* may have developed competitive interactions with the other Gram-negative bacteria suppressing their biomass and leading to the increase of the Gram$^+$/Gram$^-$ ratio. The dominance of Gram-positive bacteria when young roots are colonized with the inoculant containing the *Pseudomonas* genus has been reported by other researchers as well [59] and was attributed to the vulnerability of r-strategy microbes to the environmental perturbation caused by the *P. fluorescens* inoculant. These negative relationships were not developed in the organically managed pots probably due to the greater presence of resources. Specifically, in the organic system, the incorporation of *B. subtilis* seemed to result in synergistic relationships with the other Gram-positive bacteria.

Apart from the ratio among bacterial groups, dissimilarities in the structure of microbial communities were revealed. Nutrient adequacy in the organically managed pots facilitated the development of distinct microbial assemblages; community structure in the Ps inoculated pots differed significantly from the community in the Ba pots and both differed from the control. Moreover, the microbial community in the BaPs organic pots differed from the control. This result looks similar to the results of Nunan et al. [60] who demonstrated that the intrinsic differences in microbial communities of different structures and/or diversities were revealed under conditions of high nutrient availability. In contrast, the limited nutrient availability in the conventional system together with the pesticide residues may have weakened the effect of the inoculants. In the conventionally managed pots, only in the case of co-inoculation did a significantly different microbial community structure emerge.

Although microbial biomasses and community structure were affected by the interaction of management type × inoculum, the soil enzyme activity was affected separately by each one of the independent variables. For all enzymes, a higher activity was recorded in the organically managed pots. Extracellular enzymes can be bound on clay minerals and stabilized by soil organic matter via the formation of enzyme-clay or enzyme-humus complexes [61]. This binding protects enzymes from decomposition by microbial proteases. For acid phosphatase and urease, the lowest and highest activity was recorded in the control and BaPs inoculated pots, respectively. Kumar et al. [23] also reported that the co-inoculation with *Bacillus* and *Pseudomonas* strains showed a synergistic effect, resulting in increased enzymatic activity. The increased urease activity in treatments that involved inoculation with *P. fluorescens* (Ps and BaPs) could be explained by the increased values of Gram-negative bacteria (at least in the organically managed pots) since most urease-producing bacteria are Gram-negative [62]. In contrast, the activity of β-glucosidase was not affected by any of the inoculants indicating the independence of the carbon cycle on inoculation.

5. Conclusions

This study confirmed the central role of the soil management system for microbial inoculation. The management system exerted the only significant effect on plant growth and an independent significant effect on enzyme activity. However, it mediated the effects of inoculants on soil rhizosphere microbial biomass and community structure confirming partly our initial hypothesis. Specifically, *P. fluorescens* inoculation and co-inoculation with *B. subtilis* resulted in increased microbial biomasses only in the organic system. Further, in the organic system, new microbial assemblages emerged because of the inoculants. To conclude, we suggest that for microbial inoculation, which has been promoted as a suitable method for sustainable agriculture, to be beneficial for plant growth and soil health, it should be considered together with the management history of the soil. Our results suggest that microbial inoculants are most effective when applied to soils that were previously cultivated using more environmentally friendly methods such as organic matter incorporation, no tillage, and with limited use of pesticides.

Supplementary Materials: The following are available online at http://www.mdpi.com/2073-4395/10/9/1428/s1, Table S1: The effects of management system, inoculum type, and their interaction on plant growth parameters as revealed by two-way ANOVA. (**: $p < 0.01$, ns: non-significant); Table S2: The effects of management system, inoculum type, and their interaction on microbial groups and PLFAs ratios as revealed by Two-way ANOVA (*: $p < 0.05$, **: $p < 0.01$, ***: $p < 0.001$, ns: non-significant); Table S3: The effects of management system, inoculum type, and their interaction on microbial groups and PLFAs ratios as revealed by Two-way ANOVA (*: $p < 0.05$, **: $p < 0.01$, ***: $p < 0.001$, ns: non-significant)

Author Contributions: Conceptualization, N.M. and E.M.P.; investigation, E.A., T.D., and N.M.; data analysis, N.M. and E.M.P.; writing—original draft preparation, E.A., and N.M.; writing—review and editing, E.M.P. and N.M. All authors have read and agreed to the published version of the manuscript.

Funding: This research received no external funding.

References

1. Tang, W.; Shan, B.; Zhang, H. Phosphorus buildup and release risk associated with agricultural intensification in the estuarine sediments of Chaohu Lake Valley, Eastern China. *Clean Soil Air Water* **2010**, *38*, 336–343. [CrossRef]

2. Ju, X.T.; Xing, G.X.; Chen, X.P.; Zhang, S.L.; Zhang, L.J.; Liu, X.J.; Cui, Z.L.; Yin, B.; Christie, P.; Zhu, Z.L.; et al. Reducing environmental risk by improving N management in intensive Chinese agricultural systems. *Proc. Natl. Acad. Sci. USA* **2009**, *106*, 3041–3046. [CrossRef] [PubMed]

3. Edenhofer, O.; Pichs-Madruga, R.; Sokona, Y.; Farahani, E.; Kadner, S.; Seyboth, K.; Adier, A.; Baum, I.; Brunner, S.; Eickemeier, P. *Climate Change 2014: Mitigation of Climate Change. Contribution of Working Group III to the Fifth Assessment Report of the Intergovernmental Panel on Climate Change*; Cambrige University Press: Cambridge, UK; New York, NY, USA, 2014.

4. Tilman, D.; Cassman, K.G.; Matson, P.A.; Naylor, R.; Polasky, S. Agricultural sustainability and intensive production practices. *Nature* **2002**, *418*, 671–677. [CrossRef] [PubMed]

5. Laetz, C.A.; Baldwin, D.H.; Collier, T.K.; Hebert, V.; Stark, J.D.; Scholz, N.L. The synergistic toxicity of pesticide mixtures: Implications for risk assessment and the conservation of endangered Pacific salmon. *Environ. Health Perspect.* **2009**, *117*, 348–353. [CrossRef] [PubMed]

6. Dou, R.; Sun, J.; Deng, F.; Wang, P.; Zhou, H.; Wei, Z.; Chen, M.; He, Z.; Lai, M.; Ye, T.; et al. Contamination of pyrethroids and atrazine in greenhouse and open-field agricultural soils in China. *Sci. Total Environ.* **2020**, *701*, 134916. [CrossRef] [PubMed]

7. Baćmaga, M.; Wyszkowska, J.; Kucharski, J. Response of soil microorganisms and enzymes to the foliar application of Helicur 250 EW fungicide on *Horderum vulgare* L. *Chemosphere* **2020**, *242*, 125163. [CrossRef]

8. Satapute, P.; Kamble, M.V.; Adhikari, S.S.; Jogaiah, S. Influence of triazole pesticides on tillage soil microbial populations and metabolic changes. *Sci. Total Environ.* **2019**, *651*, 2334–2344. [CrossRef] [PubMed]

9. Neuwirthová, N.; Trojan, M.; Svobodová, M.; Vašíčková, J.; Šimek, Z.; Hofman, J.; Bielská, L. Pesticide residues remaining in soils from previous growing season(s)—Can they accumulate in non-target organisms and contaminate the food web? *Sci. Total Environ.* **2019**, *646*, 1056–1062. [CrossRef]

10. Birkhofer, K.; Bezemer, T.M.; Bloem, J.; Bonkowski, M.; Christensen, S.; Dubois, D.; Ekelund, F.; Fließbach, A.; Gunst, L.; Hedlund, K.; et al. Long-term organic farming fosters below and aboveground biota: Implications for soil quality, biological control and productivity. *Soil Biol. Biochem.* **2008**, *40*, 2297–2308. [CrossRef]

11. Zhang, Q.; Zhou, W.; Liang, G.; Wang, X.; Sun, J.; He, P.; Li, L. Effects of different organic manures on the biochemical and microbial characteristics of albic paddy soil in a short-term experiment. *PLoS ONE* **2015**, *10*, e0124096. [CrossRef]

12. Hartmann, M.; Frey, B.; Mayer, J.; Mäder, P.; Widmer, F. Distinct soil microbial diversity under long-term organic and conventional farming. *ISME J.* **2015**, *9*, 1177–1194. [CrossRef] [PubMed]

13. Li, F.; Chen, L.; Zhang, J.; Yin, J.; Huang, S. Bacterial community structure after long-term organic and inorganic fertilization reveals important associations between soil nutrients and specific taxa involved in nutrient transformations. *Front. Microbiol.* **2017**, *8*, 187. [CrossRef] [PubMed]

14. Ierna, A.; Mauromicale, G. Sustainable and profitable nitrogen fertilization management of potato. *Agronomy* **2019**, *9*, 582. [CrossRef]

15. Bizos, G.; Papatheodorou, E.M.; Chatzistathis, T.; Ntalli, N.; Aschonitis, V.G.; Monokrousos, N. The role of microbial inoculants on plant protection, growth stimulation, and crop productivity of the olive tree (*Olea europea* L.). *Plants* **2020**, *9*, 743. [CrossRef] [PubMed]

16. Mirshad, P.P.; Puthur, J.T. Drought tolerance of bioenergy grass *Saccharum spontaneum* L. enhanced by arbuscular mycorrhizae. *Rhizosphere* **2017**, *3*, 1–8. [CrossRef]

17. Vurukonda, S.S.K.P.; Vardharajula, S.; Shrivastava, M.; SkZ, A. Multifunctional *Pseudomonas putida* strain FBKV2 from arid rhizosphere soil and its growth promotional effects on maize under drought stress. *Rhizosphere* **2016**, *1*, 4–13. [CrossRef]

18. Deshmukh, Y.; Khare, P.; Patra, D. Rhizobacteria elevate principal basmati aroma compound accumulation in rice variety. *Rhizosphere* **2016**, *1*, 53–57. [CrossRef]

19. Bhattacharyya, P.N.; Jha, D.K. Plant growth-promoting rhizobacteria (PGPR): Emergence in agriculture. *World J. Microbiol. Biotechnol.* **2012**, *28*, 1327–1350. [CrossRef]

20. Barea, J.M. Future challenges and perspectives for applying microbial biotechnology in sustainable agriculture based on a better understanding of plant-microbiome interactions. *J. Soil Sci. Plant Nutr.* **2015**, *15*, 261–282. [CrossRef]

21. Kamou, N.N.; Karasali, H.; Menexes, G.; Kasiotis, K.M.; Bon, M.C.; Papadakis, E.N.; Tzelepis, G.D.; Lotos, L.; Lagopodi, A.L. Isolation screening and characterisation of local beneficial rhizobacteria based upon their ability to suppress the growth of *Fusarium oxysporum* f. sp. radicis-lycopersici and tomato foot and root rot. *Biocontrol Sci. Technol.* **2015**, *25*, 928–949. [CrossRef]

22. Ruiu, L. Microbial biopesticides in agroecosystems. *Agronomy* **2018**, *8*, 235. [CrossRef]

23. Kumar, M.; Mishra, S.; Dixit, V.; Kumar, M.; Agarwal, L.; Chauhan, P.S.; Nautiyal, C.S. Synergistic effect of *Pseudomonas putida* and *Bacillus amyloliquefaciens* ameliorates drought stress in chickpea (*Cicer arietinum* L.). *Plant Signal. Behav.* **2016**, *11*, 1–9. [CrossRef] [PubMed]

24. Haas, D.; Défago, G. Biological control of soil-borne pathogens by fluorescent pseudomonads. *Nat. Rev. Microbiol.* **2005**, *3*, 307–319. [CrossRef] [PubMed]

25. Leveau, J.H.J.; Lindow, S.E. Utilization of the plant hormone Indole-3-Acetic Acid for growth by. *Society* **2005**, *71*, 2365–2371.

26. Srivastava, S.; Bist, V.; Srivastava, S.; Singh, P.C.; Trivedi, P.K.; Asif, M.H.; Chauhan, P.S.; Nautiyal, C.S. Unraveling aspects of *Bacillus amyloliquefaciens* mediated enhanced production of rice under biotic stress of *Rhizoctonia solani*. *Front. Plant Sci.* **2016**, *7*, 1–16. [CrossRef]

27. Palma, L.; Muñoz, D.; Berry, C.; Murillo, J.; Caballero, P.; Caballero, P. *Bacillus thuringiensis* toxins: An overview of their biocidal activity. *Toxins* **2014**, *6*, 3296–3325. [CrossRef]

28. Choudhary, D.K.; Johri, B.N. Interactions of *Bacillus* spp. and plants—With special reference to induced systemic resistance (ISR). *Microbiol. Res.* **2009**, *164*, 493–513. [CrossRef]

29. Alori, E.T.; Glick, B.R.; Babalola, O.O. Microbial phosphorus solubilization and its potential for use in sustainable agriculture. *Front. Microbiol.* **2017**, *8*, 1–8. [CrossRef]

30. Aini, N.; Yamika, W.S.D.; Ulum, B. Effect of nutrient concentration, PGPR and AMF on plant growth, yield and nutrient uptake of hydroponic lettuce. *Int. J. Agric. Biol.* **2019**, *21*, 175–183.

31. Hsu, C.K.; Micallef, S.A. Plant-mediated restriction of Salmonella enterica on tomato and spinach leaves colonized with *Pseudomonas* plant growth-promoting rhizobacteria. *Int. J. Food Microbiol.* **2017**, *259*, 1–6. [CrossRef]

32. Chowdhury, S.P.; Dietel, K.; Rändler, M.; Schmid, M.; Junge, H.; Borriss, R.; Hartmann, A.; Grosch, R. Effects of *Bacillus amyloliquefaciens* FZB42 on lettuce growth and health under pathogen pressure and its impact on the rhizosphere bacterial community. *PLoS ONE* **2013**, *8*, 1–10. [CrossRef]

33. Cipriano, M.A.P.; Lupatini, M.; Lopes-Santos, L.; da Silva, M.J.; Roesch, L.F.W.; Destéfano, S.A.L.; Freitas, S.S.; Kuramae, E.E. Lettuce and rhizosphere microbiome responses to growth promoting *Pseudomonas* species under field conditions. *FEMS Microbiol. Ecol.* **2016**, *92*. [CrossRef] [PubMed]

34. Cucu, M.A.; Gilardi, G.; Pugliese, M.; Matić, S.; Gisi, U.; Gullino, M.L.; Garibaldi, A. Influence of different biological control agents and compost on total and nitrification-driven microbial communities at rhizosphere and soil level in a lettuce—*Fusarium oxysporum* f. sp. lactucae pathosystem. *J. Appl. Microbiol.* **2019**, *126*, 905–918. [CrossRef] [PubMed]

35. Castro-Sowinski, S.; Herschkovitz, Y.; Okon, Y.; Jurkevitch, E. Effects of inoculation with plant growth-promoting rhizobacteria on resident rhizosphere microorganisms. *FEMS Microbiol. Lett.* **2007**, *276*, 1–11. [CrossRef] [PubMed]

36. Troelstra, S.R.; Wagenaar, R.; Smant, W.; Peters, B.A.M. Interpretation of bioassays in the study of interactions between soil organisms and plants: Involvement of nutrient factors. *New Phytol.* **2001**, *150*, 697–706. [CrossRef]

37. Kapali, S.; Gade, R.M.; Shitole, A.V.; Swathi, S. Isolation and characterization of *Pseudomonas fluorescens* and *Bacillus subtilis* and their in vitro evaluation. *Adv. Life Sci.* **2016**, *5*, 5856–5859.

38. Versalovic, J.; Schneider, M.; de Bruijn, F.J.; Lupski, J.R. Versalovic_MMCB 1994_Genomic Fingerprinting.pdf. *Methods Mol. Cell. Biol.* **1994**, *5*, 25–40.

39. Allen, E. *Chemical Analysis of Ecological Materials*; Blackwell Scientific Publishing: Oxford, UK, 1976.

40. Ntalli, N.; Monokrousos, N.; Rumbos, C.; Kontea, D.; Zioga, D.; Argyropoulou, M.D.; Menkissoglu-Spiroudi, U.; Tsiropoulos, N.G. Greenhouse biofumigation with *Melia azedarach* controls *Meloidogyne* spp. and enhances soil biological activity. *J. Pest Sci.* **2018**, *91*, 29–40. [CrossRef]

41. Ntalli, N.; Menkissoglu-Spiroudi, U.; Doitsinis, K.; Kalomoiris, M.; Papadakis, E.N.; Boutsis, G.; Dimou, M.; Monokrousos, N. Mode of action and ecotoxicity of hexanoic and acetic acids on *Meloidogyne javanica*. *J. Pest Sci.* **2020**, *93*, 867–877. [CrossRef]

42. Panayiotou, E.; Dimou, M.; Monokrousos, N. The effects of grazing intensity on soil processes in a Mediterranean protected area. *Environ. Monit. Assess.* **2017**, *189*, 441. [CrossRef]

43. Allison, S.D.; Jastrow, J.D. Activities of extracellular enzymes in physically isolated fractions of restored grassland soils. *Soil Biol. Biochem.* **2006**, *38*, 3245–3256. [CrossRef]

44. Sinsabaugh, R.L.; Reynolds, H.; Long, T.M. Rapid assay for amidohydrolase (urease) activity in environmental samples. *Soil Biol. Biochem.* **2000**, *32*, 2095–2097. [CrossRef]

45. Poorter, H.; Nagel, O. Poorter and Nagel 1999 The role of biomass allocation in the growth response of plants to different levels of light, CO_2, nutrients and water: A quantitative review. *Aust. J. Plant Physiol.* **2000**, *27*, 1191.

46. Khosravi, A.; Zarei, M.; Ronaghi, A. Effect of PGPR, Phosphate sources and vermicompost on growth and nutrients uptake by lettuce in a calcareous soil. *J. Plant Nutr.* **2018**, *41*, 80–89. [CrossRef]

47. Nguyen, M.L.; Glaes, J.; Spaepen, S.; Bodson, B.; du Jardin, P.; Delaplace, P. Biostimulant effects of *Bacillus* strains on wheat from in vitro towards field conditions are modulated by nitrogen supply. *J. Plant Nutr. Soil Sci.* **2019**, *182*, 325–334. [CrossRef]

48. Mohite, B. Isolation and characterization of indole acetic acid (IAA) producing bacteria from rhizospheric soil and its effect on plant growth. *J. Soil Sci. Plant Nutr.* **2013**, *13*, 638–649. [CrossRef]

49. Lynch, W.H. Effect of temperature on *Pseudomonas fluorescens* chemotaxis. *J. Bacteriol.* **1980**, *143*, 338–342. [CrossRef]

50. Faucon, M.P.; Houben, D.; Lambers, H. Plant Functional Traits: Soil and Ecosystem Services. *Trends Plant Sci.* **2017**, *22*, 385–394. [CrossRef]

51. Thirup, L.; Ekelund, F.; Johnsen, K.; Jacobsen, C.S. Population dynamics of the fast-growing sub-populations of *Pseudomonas* and total bacteria, and their protozoan grazers, revealed by fenpropimorph treatment. *Soil Biol. Biochem.* **2000**, *32*, 1615–1623. [CrossRef]

52. Ke, X.; Feng, S.; Wang, J.; Lu, W.; Zhang, W.; Chen, M.; Lin, M. Effect of inoculation with nitrogen-fixing bacterium *Pseudomonas stutzeri* A1501 on maize plant growth and the microbiome indigenous to the rhizosphere. *Syst. Appl. Microbiol.* **2019**, *42*, 248–260. [CrossRef]

53. Kozdrój, J.; Trevors, J.T.; Van Elsas, J.D. Influence of introduced potential biocontrol agents on maize seedling growth and bacterial community structure in the rhizosphere. *Soil Biol. Biochem.* **2004**, *36*, 1775–1784. [CrossRef]

54. Han, L.; Wang, Z.; Li, N.; Wang, Y.; Feng, J.; Zhang, X. Bacillus amyloliquefaciens B1408 suppresses Fusarium wilt in cucumber by regulating the rhizosphere microbial community. *Appl. Soil Ecol.* **2019**, *136*, 55–66. [CrossRef]

55. Zhang, L.; Lueders, T. Micropredator niche differentiation between bulk soil and rhizosphere of an agricultural soil depends on bacterial prey. *FEMS Microbiol. Ecol.* **2017**, *93*, 1–11. [CrossRef] [PubMed]

56. Viollet, A.; Pivato, B.; Mougel, C.; Cleyet-Marel, J.C.; Gubry-Rangin, C.; Lemanceau, P.; Mazurier, S. *Pseudomonas fluorescens* C7R12 type III secretion system impacts mycorrhization of *Medicago truncatula* and associated microbial communities. *Mycorrhiza* **2017**, *27*, 23–33. [CrossRef]

57. El Arbi, A.; Rochex, A.; Chataigné, G.; Béchet, M.; Lecouturier, D.; Arnauld, S.; Gharsallah, N.; Jacques, P. The Tunisian oasis ecosystem is a source of antagonistic *Bacillus* spp. producing diverse antifungal lipopeptides. *Res. Microbiol.* **2016**, *167*, 46–57. [CrossRef]

58. Moeskops, B.; Buchan, D.; Sleutel, S.; Herawaty, L.; Husen, E.; Saraswati, R.; Setyorini, D.; De Neve, S. Soil microbial communities and activities under intensive organic and conventional vegetable farming in West Java, Indonesia. *Appl. Soil Ecol.* **2010**, *45*, 112–120. [CrossRef]

59. Nacamulli, C.; Bevivino, A.; Dalmastri, C.; Tabacchioni, S.; Chiarini, L. Perturbation of maize rhizosphere microflora following seed bacterization with *Burkholderia cepacia* MCI 7. *FEMS Microbiol. Ecol.* **1997**, *23*, 183–193. [CrossRef]

60. Nunan, N.; Leloup, J.; Ruamps, L.S.; Pouteau, V.; Chenu, C. Effects of habitat constraints on soil microbial community function. *Sci. Rep.* **2017**, *7*, 1–10. [CrossRef]

61. Nannipieri, P.; Ascher, J.; Ceccherini, M.T.; Landi, L.; Pietramellara, G.; Renella, G. Microbial diversity and soil functions. *Eur. J. Soil Sci.* **2003**, *54*, 655–670. [CrossRef]

62. Fujita, M.; Nakashima, K.; Achal, V.; Kawasaki, S. Whole-cell evaluation of urease activity of *Pararhodobacter* sp. isolated from peripheral beachrock. *Biochem. Eng. J.* **2017**, *124*, 1–5. [CrossRef]

Article

Impact of Maize–Mushroom Intercropping on the Soil Bacterial Community Composition in Northeast China

Xiaoqin Yang [1], Yang Wang [1,*], Luying Sun [1], Xiaoning Qi [1], Fengbin Song [1] and Xiancan Zhu [1,2,*]

[1] Northeast Institute of Geography and Agroecology, Chinese Academy of Sciences, Changchun 130102, China; yangxiaoqin@iga.ac.cn (X.Y.); sunluying@iga.ac.cn (L.S.); qxn@iga.ac.cn (X.Q.); songfb@iga.ac.cn (F.S.)

[2] College of Life Sciences, Anhui Normal University, Wuhu 241000, China

* Correspondence: wangyang@iga.ac.cn (Y.W.); zhuxiancan@ahnu.edu.cn (X.Z.)

Received: 18 August 2020; Accepted: 30 September 2020; Published: 7 October 2020

Abstract: Conservative agricultural practices have been adopted to improve soil quality and maintain crop productivity. An efficient intercropping of maize with mushroom has been developed in Northeast China. The objective of this study was to evaluate and compare the effects of planting patterns on the diversity and structure of the soil bacterial communities at a 0–20 cm depth in the black soil zone of Northeast China. The experiment consisted of monoculture of maize and mushroom, and intercropping in a split-plot arrangement. The characteristics of soil microbial communities were performed by 16S rRNA gene amplicom sequencing. The results showed that intercropping increased soil bacterial richness and diversity compared with maize monoculture. The relative abundances of Acidobacteria, Chloroflexi, Saccharibacteria and Planctomycetes were significantly higher, whereas Proteobacteria and Firmicutes were lower in intercropping than maize monoculture. Redundancy analysis suggested that pH, NO_3^--N and NH_4^+-N contents had a notable effect on the structure of the bacterial communities. Moreover, intercropping significantly increased the relative abundance of carbohydrate metabolism pathway functional groups. Overall, these findings demonstrated that intercropping of maize with mushroom strongly impacts the physical and chemical properties of soil as well as the diversity and structure of the soil bacterial communities, suggesting this is a sustainable agricultural management practice in Northeast China.

Keywords: 16S rRNA; planting pattern; soil chemical properties; soil microbial community

1. Introduction

Maize (*Zea mays* L.) is not only one of the three major food crops in the world, but also a high-quality animal feed and an important industrial raw material. Maize production in China accounts for 21.9% of total maize output in the world. The northeast area is the largest maize producer in China, particularly in the Jilin province, which accounts for 1/8th of maize production in China [1]. However, the maize planting pattern in Northeast China was mainly based upon conventional agricultural practices (e.g., monoculture cropping, tillage and removal of crop residues) that have caused soil erosion, fertility decline, and loss of soil biodiversity [2]. Therefore, conservation and sustainable agricultural practices, such as intercropping, no tillage and mulching, have been adopted to improve soil quality and increase crop productivity.

Intercropping, a method of simultaneously planting two or more crops on the same field, has been practiced in many countries for several decades [3]. It increases the productivity per unit of land through better utilization of resources, control of disease and reducing soil erosion [4]. Soil microbial activity, nutrient cycling, decomposition of organic matter, and physical and chemical properties can be changed in intercropping systems [5,6]. Soil organic matter, reflecting the physical, chemical and biological properties of soil, is an effective indicator of soil quality [2,4], which is defined as the capacity of a soil to function within ecosystem and land-use boundaries to sustain biological productivity, maintain environmental quality, and promote plant and animal health [7]. Mushroom compost is used as a soil conditioner for the growth of soybean, and the substrate pH for growing soybean is increased from acidity to neutrality [8]. The residual fungi and a large number of mycelium are incorporated into the soil, making the soil loose, which improves the physical and chemical properties of the soil, and increases the content of various nutrients in the soil in the maize–mushroom intercropping system [9]. Soil microorganisms play crucial roles in soil ecosystem functioning, as they are involved in soil nutrient cycling and energy flows, micro-ecology regulation and soil sustainable productivity [4,10]. In the soil ecosystem, soil bacteria are crucial decomposers that metabolize organic matter by secreting specific extracellular enzymes to break down large organic molecules into monomers, which are then available for plant uptake [11,12]. The cultivation of continuous monoculture often results in the accumulation of soil-borne pathogens, the growth inhibition of beneficial bacteria [13], and a decrease of bacterial diversity, ultimately [14]. Pear/mushroom intercropping significantly increased the number and biomass of cultivable microorganisms in the 0–40-cm soil layer and had a marked effect on the soil fertility of the pear garden and the quality of the pear fruit [15].

Oyster mushroom (class Basidiomycetes) is a fungus that is an excellent source of human nutrients (e.g., vitamins, minerals and micro nutrients) and has medicinal values [16]. It can be used as an intercropping crop to promote the efficient uptake of nutrients and resources [8]. For example, the mushroom (*Plearotus* spp.) intercropped with field-grown faba bean (*Vicia faba* L.) increased the faba bean dry seed yield and mushroom basidiocarp yield compared to sole cultivation [16]. Maize–mushroom intercropping systems have been previously studied and were proved to improve maize yield and land equivalent ratio [9,17]. However, few studies have attempted to investigate the effects of maize–mushroom intercropping on soil physical and chemical properties and the diversity and composition of soil microbial community. Therefore, the objective of this study was to explore the effect of maize–mushroom intercropping on the soil bacterial community in Northeast China. Recently, we developed an integrated agricultural management practice involving wide–narrow-row spacing alternation (160 + 40 cm), adoption of no tillage, and mulching by crop residues in maize cropping system in the Northeast China plain, resulting in an improvement of soil quality, reduced soil erosion, and a more sustainable use of cultivated land [18]. Furthermore, in order to use land resources with high efficiency, control disease, and increase the income of farmers, maize–mushroom intercropping was developed based on the integrated agricultural practice. We hypothesized that the intercropping could lead to (i) a marked effect on the structure and diversity of the soil bacterial community and (ii) an improvement of soil quality compared with monocultures.

2. Materials and Methods

2.1. Site Description and Experimental Design

The integrated agricultural management practice was initiated in spring 2012 and the maize–mushroom intercropping under integrated agricultural practice experiment was initiated in spring 2014 at Changchun Agricultural Experimental Field of the Northeast Institute of Geography and Agroecology, Jilin Province, China (43°59′54″ N, 125°23′57″ E). The climate is a north temperate continental monsoon. The mean annual air temperature is 7.2 °C. The mean annual precipitation is 530.5 mm. The frost-free period is 138 days. Snow cover usually occurs from November to April. The soil is a typical silty clay loam (classified as Mollisols), which is highly fertile, inherently productive and suitable soil for cultivation in the world [19], with organic matter 3.1%, total nitrogen 1.49 g kg^{-1}, total phosphorus 0.59 g kg^{-1}, total potassium 21.53 g kg^{-1}, available nitrogen 204.54 mg kg^{-1}, available phosphorus 9.43 mg kg^{-1}, effective potassium 125.84 mg kg^{-1}, soil bulk density 1.19 g cm^{-3}, cation exchange capacity 23.3 cmol kg^{-1}, and pH 6.71.

The experiment was conducted in a plot comparison experiment with three treatments including (1) maize–mushroom intercropping, (2) maize monocultures, and (3) mushroom monocultures. Each plot was 20 m long and 8 m wide with four plot replicates for each treatment. The cultivars were maize Liangyu 99 and oyster mushroom (provided by Liaoning Sanyou Agricultural Biotechnology Company). The maize seeds were sown in late April and harvested in late September, and mushroom sticks were planted in early July and harvested in late September every year. In the maize monoculture treatment, the wide–narrow row spacing was 160 and 40 cm, respectively; no tillage was implemented; the maize stalks were cut at 35 cm above the ground, and then maize stalks were left in the field. Maize was planted on a narrow line of 40-cm spacing, with a plant spacing of 15 cm and a planting density of 6.5×10^4 plants ha^{-1}. The intercropping of maize with mushroom was based on the integrated agricultural management practice. In the intercropping and maize monoculture, maize seeds were sown in 40-cm, narrow rows using a small jukebox in late April. During the maize spinning period, a 10-cm-deep and 40-cm-wide sulcus was dug in advance to place mushroom sticks in the intercropping. Then, mushroom sticks were evenly placed in the sulcus, covered in 2–3 cm soil (Figure 1). The distance between mushroom and maize, mushroom and high stubble both were 27 cm. The mushroom row spacing of mushroom monoculture treatment was 160 cm. A based controlled release compound fertilizer (resin coating) of 600 kg ha^{-1} (total nutrient $\geq$ 53%, Zn $\geq$ 2%) was applied to all treatments (as the base of the fertilizer) before rotating the soil (using the small rotating machine). The type of controlled release nutrient was nitrogen, and the amount of controlled release nutrient was greater or equal to 8%. In mid-May, the compound fertilizer was added once. Throughout the growth period of mushroom, intermittent spray irrigation was used to maintain the optimum moisture for mushroom production.

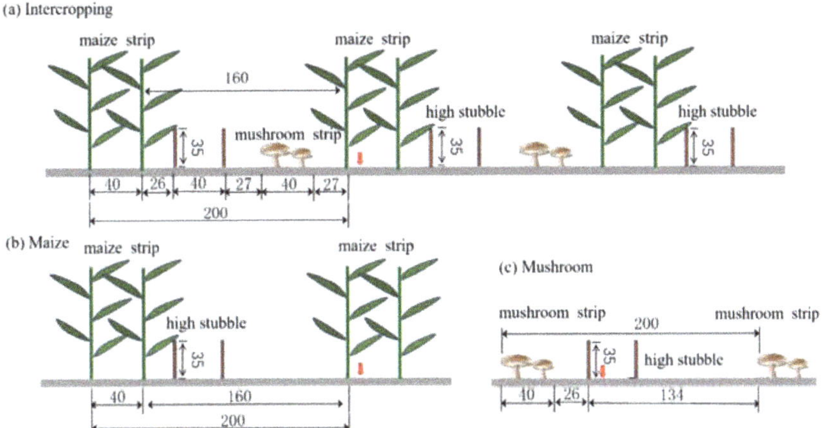

Figure 1. Schematic illustration of row placement of (**a**) maize–mushroom intercropping, (**b**) maize monoculture and (**c**) mushroom monoculture. Red arrows indicate the sampling point. The unit of distances is the centimeter (cm).

2.2. Soil Sampling

On September 20 2017 (After the mushrooms had been planted for 70 days), soil core samples (5.5 cm in diameter, 20 cm in depth, repeated four times) from each treatment were taken randomly from the maize rows (10 cm from the maize rows) of maize monoculture and intercropping treatment, and high stubble rows (10 cm from high stubble rows) of mushroom treatment. In total, there were 12 soil samples used for the analysis. All samples were sieved by a 2-mm sieve to remove rocks and were then separated into two parts: one part air dried to determine soil properties, and another stored at $-80\,°C$ for DNA extraction.

2.3. Soil Chemical Assays

Total nitrogen (TN), nitrate nitrogen (NO_3^--N), ammonium nitrogen (NH_4^+-N), were determined by a continuous flow analyzer (San++, Skalar, Breda, Holland). Soil organic matter (SOM) was measured by the method of potassium dichromate heating oxidation-volumetric based on the standard LY/T1237-1999 [20]. Briefly, 0.2 g of air-dried soil was added into 5 mL $K_2Cr_2O_7$ (0.8 M) and 5 mL concentrated H_2SO_4 solution and heated on a hot plate (300 °C) for 5 min. Three to four drops of phenanthroline indicator were then added, and titrated with $FeSO_4$ (0.4 M).

TN, available nitrogen (AN), NH_4^+-N, and NO_3^--N were measured by standard methods based on LY/T1228-2015. For TN, 1 g soil was added to 1.8 g catalyst (selenium powder: copper sulfate: potassium sulfate = 1:10:100), mixed with 4 mL concentrated sulfuric acid, and removed on the electric heating plate until the soil became grayish white. The resulting solution was green and was transferred to a 100-mL volumetric flask after cooling. The solution was then diluted to volume with distilled water and shaken for testing. For AN, 2 g soil was mixed with 3 mL of 2% H_3BO_3, 10 mL of 1 M NaOH solution, and then incubated at 40 °C for 24 h. After cooling, the sample was titrated with HCl (0.012 M) until the color changed from blue-green to purple. For NH_4^+-N and NO_3^--N, 5 g soil was mixed with 25 mL of KCl (2 M), shaken for 30 min, and filtered for testing.

To measure soil pH, 10 g soil was placed in a 100-mL beaker. Then, 50 mL of distilled water (water:soil = 5:1) was added, shaken for 30 min, and measured with a calibrated pH meter (Mettler-Toledo FE 20, Zurich, Switzerland).

2.4. DNA Extraction and MiSeq Sequencing

DNA was extracted from 0.25 g fresh soil using E.Z.N.A Mag-Bind Soil DNA Kits (OMEGA, Irving, TX, USA) according to the manufacturer's instructions. The DNA extraction quality was measured by 0.8% agarose gel electrophoresis, and the DNA was quantified by an ultraviolet spectrophotometer. The PCR amplification used Q5 high-fidelity DNA polymerase (NEB, Ipswich, MA, USA), and the V3-V4 region of the 16S rDNA genes were amplified using the primers 338F (ACTCCTACGGGAGGCAGCA) and 806R (GGACTACHVGGGTWTCTAAT) [21]. Referring to the preliminary quantitative results of gel electrophoresis, the PCR amplified product was subjected to fluorescence quantification, the fluorescent reagent was Quant-iTPicoGreen dsDNA Assay Kit, and the quantitative instrument was Microplate reader (BioTek, FLx800). A total of 20 pM DNA for each sample were pooled and sequenced in an Illumina MiSeq platform (Illumina, SanDiego, CA, USA) with a 600-cycle kit (2 × 300 bp paired ends). The Miseq sequencing raw data were deposited in the NCBI Sequence Read Archive database, and the project ID is PRJNA432129 and BioSample is SAMN10362720.

2.5. Data Analysis

The original paired-end sequencing data were exported in a FASTQ format. Sequences were removed if the read length was <150 bp, with a mean quality score <20 [22]. Sequence analysis was performed using QIIME software (Version 1.8.0) [23]. Sequences with ≥97% similarity were assigned to the same operational taxonomic units (OTUs) using UCLUST [24]. In detail, the sequences were merged according to ≥97% similarity into OTUs, and the most abundant sequence in each OTU was chose as the representative OTU sequence. Then, according to the number of sequences of each OTU in each sample, a matrix file of OTU abundance in each sample was constructed (i.e., OTU table). The relative abundance of OTUs with <0.001% of the total reads of all samples were removed [25]. The Greengenes database (Release 13.8) was used to annotate taxonomic information [26].

Data were compared using analysis of variance (ANOVA) in IBM SPSS 23.0 software (SPSS Inc., USA). Alpha diversity (Chao1 and Shannon index) was calculated with QIIME (Version 1.8.0).

Pearson's correlation coefficients between soil properties, OTU richness (Chao1 index) and diversity (Shannon index), and bacterial phyla were computed.

Differences of soil bacterial communities based on OTUs between treatments were analyzed using LEfSe [27]. Cluster analysis of soil bacterial communities based on the non-metric multidimensional scaling (NMDS) dissimilarity matrix was performed using QIIME. Permutational multivariate analysis of variance (PERMANOVA; function 'adonis') was adopted to compare community composition on three treatments with QIIME. Redundancy analysis (RDA) was carried out to explore the relationship between soil properties and microbial community composition using the 'vegan' package in the R program (Version 3.3.1). Function predictions were classified into KEGG pathways using PICRUSt (Version 1.1.4) method [28]. In detail, firstly, OTU table was standardized by copy number; the full-length 16S rRNA gene sequence of the tested microbial genome was used to infer the gene function spectrum of their common ancestor; then, the gene function profiles of other untested species in the Greengenes 16S rRNA gene full-length sequence database was inferred and constructed the gene function prediction profiles of the entire lineage of archaea and bacteria; thirdly, the 16S rRNA gene sequence data obtained by sequencing with the Greengenes database was compared to find the "nearest neighbor of the reference sequence" of each sequenced sequence and classified it as a reference OTU; the obtained OTU abundance matrix according to the copy number of the rRNA gene of the nearest neighbor of the reference sequence was corrected; finally, the microbial composition data to the known gene function profile database was "mapped" to realize the prediction of the metabolic function of the microbial communities. The KEGG pathways statistical analysis was implemented using SPSS.

3. Results

3.1. Soil Chemical Properties

The planting pattern was found to significantly affect soil pH and the contents of AN, NO_3^--N, NH_4^+-N, and SOM (Table 1). The soil pH and SOM were markedly increased in intercropping, whereas the contents of AN, NO_3^--N and NH_4^+-N were decreased, compared with maize monoculture. Soil pH was higher, whereas AN, NO_3^--N and NH_4^+-N contents were lower in mushroom monoculture than maize monoculture.

Table 1. Soil chemical properties under maize monoculture, mushroom monoculture and maize–mushroom intercropping.

Planting Pattern	pH	TN (g kg^{-1})	AN (mg kg^{-1})	NO_3^--N (mg kg^{-1})	NH_4^+-N (mg kg^{-1})	SOM%
Maize	6.37 ± 0.10 b	1.28 ± 1.49 a	257 ± 33.6 a	191 ± 58 a	51.3 ± 15.3 a	3.12 ± 0.15 b
Mushroom	7.49 ± 0.05 a	1.39 ± 0.60 a	169 ± 10.4 b	2.19 ± 0.27 b	4.15 ± 0.46 b	3.23 ± 0.05 ab
Intercropping	7.40 ± 0.04 a	1.51 ± 1.05 a	170 ± 8.65 b	19.8 ± 3.39 b	8.41 ± 0.69 b	3.59 ± 0.10 a

Values are means ± standard errors. Mean values in each column followed by the different letters are significantly different ($p < 0.05$) according to Tukey's test. The number of replicates per treatment is 4 ($n = 4$). TN, AN, NO_3^--N, NH_4^+-N and SOM represent total nitrogen, available nitrogen, nitrate nitrogen, ammonium nitrogen and soil organic matter, respectively.

3.2. Soil Bacterial Community Diversity

Sequencing of the 16S rRNA gene fragment from 12 soil samples produced a total of 462,813 sequences (Table S1). After filtration, alignment, pre-clustering and removal of chimeric sequences and singletons, 359,341 sequences were obtained. They were assigned into 18,519 OTUs.

Intercropping and mushroom monoculture had higher bacterial OTU richness and Shannon index compared with maize monoculture (Figure 2). There was a striking positive relationship between Shannon index and pH and SOM, whereas there was a negative association between Shannon index and the contents of AN, NO_3^--N, NH_4^+-N (Table S2).

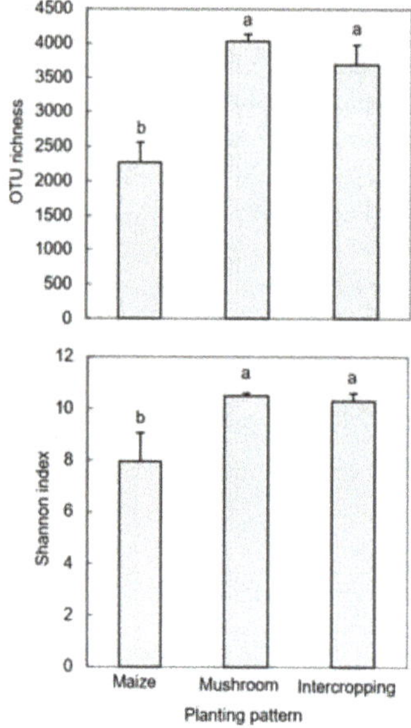

Figure 2. Operational taxonomic units (OTUs) richness and Shannon index of soil bacterial community under maize monoculture, mushroom monoculture and maize–mushroom intercropping. The error bars represent standard errors (SE). Different letters indicate significant different ($p < 0.05$) according to Tukey's test. The number of replicates per treatment is 4 ($n = 4$).

3.3. Soil Bacterial Community Structure

In order to characterize the effect of planting pattern on soil bacterial communities, the relative abundances at phylum, class, order, family and genus levels were analyzed. Then, 16S rRNA gene amplicon sequencing showed that the dominant bacterial phyla were Proteobacteria, Chloroflexi, Acidobacteria, Actinobacteria, Firmicutes, Gemmatimonadetes, Cyanobacteria, Saccharibacteria, Bacteroidetes, Nitrospirae, Planctomycetes, Verrucomicrobia and Parcubacteria (Figure 3), and these groups accounted for over 88–95% of the sequences. Moreover, soil chemical properties were closely correlated with the relative abundance of some dominant microbial phyla groups (Table S2). There was a marked positive association between soil pH and Chloroflexi, Acidobacteria, Planctomycetes, Verrucomicrobia, and Parcubacteria, whereas the relationships were negative between the contents of AN, NO_3^--N, NH_4^+-N and those dominant microbial phyla.

In total, 13 phyla, 31 classes, 64 orders, and 84 families in the bacterial community were significantly affected by planting pattern. At the phylum level, planting pattern significantly altered the relative abundance of Chloroflexi, Acidobacteria, Saccharibacteria, Planctomycetes, Armatimonadetes, Fusobacteria, Euryarchaeota, WS6, Elusimicrobia, Peregrinibacteria, WWE3 and Gracilibacteria. Intercropping had higher relative abundance of Acidobacteria, Chloroflexi, and Planctomycetes, and lower relative abundance of Proteobacteria and Firmicutes compared with maize monoculture treatment (Figure 3). Intercropping had higher relative abundance of Saccharibacteria, and lower Cyanobacteria than mushroom monoculture. The distribution of related genera varied between maize monoculture, mushroom monoculture and intercropping soils. *Pseudomonas*, *Sphingomonas*, *Lactobacillus* and *Rhodanobacter* were the most abundant genera across all soil samples, representing 5.57%, 2.83%, 0.19%, and 3.74% of all classified sequences in intercropping, 17.24%, 9.28%, 10.79%, and 2.05% in maize monoculture and 6.64%, 1.59%, 0.14%, and 1.05% in mushroom monoculture, respectively. *Gemmatimonas*, *Cystobacteraceae*, *Marmoricola*, *Bradyrhizobium*, *Streptomyces*, *Rhodoglobus*, *Nakamurella*, *Frateuria*, *Sphingobium*, *Woodsholea*, and *Oribacterium* showed an increased relative abundance in intercropping, while the relative abundance of *Streptomycetaceae* decreased in intercropping compared with maize monoculture (Figure 4).

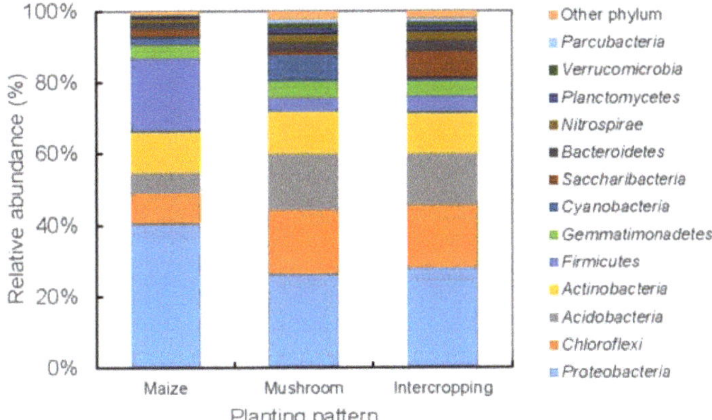

Figure 3. Relative abundances of the dominant phyla of bacteria at 0-20-cm soil depths under maize monoculture, mushroom monoculture and maize–mushroom intercropping. The number of replicates per treatment is 4 ($n = 4$).

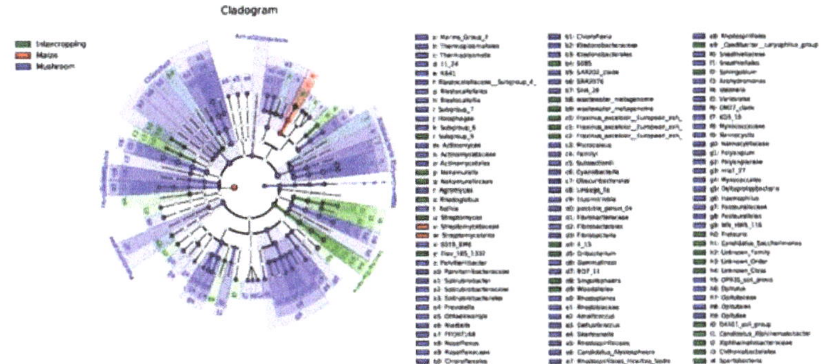

Figure 4. Cladogram of bacteria at 0–20-cm soil depths under maize monoculture, mushroom monoculture and maize–mushroom intercropping. The Cladogram shows the hierarchical relationship of all classification units from the phylum to the genus (from the inner circle to the outer circle). The node size corresponds to the average relative abundance of the classification units. The blue, green and red node represents that the difference of bacterial relative abundance between groups is significant. The letters identify the taxon name that has a significant difference between groups. The number of replicates per treatment is 4 ($n = 4$).

3.4. Comparative Analysis of Soil Bacterial Community

Based on the NMDS dissimilarity matrix, hierarchical cluster analysis for investigated the beta diversity of bacterial communities showed that soil bacterial communities were affected by planting patterns (Figure 5). PERMANOVA analysis indicated that soil bacterial community composition was significantly affected by maize monoculture, mushroom monoculture and intercropping treatments. The RDA was performed to determine the strength of the association between the soil bacterial community and soil physical, chemical properties. RDA revealed a strong difference between maize monoculture, mushroom monoculture and intercropping soils (Figure 6). The first two canonical axes are responsible for 45.73% of variance (25.42% by RDA1 axis and 20.31% by RDA2 axis). The RDA indicated that pH, NO_3^--N and NH_4^+-N contents had an extremely significant influence on the structure of the bacterial community ($p < 0.05$) (Table S3).

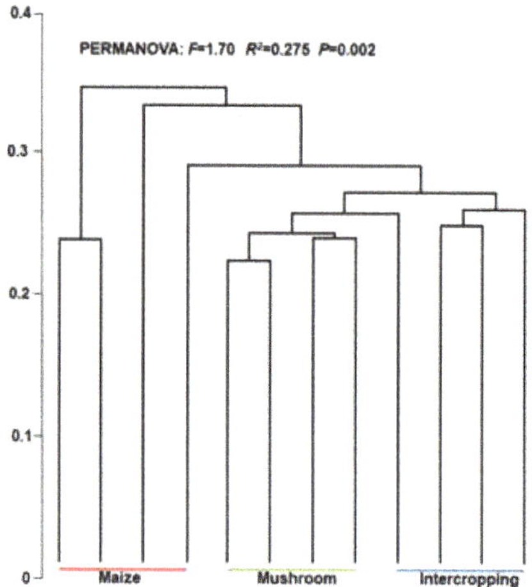

Figure 5. Hierarchical cluster analysis of soil bacterial communities based on the NMDS dissimilarity matrix among maize monoculture, mushroom monoculture and intercropping. Permutational multivariate analysis of variance (PERMANOVA) was adopted to compare community composition on three treatments. The number of replicates per treatment is 4 ($n = 4$).

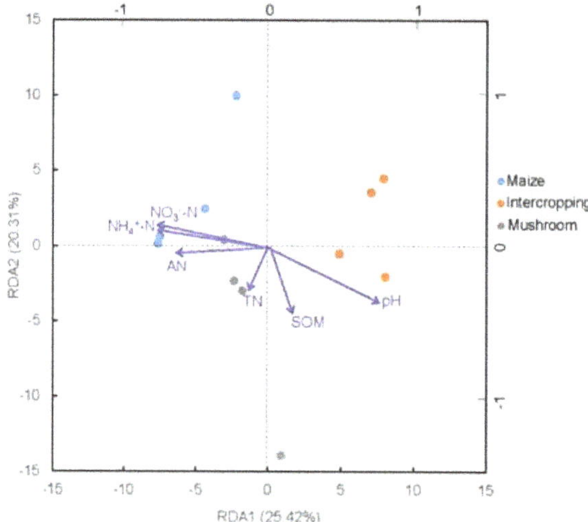

Figure 6. Redundancy analysis (RDA) of soil bacterial community structure and soil properties under maize monoculture, mushroom monoculture and intercropping. Soil factors indicated in blue text include pH, contents of organic matter (SOM), total nitrogen (TN), available nitrogen (AN), nitrate nitrogen (NO_3^--N) and ammonium nitrogen (NH_4^+-N). The circles are the RDA scores of the samples and the arrows are the scores of the soil variables by RDA. The number of replicates per treatment is 4 ($n = 4$).

3.5. Metabolism of Soil Microbial Community

Potential metabolism were assigned to predicted functional annotation of protein sequences. The KEGG metabolic pathway difference analysis of soil bacteria revealed significant changes in 11 metabolic networks between the groups of planting patterns. The analysis showed that intercropping significantly increased the relative abundance of carbohydrate metabolism pathway functional groups compared with maize monoculture (Figure 7). In addition, the relative abundance of glycan biosynthesis and other secondary metabolites functional groups in intercropping and mushroom monoculture was higher compared with maize monoculture ($p < 0.05$) (Figure 7 and Table S4).

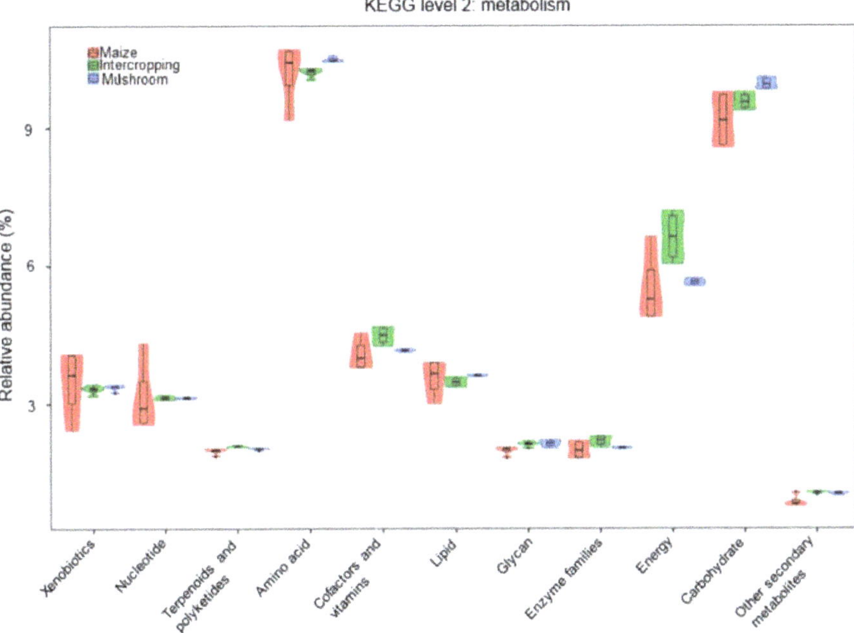

Figure 7. The KEGG metabolic pathway difference analysis of bacteria at 0–20-cm soil depths under maize monoculture, mushroom monoculture and intercropping. The number of replicates per treatment is 4 ($n = 4$).

4. Discussion

4.1. Soil Physicochemical Properties

Nutrients such as carbon, nitrogen, phosphorus and potassium, are essential for proper plant growth. Soil carbon and nitrogen contents are the most sensitive indicators of soil quality [29], and it has been suggested that below ground interspecific interactions improved soil nitrogen supply capacity [30], mobilization [31] and increased TN in intercropping [32]. The results of the current study have shown that contents of AN, NO_3^--N, NH_4^+-N in intercropping treatment were no higher than maize monoculture treatment. Below ground interactions through intercropping could affect N-cycling [30,33]. Moreover, soil nitrate nitrogen is dynamic, and is influenced by soil particle distribution, soil depth, and precipitation, and varies during crop growth and development [34]. Soil nitrate would likely decline faster in intercropping and mushroom monoculture treatments in comparison with maize monoculture, mostly due to decomposition of mineral nutrients by saprotrophs and leaching losses [35], using intermittent spray irrigation to maintain the optimum moisture for mushroom production. Vieira et al. [36] also showed that nitrogen losses after 10 and 15 days impacted by mushroom yield. In this study, intercropping treatment significantly increased the contents of SOM compared with maize monoculture. This is consistent with previous studies that demonstrated that soil organic carbon fraction, carbon pool management index and soil carbon sequestration were improved under intercropping [4,37,38].

4.2. Bacterial Community Diversity

The soil microbial community, a biomarker indicator of soil quality and ecosystem processes [39], is very sensitive to vegetation. On the other hand, it can also strongly affect plant growth and yield formation [40]. Our study revealed that intercropping significantly increased the OTU richness and diversity of the bacterial community compared to maize monoculture treatment. Consistent with our findings, Fu et al. [4] reported that maize–soybean intercropping had higher Shannon index compared with monocultures. Qin et al. [41] demonstrated that maize–potato intercropping increased the carbon source utilization rate and diversity of the microbial community. In addition, the bacterial Shannon index was correlated with contents of N, SOM and pH, suggesting that the improvements in C and N source utilization were beneficial to increased soil bacterial diversity, such as species richness [12]. Moreover, soil bacterial community could be changed by pH with a higher bacterial diversity in neutral soil than acidic soil, which a significant positive association between Shannon index and pH was found in this study.

4.3. Bacterial Community Structure

Microbial community composition has large effects on organic matter dynamics and nutrient cycling, and can influence soil function and ecosystem sustainability [42]. In total, 46 phyla, 168 classes, 356 orders, 605 families and 1207 genera of bacterial communities were obtained in our samples. Within the thirteen dominant bacterial phyla, Proteobacteria was the most abundant bacterial phylum, which was consistent with the results of mulberry and alfalfa intercropping system [43]. The major microbial phyla, such as Chloroflexi, Acidobacteria, Actinobacteria, identified in this study are often observed in other soils, though the relative abundance was different [44,45].

Previous studies have indicated that the dominant bacterial phyla could be changed by manipulating planting patterns and plant species [4,12,43]. In the present study, the bacterial community structures among intercropping and maize and mushroom monoculture treatments were significantly different. Planting patterns significantly affected the four dominant and eight other bacterial phyla. The relative abundance of Chloroflexi, Acidobacteria, Planctomycetes and Saccharibacteria were higher in intercropping than maize or mushroom monoculture treatments. Anaerolineae, a dominant class of Chloroflexi, has been thought to be ubiquitous and to play important roles in ecosystems [46]. The greater abundance of Chloroflexi in the intercropping than maize monoculture treatment likely indicated that intercropping would better coordinate soil ecosystems [47]. Acidobacteria is an acidophilic and oligotrophic chemoorganotrophic bacterium; Planctomycetes play possible role in the evolution of the methane cycle [48]. Saccharibacteria play a role in the degradation of various organic compounds as well as sugar compounds under aerobic, nitrate reducing, and anaerobic conditions [49]. Although the phylum Proteobacteria was not affected by planting pattern, Gammaproteobacteria, the most dominant class of Proteobacteria in our study, had a prominent higher abundance under intercropping than maize monoculture treatment, stimulated by higher SOM and lower nitrate nitrogen contents of intercropping treatment [50]. *Pseudomonas* is one of the widely distributed plant growth promoting rhizobacteria [51]. The result revealed that the soil *Pseudomonas* community was affected by planting pattern. Consistent with previous studies, the altered plant species affected the microbial community composition [52].

In this study, the relative abundance of Actinobacteria in intercropping and maize monoculture treatment were significant different at class level. At genus level, *Gemmatimonas*, *Streptomyces*, *Nakamurella* and *Frateuria* had greater abundance in intercropping soils. *Actinobacteria* have a critical role in decomposition of soil organic materials, such as cellulose, chitin and polysaccharides [53]. Streptomyces could produce bioactive secondary metabolites which show antifungals and antivirals biological activities [54], and have more efficient secretion mechanisms which could promote protein solubilization [55]; *Nakamurella* is able to accumulate polysaccharides [56]; *Frateuria* is able to enhance potassium uptake efficiently in plants and has been found to increase biomass and nutrient content [57]. In addition, *Gemmatimonas*, the dominant genus of Gemmatimonadetes, is involved in modulating carbon and nitrogen intake, decomposing polyaromatic carbon and promoting plant development [58]. The greater abundance of *Actinobacteria*, *Streptomyces*, *Nakamurella*, *Frateuria* and *Gemmatimonas* likely be attributed to the effects of secondary metabolites such as carbohydrates, free amino acids, and nucleotides produced by mushroom during metabolism [59]. Our study showed that the relative abundance of glycan and secondary metabolites functional groups in intercropping and mushroom monoculture significantly increased compared with maize monoculture. This is consistent with the other research that intercropping lead to changes in plants accumulation of minerals and secondary metabolites [60].

To investigate the relationships between soil microbial community structure and measured soil variables in the maize monoculture, monoculture mushroom and intercropping systems, we analyzed the dominant bacterial phyla and OTUs data using Pearson's correlation and RDA. The soil variables have a substantial impact on the dominant bacterial phyla. In this study, SOM, TN, AN, $NO_3{}^--N$, $NH_4{}^+-N$, and pH had positive/negative correlations with the dominant bacterial phyla. For example, the abundance of Chloroflexi and Planctomycetes were positively correlated with SOM content, which indicated that SOM correlated with the relative abundance of these bacteria [4]. SOM may play non-negligible roles in influencing on the soil microbial community structure by affecting the metabolism of soil microbes [61]. The output of RDA with soil variation in bacterial community indicated that pH plays critical roles in the structure of the bacterial community. This was in accordance with other studies that indicated that soil pH was a major factor in determining the structure of the soil bacterial community [12,61].

5. Conclusions

In total, 13 phyla, 31 classes, 64 orders, 84 families in the bacterial community were significantly affected by planting pattern. The results revealed that SOM, TN contents, Shannon index and the relative abundance of Chloroflexi, Acidobacteria, Saccharibacteria, Planctomycetes and carbohydrate metabolism pathway functional groups were significantly increased in intercropping compared with maize monoculture treatment. Moreover, soil chemical properties closely correlated with OTU richness, Shannon index and the relative abundance of Chloroflexi, Acidobacteria, Planctomycetes, Verrucomicrobia, Parcubacteria phyla groups. Our study demonstrates that intercropping of maize with mushroom affected the physical and chemical properties of the soil, and altered the structure and diversity of the soil microbial community. These results suggest that this crop production system could be a sustainable efficient agricultural management practice in Northeast China.

Supplementary Materials: The following are available online at http://www.mdpi.com/2073-4395/10/10/1526/s1, Table S1: Numbers of bacterial sequences and OTUs identified by 16S rRNA gene sequencing, Table S2: Pearson's correlation coefficients between relative abundances of dominant bacterial phyla and soil properties, Table S3: The environmental vectors onto two ordination of redundancy analysis, Table S4: The analysis of variance of the KEGG metabolic pathway.

Author Contributions: Conceptualization, F.S. and X.Z.; methodology, X.Y.; software, X.Y. and X.Z.; investigation, X.Y., L.S. and X.Q.; data curation, X.Y.; writing—original draft preparation, X.Y.; writing—review and editing, Y.W. and X.Z.; supervision, X.Q.; project administration, Y.W. and X.Z. All authors have read and agreed to the published version of the manuscript.

Funding: This research was funded by the "One-Three-Five" Strategic Planning Program of Chinese Academy of Sciences, grant number IGA-135-04.

Conflicts of Interest: The authors declare no conflict of interest.

References

1. NBSC (National Bureau of Statistics of China). *National Statistical Yearbook*; China Statistics Press: Beijing, China, 2017.
2. Zhu, X.; Sun, L.; Song, F.; Liu, S.; Liu, F.; Li, X. Soil microbial community and activity are affected by integrated agricultural practices in China. *Eur. J. Soil Sci.* **2018**, *69*, 924–935. [CrossRef]
3. Zaeem, M.; Nadeem, M.; Pham, T.H.; Ashiq, W.; Ali, W.; Gilani, S.S.M.; Elavarthi, S.; Kavanagh, V.; Cheema, M.; Galagedara, L.; et al. The potential of corn-soybean intercropping to improve the soil health status and biomass production in cool climate boreal ecosystems. *Sci. Rep.* **2019**, *9*, 13148. [CrossRef] [PubMed]
4. Fu, Z.; Zhou, L.; Chen, P.; Du, Q.; Pang, T.; Song, C.; Wang, X.; Liu, W.; Yang, W.; Yong, T. Effects of maize-soybean relay intercropping on crop nutrient uptake and soil bacterial community. *J. Integr. Agr.* **2019**, *18*, 2006–2018. [CrossRef]
5. De Freitas, M.A.M.; Silva, D.V.; Guimaraes, F.R.; Leal, P.L.; Moreira, F.M.D.; da Silva, A.A.; Souza, M.D. Biological attributes of soil cultivated with corn intercropped with *Urochloa brizantha* in different plant arrangements with and without herbicide application. *Agr. Ecosyst. Environ.* **2018**, *254*, 35–40. [CrossRef]
6. Nyawade, S.O.; Karanja, N.N.; Gachene, C.K.K.; Gitari, H.I.; Schulte-Geldermann, E.; Parker, M.L. Short-term dynamics of soil organic matter fractions and microbial activity in smallholder potato-legume intercropping systems. *Appl. Soil Ecol.* **2019**, *142*, 123–135. [CrossRef]
7. Bünemann, E.K.; Bongiorno, G.; Bai, Z.; Creamer, R.E.; De Deyn, G.; de Goede, R.; Fleskens, L.; Geissen, V.; Kuyper, T.W.; Mäder, P.; et al. Soil quality—A critical review. *Soil Biol. Biochem.* **2018**, *120*, 105–125. [CrossRef]
8. Jonathan, S.; Oyetunji, O.; Asemoloye, M.J.N. Asemoloye, Influence of spent mushroom compost (SMC) of *Pleurotus ostreatus* on the yield and nutrient compositions of *Telfairia occidentalis* Hook.F.A. (Pumpkin), a Nigerian leafy vegetable. *Nat. Sci.* **2012**, *10*, 149–156.
9. Wang, B. New mushroom cultivation technology between field crop maize rows. *J. Shanxi Agric. Sci.* **2009**, *37*, 91–92.
10. Griffiths, B.S.; Philippot, L. Insights into the resistance and resilience of the soil microbial community. *FEMS Microbiol. Rev.* **2013**, *37*, 112–129. [CrossRef]
11. Zheng, W.; Gong, Q.; Zhao, Z.; Liu, J.; Zhai, B.; Wang, Z.; Li, Z. Changes in the soil bacterial community structure and enzyme activities after intercrop mulch with cover crop for eight years in an orchard. *Eur. J. Soil Biol.* **2018**, *86*, 34–41. [CrossRef]
12. Gong, X.; Liu, C.; Li, J.; Luo, Y.; Yang, Q.; Zhang, W.; Yang, P.; Feng, B. Responses of rhizosphere soil properties, enzyme activities and microbial diversity to intercropping patterns on the *Loess Plateau* of China. *Soil Till. Res.* **2019**, *195*, 104355. [CrossRef]
13. Monneveux, P.; Quillerou, E.; Sanchez, C.; Lopez-Cesati, J. Effect of zero tillage and residues conservation on continuous maize cropping in a subtropical environment (Mexico). *Plant Soil* **2006**, *279*, 95–105. [CrossRef]
14. She, S.; Niu, J.; Zhang, C.; Xiao, Y.; Chen, W.; Dai, L.; Liu, X.; Yin, H.J.A.o.M. Significant relationship between soil bacterial community structure and incidence of bacterial wilt disease under continuous cropping system. *Arch. Microbiol.* **2017**, *199*, 267–275. [CrossRef] [PubMed]
15. Chen, S.; Hou, D.; Wu, W.; Sun, W.; Qiu, L. Influence of interplanting *Pleurotus ostreatus* on soil biological activity and fruit quality in pear orchard. *J. Fruit Sci.* **2012**, *29*, 583–588.
16. Mohamed, M.F.; Nassef, D.M.T.; Waly, E.A.; Kotb, A.M. Production of oyster mushroom (*Pleurotus* spp.) intercropped with field grown faba bean (*Vicia faba* L.). *Asian. J. Crop Sci.* **2014**, *6*, 27–37. [CrossRef]
17. Yang, X.; Wang, Y.; Qi, X.; Sun, L.; Zhang, M.; Song, F.; Liu, S.; Li, X.; Zhu, X. Effects of maize/mushroom intercropping on photosynthetic characteristics of maize ear leaf under high photosynthetic efficiency planting pattern. *Soil Crops* **2020**, *9*, 166–177.
18. Sun, L.; Song, F.; Liu, S.; Cao, Q.; Liu, F.; Zhu, X. Integrated agricultural management practice improves soil quality in Northeast China. *Arch. Agron. Soil Sci.* **2018**, *64*, 1932–1943. [CrossRef]
19. Zhu, X.; Yang, W.; Song, F.; Li, X. Diversity and composition of arbuscular mycorrhizal fungal communities in the cropland black soils of China. *Glob. Ecol. Conserv.* **2020**, *22*, e00964. [CrossRef]
20. Gu, Z.; Li, G.; Du, Y.; Li, J.; Cai, F.; Wang, H. Investigation on parameters affecting the determination of organic matter in forest soil by potassium dichromate heating oxidation-volumetric method. *Hubei For. Sci. Technol.* **2014**, *43*, 24–26.

21. Lei, Y.; Xiao, Y.; Li, L.; Jiang, C.; Zu, C.; Li, T.; Cao, H. Impact of tillage practices on soil bacterial diversity and composition under the tobacco-rice rotation in China. *J. Microbiol.* **2017**, *55*, 349–356. [CrossRef]
22. Magoč, T.; Salzberg, S.L. FLASH: Fast length adjustment of short reads to improve genome assemblies. *Bioinformatics* **2011**, *27*, 2957–2963. [CrossRef] [PubMed]
23. Caporaso, J.G.; Kuczynski, J.; Stombaugh, J.; Bittinger, K.; Bushman, F.D.; Costello, E.K.; Fierer, N.; Peña, A.G.; Goodrich, J.K.; Gordon, J.I.J.N.M. QIIME allows analysis of high-throughput community sequencing data. *Nat. Methods* **2010**, *7*, 335–336. [CrossRef] [PubMed]
24. Edgar, R.C. Search and clustering orders of magnitude faster than BLAST. *Bioinformatics* **2010**, *26*, 2460–2461. [CrossRef] [PubMed]
25. Bokulich, N.A.; Subramanian, S.; Faith, J.J.; Gevers, D.; Gordon, J.I.; Knight, R.; Mills, D.A.; Caporaso, J.G. Quality-filtering vastly improves diversity estimates from Illumina amplicon sequencing. *Nat. Methods* **2013**, *10*, 57. [CrossRef]
26. DeSantis, T.Z.; Hugenholtz, P.; Larsen, N.; Rojas, M.; Brodie, E.L.; Keller, K.; Huber, T.; Dalevi, D.; Hu, P.; Andersen, G.L. Greengenes, a chimera-checked 16S rRNA gene database and workbench compatible with ARB. *Appl. Environ. Microbiol.* **2006**, *72*, 5069–5072. [CrossRef]
27. Segata, N.; Izard, J.; Waldron, L.; Gevers, D.; Miropolsky, L.; Garrett, W.S.; Huttenhower, C. Metagenomic biomarker discovery and explanation. *Genome Biol.* **2011**, *12*, R60. [CrossRef]
28. Langille, M.G.I.; Zaneveld, J.; Caporaso, J.G.; McDonald, D.; Knights, D.; Reyes, J.A.; Clemente, J.C.; Burkepile, D.E.; Vega Thurber, R.L.; Knight, R.; et al. Predictive functional profiling of microbial communities using 16S rRNA marker gene sequences. *Nat. Biotechnol.* **2013**, *31*, 814–821. [CrossRef]
29. Fuentes, M.; Govaerts, B.; De León, F.; Hidalgo, C.; Dendooven, L.; Sayre, K.D.; Etchevers, J. Fourteen years of applying zero and conventional tillage, crop rotation and residue management systems and its effect on physical and chemical soil quality. *Eur. J. Agron.* **2009**, *30*, 228–237. [CrossRef]
30. Li, Q.; Chen, J.; Wu, L.; Luo, X.; Li, N.; Arafat, Y.; Lin, S.; Wang, X. Belowground interactions impact the soil bacterial community, soil fertility, and crop yield in maize/peanut intercropping systems. *Int. J. Mol. Sci.* **2018**, *19*, 622. [CrossRef]
31. Sun, Y.; Zhang, N.; Wang, E.; Yuan, H.; Yang, J.; Chen, W. Influence of intercropping and intercropping plus rhizobial inoculation on microbial activity and community composition in rhizosphere of alfalfa (*Medicago sativa* L.) and Siberian wild rye (*Elymus sibiricus* L.). *FEMS Microbiol. Ecol.* **2010**, *70*, 62–70.
32. Du, B.; Pang, J.; Hu, B.; Allen, D.E.; Bell, T.L.; Pfautsch, S.; Netzer, F.; Dannenmann, M.; Zhang, S.; Rennenberg, H. N$_2$-fixing black locust intercropping improves ecosystem nutrition at the vulnerable semi-arid Loess Plateau region, China. *Sci. Total Environ.* **2019**, *688*, 333–345. [CrossRef] [PubMed]
33. de Oliveira, A.B.; Cantarel, A.A.M.; Seiller, M.; Florio, A.; Bérard, A.; Hinsinger, P.; Cadre, E.L. Short-term plant legacy alters the resistance and resilience of soil microbial communities exposed to heat disturbance in a *Mediterranean calcareous* soil. *Ecol. Indic.* **2020**, *108*, 105740. [CrossRef]
34. Kebeney, S.J.; Semoka, J.M.R.; Msanya, B.M.; Ng'Etich, W.K.; Chemei, D.K. Effects of nitrogen fertilizer rates and soybean residue management on nitrate nitrogen in sorghum-soybean intercropping system. *Int. J. Plant Soil Sci.* **2014**, *4*, 212–229. [CrossRef]
35. Phiri, A.T.; Weil, R.R.; Yobe, G.; Msaky, J.J.; Mrema, J.; Grossman, J.; Harawa, R. Insitu assessment of soil nitrate-nitrogen in the pigeon pea-groundnut intercropping-maize rotation system: Implications on nitrogen management for increased maize productivity. *Int. Res. J. Agric. Sci. Soil Sci.* **2014**, *2*, 13–29.
36. Vieira, F.R.; Pecchia, J.A.; Segato, F.; Polikarpov, I. Exploring oyster mushroom (*Pleurotus ostreatus*) substrate preparation by varying phase I composting time: Changes in bacterial communities and physicochemical composition of biomass impacting mushroom yields. *J. Appl. Microbiol.* **2019**, *126*, 931–944. [CrossRef]
37. Kim, D.G. Estimation of net gain of soil carbon in a nitrogen-fixing tree and crop intercropping system in sub-Saharan Africa: Results from re-examining a study. *Agroforest Syst.* **2012**, *86*, 175–184. [CrossRef]
38. Dos Santos Soares, D.; Ramos, M.L.G.; Marchão, R.L.; Maciel, G.A.; de Oliveira, A.D.; Malaquias, J.V.; de Carvalho, A.M. How diversity of crop residues in long-term no-tillage systems affect chemical and microbiological soil properties. *Soil Tillage Res.* **2019**, *194*, 104316. [CrossRef]
39. Trivedi, P.; Delgadobaquerizo, M.; Anderson, I.C.; Singh, B.K. Response of soil properties and microbial communities to agriculture: Implications for primary productivity and soil health indicators. *Front. Plant Sci.* **2016**, *7*, 990. [CrossRef]

40. Yang, G.; Ma, K.; Lu, F.; Wei, C.; Dai, X. Effect of continuous cropping of potato on allelochemicals and soil microbial community. *J. Ecol. Rural Environ.* **2015**, *31*, 711–717.

41. Qin, X.; Zheng, Y.; Tang, L.; Long, G. Effects of maize and potato intercropping on rhizosphere microbial community structure and diversity. *Acta Agron. Sin.* **2015**, *41*, 919–928. [CrossRef]

42. Acosta-Martínez, V.; Acosta-Mercado, D.; Sotomayor-Ramírez, D.; Cruz-Rodríguez, L. Microbial communities and enzymatic activities under different management in semiarid soils. *Appl. Soil Ecol.* **2008**, *38*, 249–260. [CrossRef]

43. Zhang, M.; Wang, N.; Hu, Y.; Sun, G. Changes in soil physicochemical properties and soil bacterial community in mulberry (*Morus alba* L.)/alfalfa (*Medicago sativa* L.) intercropping system. *Microbiol. Open* **2018**, *7*, e00555. [CrossRef] [PubMed]

44. Lupwayi, N.Z.; May, W.E.; Kanashiro, D.A.; Petri, R.M. Soil bacterial community responses to black medic cover crop and fertilizer N under no-till. *Appl. Soil Ecol.* **2018**, *124*, 95–103. [CrossRef]

45. Yu, H.; Chen, S.; Zhang, X.; Zhou, X.; Wu, F. Rhizosphere bacterial community in watermelon-wheat intercropping was more stable than in watermelon monoculture system under *Fusarium oxysporum* f. sp. niveum invasion. *Plant Soil* **2019**, *445*, 369–381. [CrossRef]

46. Hugenholtz, P.; Goebel, B.M.; Pace, N.R. Impact of culture-independent studies on the emerging phylogenetic view of bacterial diversity. *J. Bacteriol.* **1998**, *180*, 4765–4774. [CrossRef]

47. Thiel, V.; Fukushima, S.I.; Kanno, N.; Hanada, S. Chloroflexi. In *Encyclopedia of Microbiology*, 4th ed.; Schmidt, T.M., Ed.; Academic Press: Cambridge, MA, USA, 2019; pp. 651–662.

48. Chistoserdova, L.; Jenkins, C.; Kalyuzhnaya, M.G.; Marx, C.J.; Lapidus, A.; Vorholt, J.A.; Staley, J.T.; Lidstrom, M.E. The enigmatic planctomycetes may hold a key to the origins of methanogenesis and methylotrophy. *Mol. Biol. Evol.* **2004**, *21*, 1234. [CrossRef]

49. Kindaichi, T.; Yamaoka, S.; Uehara, R.; Ozaki, N.; Ohashi, A.; Albertsen, M.; Nielsen, P.H.; Nielsen, J.L. Phylogenetic diversity and ecophysiology of candidate phylum Saccharibacteria in activated sludge. *FEMS Microbiol. Ecol.* **2016**, *92*, fiw078. [CrossRef]

50. Ahn, J.H.; Lee, S.A.; Kim, J.M.; Kim, M.S.; Song, J.; Weon, H.Y. Dynamics of bacterial communities in rice field soils as affected by different long-term fertilization practices. *J. Microbiol.* **2016**, *54*, 724–731. [CrossRef]

51. Wackett, L.P. *Pseudomonas* putida-a versatile biocatalyst. *Nat. Biotechnol.* **2003**, *21*, 136–138. [CrossRef]

52. Bever, J.D.; Dickie, I.A.; Facelli, E.; Facelli, J.M.; Klironomos, J.; Moora, M.; Rillig, M.C.; Stock, W.D.; Tibbett, M.; Zobel, M. Rooting theories of plant community ecology in microbial interactions. *Trends Ecol. Evol.* **2010**, *25*, 468–478. [CrossRef]

53. Lladó, S.; Žifčáková, L.; Větrovský, T.; Eichlerová, I.; Baldrian, P. Functional screening of abundant bacteria from acidic forest soil indicates the metabolic potential of Acidobacteria subdivision 1 for polysaccharide decomposition. *Biol. Fertil. Soils* **2016**, *52*, 251–260. [CrossRef]

54. Rashad, F.M.; Fathy, H.M.; El-Zayat, A.S.; Elghonaimy, A.M. Isolation and characterization of multifunctional Streptomyces species with antimicrobial, nematicidal and phytohormone activities from marine environments in Egypt. *Microbiol. Res.* **2015**, *175*, 34–47. [CrossRef] [PubMed]

55. Binnie, C.; Cossar, J.D.; Stewart, D.I.H. Heterologous biopharmaceutical protein expression in Streptomyces. *Trends Biotechnol.* **1997**, *15*, 315–320. [CrossRef]

56. Tao, T.; Yue, Y.; Chen, W.; Chen, W. Proposal of Nakamurella gen. nov. as a substitute for the bacterial genus Microsphaera Yoshimi et al. 1996 and Nakamurellaceae fam. nov. as a substitute for the illegitimate bacterial family Microsphaeraceae Rainey et al. 1997. *Int. J. Syst. Evol. Microbiol.* **2004**, *54*, 999–1000. [CrossRef] [PubMed]

57. Subhashini, D. Growth promotion and increased potassium uptake of tobacco by potassium-mobilizing bacterium Frateuria aurantia grown at different potassium levels in vertisols. *Commun. Soil Sci. Plant Anal.* **2015**, *46*, 210–220. [CrossRef]

58. Li, F.; Chen, L.; Zhang, J.; Yin, J.; Huang, S. Bacterial community structure after long-term organic and inorganic fertilization reveals important associations between soil nutrients and specific taxa involved in nutrient transformations. *Front. Microbiol.* **2017**, *8*, 187. [CrossRef]

59. Kojima, M.; Kimura, N.; Miura, R. Regulation of primary metabolic pathways in oyster mushroom mycelia induced by blue light stimulation: Accumulation of shikimic acid. *Sci. Rep.* **2015**, *5*, 8630. [CrossRef]

60. Ngwene, B.; Neugart, S.; Baldermann, S.; Ravi, B.; Schreiner, M. Intercropping induces changes in specific secondary metabolite concentration in Ethiopian Kale (*Brassica carinata*) and African Nightshade (*Solanum scabrum*) under controlled conditions. *Front. Plant Sci.* **2017**, *8*, 1700. [CrossRef]
61. Zhang, X.; Gao, G.; Wu, Z.; Wen, X.; Zhong, H.; Zhong, Z.; Bian, F.; Gai, X. Agroforestry alters the rhizosphere soil bacterial and fungal communities of moso bamboo plantations in subtropical China. *Appl. Soil Ecol.* **2019**, *143*, 192–200. [CrossRef]

 **agronomy**

Article

Compost as an Option for Sustainable Crop Production at Low Stocking Rates in Organic Farming

Christopher Brock [1,*], Meike Oltmanns [1], Christoph Matthes [2], Ben Schmehe [2], Harald Schaaf [3], Detlef Burghardt [3], Hartmut Horst [3] and Hartmut Spieß [2]

[1] Forschungsring e.V., Brandschneise 5, 64295 Darmstadt, Germany; oltmanns@forschungsring.de
[2] Forschung & Züchtung Dottenfelderhof, Dottenfelderhof 1, 61118 Bad Vilbel, Germany;
     christoph.matthes@dottenfelderhof.de (C.M.); ben.schmehe@dottenfelderhof.de (B.S.);
     h.spiess@dottenfelderhof.de (H.S.)
[3] Landesbetrieb Hessisches Landeslabor, Am Versuchsfeld 13, 34128 Kassel, Germany;
     harald.schaaf@outlook.de (H.S.); detlef.burghardt@lhl.hessen.de (D.B.); apiverdi@gmx.de (H.H.)
*   Correspondence: brock@forschungsring.de; Tel.: +49-6155-8421-23

Citation: Brock, C.; Oltmanns, M.; Matthes, C.; Schmehe, B.; Schaaf, H.; Burghardt, D.; Horst, H.; Spieß, H. Compost as an Option for Sustainable Crop Production at Low Stocking Rates in Organic Farming. *Agronomy* **2021**, *11*, 1078. https://doi.org/10.3390/agronomy11061078

Academic Editors: Nikolaos Monokrousos and Efimia M. Papatheodorou

Received: 14 April 2021
Accepted: 17 May 2021
Published: 27 May 2021

Publisher's Note: MDPI stays neutral with regard to jurisdictional claims in published maps and institutional affiliations.

Abstract: Mixed-crop-livestock farms offer the best conditions for sustainable nutrient management in organic farming. However, if stocking rates are too low, sustainability might be threatened. Therefore, we studied the development of soil organic matter and nutrients as well as crop yields over the first course of a new long-term field experiment with a mimicked cattle stocking rate of 0.6 LU ha^{-1}, which is the actual average stocking rate for organic farms in Germany. In the experiment, we tested the effects of additional compost application to improve organic matter supply to soils, and further, potassium sulfate fertilization for an improved nutrition of fodder legumes. Compost was made from internal resources of the farm (woody material from hedge-cutting). Soil organic matter and nutrient stocks decreased in the control treatment, even though yield levels, and thus nutrient exports, were comparably low. With compost application, soil organic matter and nutrient exports could be compensated for. At the same time, the yields increased but stayed at a moderate level. Potassium sulfate fertilization further improved N yields. We conclude that compost from internal resources is a viable solution to facilitate sustainable organic crop production at low stocking rates. However, we are aware that this option does not solve the basic problem of open nutrient cycles on the farm gate level.

Keywords: long term field experiment; sustainable crop production; nutrient balances; legume nutrition

1. Introduction

Soil organic matter is recognized as a key factor of soil fertility [1]. For this reason, the supply of soils with organic matter was always a major concern in organic agriculture. Meanwhile, it was shown that organic farming in fact leads to higher soil organic matter levels than conventional management [2]. However, a sufficient supply of soils with organic matter is not an effect of organic farming per se, but of the specific structure of organic systems. Leithold et al. [3] emphasized that fodder legumes and cattle manure are the basic factors for a sufficient supply of soils with organic matter. These factors must balance the loss of soil organic matter in turnover. If the supply of soil with organic matter is too low to meet the specific requirements, SOM levels might decrease even under organic management. This situation was observed in the OAFEG long-term field experiment in Germany that is designed to study the effects of mixed, as compared to stockless organic farming [4]. Under the conditions of this experiment, SOM stocks increased under the mixed farming treatment, but stayed unchanged or even decreased under the two stockless treatments. In a modeling study, Brock et al. [5] calculated that the actual average soil organic matter balance of organic farming in Germany was slightly negative, as the mean animal stocking rate was only 0.63 LU per ha at that time. Even though this result should not be overrated due to the high uncertainty of the calculation, it seems necessary to further study soil organic matter changes under organic management with low stocking rates or even stockless systems. If manure availability is too low, farmers will need to utilize further sources of organic matter. Here, green manure and compost are the most important options.

The demand for organic matter to maintain or even increase soil organic matter stocks is dependent on site conditions, management history, and actual management [6]. Organic inputs of plant roots and residues, animal manure, and other material must balance the loss of organic matter in turnover. As organic matter supply and turnover are directly linked to N supply in organic farming, the demand for organic matter is greater with higher yield levels (of non-legumes), due to the export of mineralized N [3].

Compost is reported as a viable option to increase soil organic matter and soil health [7,8]. In principle, composting is the biological decomposition of organic residues [9]. Compost can be made from different substrates, e.g., municipal waste, sewage sludges, plant residues/green waste, farmyard manure, or biogas production residues. Farm compost, as applied in the field experiment reported in this study, is carried out individually in farms, depending on the available materials. However, Lehtinen et al. [10] found that impacts on soil properties and crop yields were not significantly different between the composts made from municipal waste, sewage sludge, green waste, and farmyard manure in a long-term field experiment, even though the macronutrient inputs differed. Further, microbial biomass and the composition of the microbial community differed between the treatments [11].

In general, compost application builds up soil organic matter [12] and enhances crop yields moderately in the short run [13]. In the long run, the build-up of soil organic matter further improves the growing conditions for arable crops and thereby further increases yields [14,15].

The biological N fixation (BNF) is an important source of N for organic crop rotations because mineral sources of N fertilizers that are allowed for organic farming are limited. Especially in organic agriculture, BNF is preferred due to different advantages, as compared to mineral N sources like higher N use efficiency of the plants and decreased volatilization, denitrification, and leaching [16]. Therefore, nitrogen fixing legumes like clover and lucerne are usually placed at the beginning of organic crop rotations and act as drivers for the subsequent crops. However, clover and lucerne react particularly sensitively to the deficiencies of P, K, and S. Although several processes and mechanisms about the dependency of legume growth to the listed elements remain unclear [17] it is evident that a good supply improves crop growth and health. It is also known that legumes that acquire N by BNF have a higher demand of P, K, and S, as compared to those that rely on soil N only [18,19]. It is generally accepted that when the host plant growth is reduced due to deficiencies of P, K, or S, an N-feedback is triggered so that the nodule development and activity is reduced. This mechanism can also be induced by plant diseases and pathogens, as well as abiotic stresses like drought, toxic levels of salt, or heavy metals [20–22].

In this study, we showed the development of crop yields and soil nutrients and organic matter over the first crop rotation in a long-term field experiment, under conditions of organic farming (more specifically—biodynamic farming). The experiment mimicked a mixed farm with a stocking rate of 0.6 LU cattle per hectare, which corresponded to the average stocking rate of organic farms in Germany. In this experiment, we compared a fertilization regime that was based on the available cattle manure with a regime that additionally utilized a farm compost made from available plant residues on the farm. Further, we examined the effect of potassium sulfate application, which was owed to the fact that the experiment was located on a potassium-fixing soil.

As the field experiment is still in an early stage, we can only study the short-term effects and development factors, rather than development trends. In this stage, we expect the positive effects of compost application on crop yields, and increased biological N fixation rates in legumes with potassium sulfate application. Further, we want to study the impact of the fertilization regimes on soil nutrient and organic matter balances. This is of high relevance in organic farming, as crop production is largely dependent on soil fertility.

2. Materials and Methods

We analyzed crop yields and the development of nutrient (N, P, K, S) and organic matter stocks in the soils under the four treatments, in a long-term field experiment on a luvisol, under conditions of biodynamic farming. Further, we calculated nutrient and soil organic matter balances to support the assessment of factor treatment effects, and modelled opportunities to improve organic matter supply to soils.

2.1. Experimental Site and Trial Design

The long-term field trial was initiated in 2010 Germany, Hesse (50°11′39.0″ N 8°45′09.5″ E) at 120 m above the sea level. It is maintained by the on-farm research and breeding department Dottenfelderhof. The soil type is a Haplic Luvisol with Silt loam from loess [23]. The average precipitation is 630 mm per year with an average temperature of 9.4 °C.

The farm was converted from conventional to biodynamic agriculture in 1968. In the time of conventional practice, sugar beet was cropped as a monoculture for many years. Since the conversion, the crop rotation consisted of a two times six year rotation with a legume/grass mixture in year one and two; winter wheat in year three; winter rye in year four; root crops in year five; and a spring cereal in year six. The legume/grass mixture alternated between clover/grass and alfalfa/grass from one six-year cycle to the next. Root crops varied widely and could be maize, potatoes, carrots, or other. The spring cereals are usually oats or spring wheat. In the rotation under study, it is important to notice that fodder maize was planted instead of winter rye in 2015 and clover/grass was ploughed and reseeded in 2013, because of drought and winter damage.

All treatments receive the same biodynamic preparations [24], i.e., BD 500 and BD 501 spray, at least once a year each. The compost used for the experiment was prepared with the usual biodynamic compost preparations and was made on site.

The trial was initiated in spring 2010 as a one factorial Latin square design with four treatments on plots of 48 m^2 gross area (6 × 8 m) and 29.25 m^2 net area (4.5 × 6.5 m). On all plots, an equivalent livestock unit (LU) of 0.6 cattle deep litter (06M) was applied. Treatments 2 and 4 were treated with potassium sulfate (K), and treatment 3 and 4 with biodynamic compost (BD).

1. Control (06M).
2. Potassium sulfate (06M + K).
3. Biodynamic plant-based compost (06M + BD).
4. Biodynamic plant-based compost + potassium sulfate (06M + BD + K).

The cattle deep litter was a fermented manure from the farms' dairy cow herd. A total of 70% of the cow manure was distributed evenly, daily in the stable, and covered with straw. Cow pat pit preparation was added daily and compost preparations were applied once a month. The deep litter was harvested after the rye harvest and worked into the soil before the root crops were planted.

Potassium sulfate was produced by the fertilizer company K + S, under the tradename "Kalisop" and consisted of 50% water-soluble potassium oxide (K$_2$O) and 45% water-soluble sulfur trioxide (SO$_3$).

The biodynamic compost consisted of 85–90% green chop, 5–8% cow manure, and 5–7% soil. To speed up the process, the material was mixed daily in the first week and prepared with the cow pat pit during this time. After that, the single biodynamic compost preparations were added for the first time. Whey from the farm dairy or water was added to keep the right moisture content, which should be over 60% to avoid overheating and thus losses of nutrients, because the initial material was usually too dry. To protect the compost from rain, it was covered with a compost membrane. After the initial week and during the following half year process of composting, the compost pile was turned three to four times. After three months, the biodynamic compost preparations were added a second time.

Table 1 shows that the climatic water balance according to Haude [25] was negative from 2012 until 2015, and was positive in 2016 and 2017.

Table 1. Mean annual temperature, annual precipitation, and climatic water balance [25] during the investigation period.

Year	Mean Annual Temperature (°C)	Total Annual Precipitation (mm)	Annual Climatic Water Balance (mm)
2012	10.6	572.8	−153.8
2013	10.5	664.4	−10.2
2014	12.0	669.7	−72.5
2015	11.7	434.7	−408.4
2016	10.7	729.0	159.8
2017	11.0	717.9	157.3

2.2. Fertilizer and Manure Application

The applied amounts of manure and fertilizer are shown in Table 2, except an application of 2 Mg ha^{-1} lime ($CaCO_3$ with 56% CaO) on all treatments in November 2009, because the pH was too low at the start of the experiment. The cattle deep litter was applied on all treatments before planting of root crops once in a 6-year rotation.

Table 2. Crop rotation, amounts of organic amendment, and total nutrient amounts (kg ha^{-1}) applied with all fertilizers (cattle deep litter, compost, and K_2SO_4).

Year	Crop	Treatment	Cattle Manure [1]	Compost [2]	N	P	K [3]	S [3]
			Mg ha^{-1} fresh matter		kg ha^{-1} dry matter			
2010	potato	06M	40.0		208.0	120.6	271.9	45.0
		06M + K	40.0		208.0	120.6	671.9	218.5
		06M + BD	40.0	30.0	344.2	150.2	383.8	67.3
		06M + BD + K	40.0	30.0	344.2	150.2	656.6	185.6
2011	oat/clover grass	06M						
		06M + K					400.0	173.5
		06M + BD		15.0	51.7	17.2	50.7	8.5
		06M + BD + K		15.0	51.7	17.2	400.0	160.1
2012	clover grass							
2013	clover grass							
2014	winter wheat	06M						
		06M + K						
		06M + BD		30.0	87.9	26.6	112.3	13.5
		06M + BD + K		30.0	87.9	26.6	112.3	13.5
2015	fodder maize	06M						
		06M + K					300	130.3
		06M + BD		30.0	98.4	59.3	269.6	29.5
		06M + BD + K		30.0	98.4	59.3	419.6	94.6
2016	red beet	06M	35.0		168.3	41.3	231.5	19.7
		06M + K	35.0		168.3	41.3	431.5	106.6
		06M + BD	35.0	30.0	294.9	92.4	426.1	40.4
		06M + BD + K	35.0	30.0	294.9	92.4	476.1	62.1
2017	spring wheat	06M						
		06M + K					400	173.5
		06M + BD		30.0	123.6	39.3	160.4	19.2
		06M + BD + K		30.0	123.6	39.3	480.4	158

[1] cattle manure from deep litter; [2] compost = plant-based compost; [3] K and S was applied at the same time as the compost in 2010, 2011, 2015, 2016, and 2017 as K_2SO_4.

The amount was calculated to represent 0.6 LU ha^{-1} and was applied in spring 2010, before planting of potatoes (40 Mg ha^{-1}) and in spring 2016 before planting of red beet (35 Mg ha^{-1}). The same amount of compost (30 Mg ha^{-1}) was applied on the 06M + BD and 06M + BD + K treatment in 2010 and from 2014 to 2017, after calculating the maximum allowed N amount by the German fertilizer regulation. In 2011, the applied amount of compost was 15 Mg ha^{-1}. Potassium sulfate was applied on the 06M + K treatment in three subsequent years from 2015 to 2017, in an amount that was derived from previous dosing tests.

2.3. Soil Samples and Chemical Analyses

Soil samples were taken every year after harvest or in autumn, for clover grass, from a soil depth of 0–30 cm. These were then mixed and sent to the laboratory "Hessisches Landeslabor" (LHL).

Soil organic carbon (SOC) were analyzed by combustion at 550 °C under O_2, using Leco$^®$ RC612 carbon analyzer. Total N were measured by the dry combustion method until 2012, according to DIN ISO 13878 [26], and afterwards according to DIN EN 16168 [27]. Total K, S, and P were determined by inductively coupled plasma optical emission spectrometry [28]. Soil pH was measured 1:10 in 0.01 M $CaCl_2$ [29]. Soil bulk density was calculated as the dry weight of soil divided by its volume and as a mean of replications at the end of the rotation [30].

2.4. Yield and Samples for Crop Nutrients

Clover grass was cut three times during the vegetation period at 12 June, 1 August, and 10 October 2012, and two times in 2013 at 19 June and 24 September. The harvest from the net plots was weighed to determine fresh matter yield. A 5 kg mixed sample of harvest was chopped and from this material 2 × 1 kg was dried at 105 °C in an oven, to determine dry weight yield. Samples for the analyses of nutrient content were taken from the chopped material.

In 2014, winter wheat cv. Butaro was harvested with a Hege 125 combine. Grain and straw were weighed separately for fresh matter yield. The straw was processed in an analogous manner to clover grass, for determination of dry weight and laboratory samples.

From maize cv. Colisee in 2015, grain was harvested on 9 September by hand. The straw was harvested one day later with a maize chopper. Maize straw was processed in an analogous manner to clover grass. Red beet cv. Robuschka was harvested on 14 September by hand. Stem and leaves were separated from the bulbs, and fresh matter yield was determined separately. From both portions, a mixed sample of 2 kg was taken and sent to the laboratory. Sping wheat cv. Heliaro was harvested on 4 August and processed in an analogous manner to winter wheat.

Crop nutrients (P, K, and S) were measured with X-ray fluorescence spectroscopy, according to VDLUFA Volume III [31]. Dry matter and N were determined according to ISO 12099 [32].

2.5. Soil Surface Nutrient Balance

The nutrient balances were calculated from 2012 until 2017, because this was a full cycle of the crop rotation, beginning with the legume-grass mixture, until the spring cereal.

The N, P, K, and S balances were annually estimated as the difference between nutrient input and nutrient output (kg ha^{-1} year^{-1}):

$$\text{Nutrient budget} = \text{nutrient input} - \text{nutrient output} \tag{1}$$

Where the nutrient inputs included fertilization (deep litter manure, plant-based compost, and potassium sulfate) and crop seeds, the outputs included harvested aboveground biomass (main and side product).

For the N balance, the N inputs were extended by atmospheric N depositions, asymbiotic N fixation, and symbiotic N fixation. The N atmospheric deposition were estimated at 15 kg ha^{-1} year^{-1}, and the asymbiotic nitrogen fixation were 5 kg ha^{-1} year^{-1}.

The symbiotic N fixation was estimated according to the Stein-Bachinger [33]:

$$\text{Symbiotic N fixation} = (N_{shoot} + N_{root} + \text{stubble}) \times Leg_{share} \times Ndfa \tag{2}$$

where N_{shoot} was calculated as the product of grass-clover biomass and the N concentrations. $N_{root+stubble}$ was calculated as the product of grass-clover biomass and the fix value of 0.75 for the root and stubble biomass, and the totally fixed root and stubble N (1.5%). For the Leg_{share}, we assumed the fix value 0.7 and for the Ndfa (nitrogen derived from the atmosphere) it was 0.8, respectively.

2.6. Soil Organic Matter Balance (HU-MOD)

The HU-MOD model [34,35] was developed as a decision support tool for application in farming practice. Unlike most other so-called humus balance methods, this model was conceptually able to analyze and predict soil organic matter changes [36]. The estimation of soil organic matter changes was based on the calculation of a coupled C and N balance in the soil–plant system. In principle, the model assumed that N in plant biomass could be used as a proxy for soil organic matter mineralization, if the N was supplied from other sources (here, atmospheric deposition, fertilizers, and—for legumes—biological nitrogen fixation) were considered. Thus, soil organic matter loss was calculated according to:

$$\text{SOMLOSS (kg SOM-N ha}^{-1}) = NPB - NFIX - NDEP - NFTLZ \tag{3}$$

NPB = N in total plant biomass (including roots), NFIX = N from biological fixation (legumes only), NDEP = N from atmospheric deposition, and NFTLZ = N from organic and mineral fertilizers.

SOM-N was transferred to SOM-C, based on the C:N ratio of the soil under assessment. Regarding the formation of new soil organic matter, the model applied a stoichiometric assumption, where the build-up of soil organic matter could be limited both by C and N availability. Again, the C:N ratio of the soil at the site under assessment was taken as a reference. Soil organic matter gain was therefore calculated according to:

$$\text{SOMGAIN (kg SOM-C ha}^{-1}) = MIN(CREM; NREM \times SITECN) \tag{4}$$

CREM = C from organic material (including plant roots), NREM = remaining N in the soil from organic material (including plat roots) and other inputs after consideration of losses, SITECN = reference C:N ratio of the soil at the site under assessment (topsoil C:N ratio was used as a proxy).

In the calculation of remaining C and N for the soil organic matter build-up, organic C and N inputs as well as mineral N inputs were considered. Losses of N in turnover were accounted for.

The model was successfully evaluated in several long-term and even in short-term field experiments [34,35,37].

2.7. Statistical Analysis

Data were analyzed using analysis of variance (ANOVA) for a Latin square design using SAS$^{®}$ Studio 3.8. Data normality was tested using the Shapiro–Wilk test ($p < 0.05$). Tukey's honestly significance difference (HSD) was used as a post-hoc mean separation test ($p < 0.05$), where the ANOVA performed significant. N stocks of 2014, 2016, and 2017 were reciprocally transformed.

3. Results

3.1. Development of Soil Organic Matter and Nutrient Levels in the Soil

In the course of the experiment, we observed an oscillating development of both carbon nitrogen stocks in soils, which were more pronounced with C (Figure 1a,b). The highest C values were measured in 2012 and 2017, which was at the start and the end of the first regular crop rotation. With N, the highest values were measured in 2010 and 2016/2017. In 2014–2016, the treatments with additional application of plant-based compost (06M + BD and 06M + BD + K) showed significantly higher stocks of SOC as compared to treatments without compost application (06M and 06M + K). The application of plant-based compost also led to a significant differentiation in the soil N stocks between the treatments in 2011 and after 2016. Nevertheless, soil total N stocks decreased from 2009 to 2017 by 17.7% and 12% for 06M and 06M + K, respectively. The 06M + BD and 06M + BD + K treatments maintained the initial values.

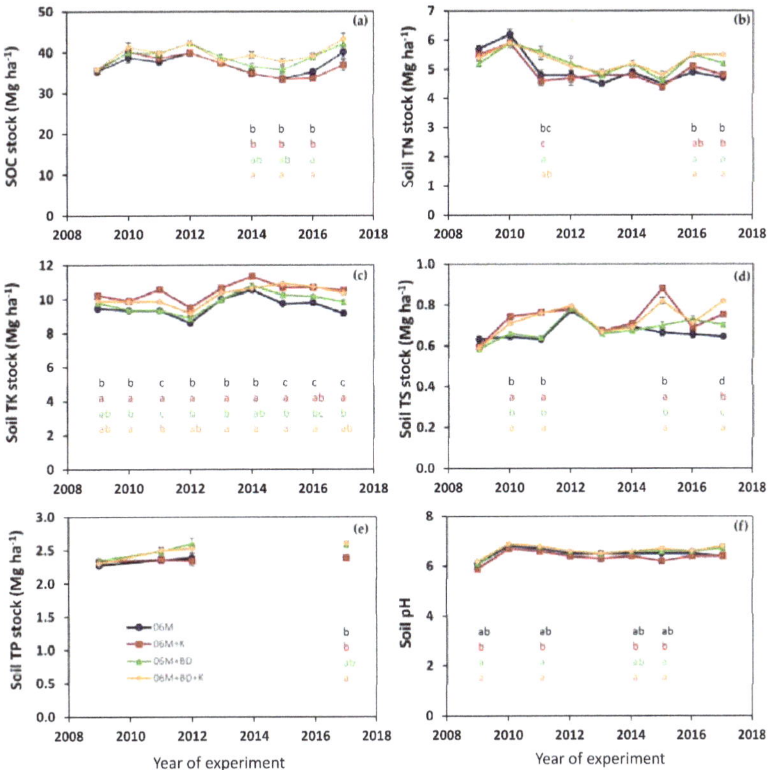

Figure 1. (**a**) Evolution of soil organic carbon stocks (Mg ha^{-1}), (**b**) soil total N stocks (Mg ha^{-1}), (**c**) soil total K stocks (Mg ha^{-1}), (**d**) soil total S stocks (Mg ha^{-1}), (**e**) soil total P stocks (Mg ha^{-1}), and soil pH (**f**) in the soil layer of 0–30 cm, over the period of 2009–2017, as affected by different fertilization treatments. Error bars represent the standard error of the mean value. Different letters within a year are significantly different at $p < 0.05$.

Potassium (K) stocks were also oscillating, but the pattern was different from that of C and N (Figure 1c). In 06M + K, the highest values were measured after the potassium sulfate fertilization events in 2011 and 2014 (cf. Table 2). In 06M + BD + K, however, these events could not be identified. As expected, potassium sulfate fertilization with and without compost application led to significantly higher soil total K stocks. As compared to

06M, all other treatments maintained or increased K stocks by 3% and 5% for 06M + K and 06M + BD + K, respectively.

Sulfur (S) stocks were higher in 06M + K and 06M + BD + K after potassium sulfate application in 2010, 2015, and 2017, but not in the other years with additional K and S fertilization in these treatments (Figure 1d). The highest increase in S stocks was observed for 06M + BD + K (+ 38.8%), followed by 06M + K (+ 26.5%) and 06M + BD (+ 20%).

Moreover, the 06M + BD + K treatment resulted in a significant increase in the soil total P stock (2.61 Mg ha^{-1}), as compared to the 06M (2.38 Mg ha^{-1}) and 06M+K (2.38 Mg ha^{-1}) treatments (Figure 1e), whereas 06M + BD were not significantly different from other treatments.

After 9 years, the soil pH in 06M, 06M + K, 06M + BD, and 06M + BD + K treatments were 0.3, 0.5, 0.6, and 0.4 units higher than the initial value in 2009 (Figure 1f). However, there were no statistically significant differences between the treatments in the last two years of the crop rotation.

3.2. Yields over the Crop Rotation 2012–2017

Depending on the crop and the year of investigation, the results varied. However, the different fertilization influenced the annual marketable yields, as shown in Table 3.

Table 3. Yields (Mg ha^{-1} dry matter), nitrogen yields (kg ha^{-1} dry matter), soil nitrate–nitrogen (mg kg^{-1} 0–90 cm) in spring of the rotation. Means followed by different letters within a row are significantly different at $p < 0.05$.

	Treatments						
	06M	**06M + K**	**06M + BD**	**06M + BD + K**	**SEM**	**Pr > F**	**LSD**
2012 clover grass							
Yield Mg ha^{-1}	10.9 [b]	11.9 [ab]	11.3 [b]	12.6 [a]	0.26	0.0138	1.3
N Yield kg ha^{-1}	306.0 [b]	338.6 [ab]	306.7 [b]	352.7 [a]	9.01	0.0237	44.1
Soil NO$_3$–N mg kg^{-1}	5.48	5.55	6.05	5.85	0.56	0.8739	2.7
2013 clover grass							
Yield Mg ha^{-1}	7.7	8.5	7.5	8	0.32	0.2386	1.5
N Yield kg ha^{-1}	172.6	198.6	164.2	175.1	8.61	0.1218	42.2
Soil NO$_3$–N mg kg^{-1}	10.63	12.1	11.3	12.2	0.55	0.245	2.7
2014 winter wheat							
Yield Mg ha^{-1}	2.5 [b]	3.0 [a]	2.8 [ab]	3.1 [a]	0.09	0.0121	0.4
N Yield kg ha^{-1}	47.2 [b]	56.8 [ab]	52.8 [ab]	58.1 [a]	1.98	0.0297	9.7
Soil NO$_3$–N mg kg^{-1}	12.28	14.05	12.6	12.1	0.53	0.1296	2.6
2015 fodder maize							
Yield Mg ha^{-1}	10.8 [b]	11.9 [ab]	12.2 [ab]	12.8 [a]	0.33	0.0244	1.6
N Yield kg ha^{-1}	164.1 [b]	180.7 [ab]	184.2 [ab]	201.4 [a]	5.42	0.0164	26.5
Soil NO$_3$–N mg kg^{-1}	19.95	21.78	20.48	21.58	0.5	0.1117	2.4
2016 red beet							
Yield Mg ha^{-1}	5.7 [ab]	5.5 [b]	6.2 [a]	6.1 [ab]	0.09	0.0339	0.7
N Yield kg ha^{-1}	91.7 [c]	86.4 [c]	97.8 [ab]	102.53 [a]	1.66	0.0021	8.1
Soil NO$_3$–N mg kg^{-1}	23.43	24.15	23.95	23.73	0.94	0.952	4.6
2017 spring wheat							
Yield Mg ha^{-1}	2.4 [b]	2.5 [b]	2.9 [a]	3.0 [a]	0.08	0.0051	0.4
N Yield kg ha^{-1}	42.3 [b]	43.7 [b]	54.9 [a]	56.5 [a]	1.64	0.0015	8
Soil NO$_3$–N mg kg^{-1}	18.1	17.43	19.73	19	0.52	0.0775	2.5

Abbreviations: SEM, standard error of the mean value; and LSD, Least Significant Difference.

The yields differed significantly in 2012, 2014, 2016, and 2017, between 06M and 06M + BD + K. Further, the application of compost plus potassium sulfate resulted in higher N yields of all treatments, as compared to 06M in all years, except for 2013. Despite the different N input, the mineral N in spring was similar in all treatments.

Fertilization resulted in significant marketable yield increases cumulated over the 6-year crop rotation, which followed the order—06M < 06M + K < 06M + BD < 06M + BD + K (Figure 2a). The significantly highest marketable yields cumulated over the 6-year crop rotation was achieved with the addition of plant-based compost, with and without potassium sulfate (26.1 Mg ha^{-1} and 25.4 Mg ha^{-1}, respectively), while the treatment with only deep litter (06M) achieved 22.7 Mg ha^{-1}, which were 13% and 10.6% less than 06M + BD + K and 06M + BD.

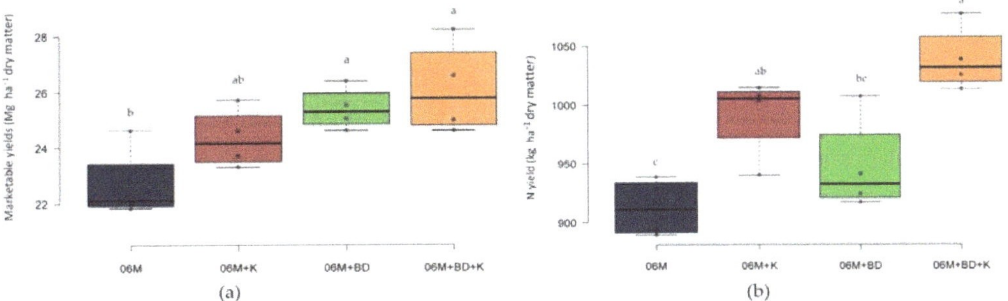

Figure 2. (**a**) Marketable yields (Mg ha^{-1} dry matter) included winter wheat, maize, red beet, and spring wheat. (**b**) Total nitrogen yields (kg ha^{-1} dry matter) included clover grass, winter wheat (grain and straw), maize, red beet (root and side product), and spring wheat (grain and straw). Results are cumulated over the 6-year crop rotation, boxplots with different letters are significantly different at $p < 0.05$.

Fertilization with potassium sulfate significantly influenced the total aboveground biomass N uptake over the crop rotation, being significantly higher in 06M + K and 06M + BD + K than in 06M (Figure 2b). Compost application (06M + BD) did not increase the N yield as compared to 06M and 06M + K, but was significantly lower than the combination of compost and potassium sulfate.

3.3. Nutrient Balance over the Crop Rotation

Nutrient inputs in the treatments varied according to the fertilization regimes, and the exports varied according to the yield levels. The N:P:K:S ratios were only different between treatments on the input side, but not for the nutrient exports.

Balances of all nutrients under study were negative with the 06M treatment (Table 4). Inputs did not compensate for nutrient export in this treatment.

Table 4. Mean total nitrogen (N), sulfur (S), potassium (K) and phosphorus (P) balance across one crop rotation (2012–2017) in kg ha^{-1} year^{-1}.

Treatments	N Balance			S Balance		
	Input kg ha^{-1} year^{-1}	Export kg ha^{-1} year^{-1}	Budget kg ha^{-1} year^{-1}	Input kg ha^{-1} year^{-1}	Export kg ha^{-1} year^{-1}	Budget kg ha^{-1} year^{-1}
06M	149	152	−3	3	12	−8
06M + K	160	165	−5	68	14	54
06M + BD	220	158	62	17	12	5
06M + BD + K	232	173	59	55	14	40
	K Balance			P Balance		
06M	39	113	−74	7	24	−17
06M + K	189	164	25	7	25	−17
06M + BD	162	127	35	37	26	11
06M + BD + K	249	169	80	37	27	10

Potassium sulfate application turned the K and S balances positive in the 06M + K treatment, while the P and N budgets became positive only with compost application in the experiment (treatments 06M + BD and 06M + BD + K).

3.4. Soil Organic Matter Balances and Modeling

The good correlation between observed and predicted C and N development (Figure 3) indicated that assumptions in the model seemed to be more or less applicable at the site, even though the undulating development of SOC was not captured in that magnitude by the model.

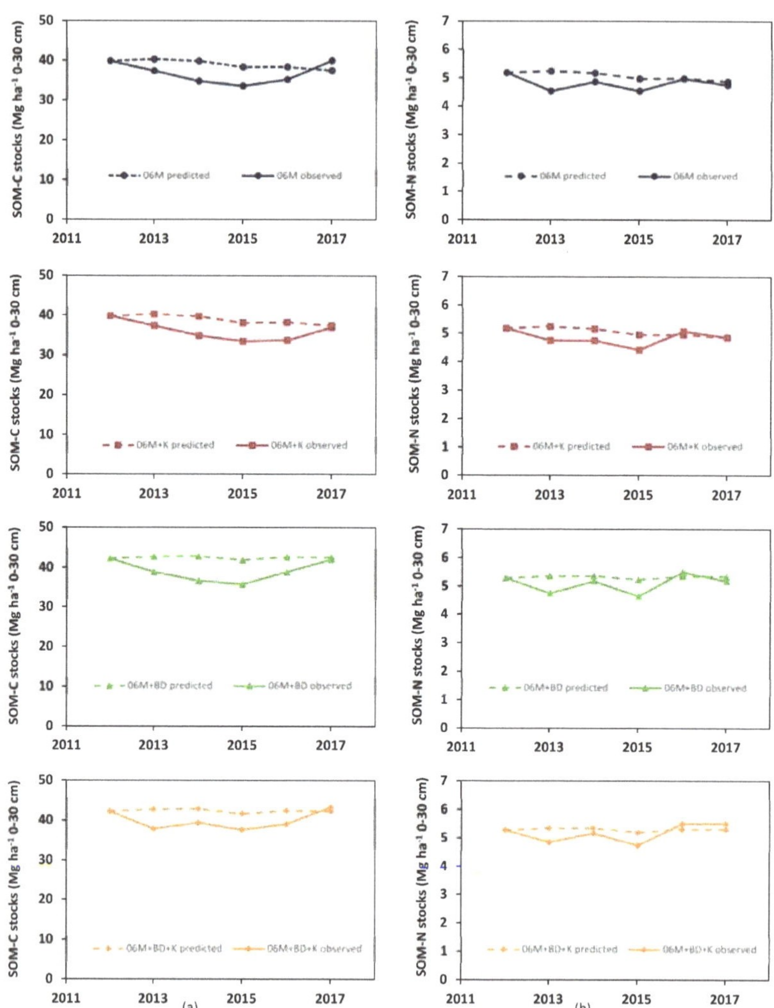

Figure 3. Observed and predicted development of soil organic C (**a**) and N (**b**) under the rotational cycle 2012–2017 in the long-term field experiment. Predicted values were calculated with the HU-MOD model.

According to the coupled C- and N-based soil organic matter balance, the supply of organic matter was too low in the 06M and 06M + K treatments to compensate for mineralization (Table 5). With compost application in 06M + BD and 06M + BD + K, the balance became slightly positive, but there was not much potential for increasing yields.

Modeling opportunities to improve organic matter supply to soils in treatments with and without compost (Table 6), we found that the inclusion of non legume catch crops would marginally improve SOM balances, but the budgets would almost not change in

06M. An optimization of the crop rotation (substitution of fodder maize by oats and new crop order) would significantly improve the SOM budget, but still the balance of the 06M treatment would stay negative. In the compost treatments, the same optimization of the crop rotation would allow for a 50% reduction of compost application, without significantly changing the budget (Table 7).

Table 5. Coupled C- and N-based soil organic matter balance with HU-MOD across one crop rotation (2012–2017) in kg ha^{-1} year^{-1}.

	06M		06M + K		06M + BD		06M + BD + K	
Crop Rotation	SOM-C kg ha^{-1} year^{-1}	SOM-N kg ha^{-1} year^{-1}	SOM-C kg ha^{-1} year^{-1}	SOM-N kg ha^{-1} year^{-1}	SOM-C kg ha^{-1} year^{-1}	SOM-N kg ha^{-1} year^{-1}	SOM-C kg ha^{-1} year^{-1}	SOM-N kg ha^{-1} year^{-1}
clover grass	290.4	37.8	354.6	46.1	356.7	44.6	411.5	51.4
clover grass	427.3	51.9	445.7	56.7	436.2	53.3	439.0	56.3
winter wheat	−453.7	−63.4	−555.8	−75.7	126.8	18.0	101.6	13.4
fodder maize	−1190.4	−160.9	−1345.0	−177.9	−692.8	−90.0	−882.7	−111.3
red beet	38.1	5.4	91.5	13.8	814.0	115.3	750.8	105.7
spring wheat	−541.2	−64.4	−480.5	−63.2	266.5	32.8	238.3	30.3
crop rotation	−238.3	−32.3	−248.3	−33.4	217.9	29.0	176.4	24.3

Table 6. Scenarios for optimized crop rotation for treatment by 06M. Coupled C- and N-based soil organic matter balance with HU-MOD across one crop rotation (2012–2017) in kg ha^{-1} year^{-1}. For the scenarios, the CN reference was standardized, therefore, the budgets of the original scenarios differed from those in Table 5.

	06M Original		06M + Catch Crops		06M Optimized		
Crop Rotation	SOM-C kg ha^{-1} year^{-1}	SOM-N kg ha^{-1} year^{-1}	SOM-C kg ha^{-1} year^{-1}	SOM-N kg ha^{-1} year^{-1}	Crop rotation	SOM-C kg ha^{-1} year^{-1}	SOM-N kg ha^{-1} year^{-1}
clover grass	302.1	37.8	302.1	37.8	clover grass	396.3	49.5
clover grass	414.9	51.9	414.9	51.9	clover grass	234.5	29.3
winter wheat	−507.0	−63.4	−507.0	−63.4	oats	−388.7	−48.6
catch crop $^{\#}$	no	no	oil radish $^{\#}$		catch crop $^{\#}$	oil radish $^{\#}$	
fodder maize	−1287.0	−160.9	−1251.7	−156.5	red beet *	552.5	69.1
red beet *	43.0	5.4	43.0	5.4	winter wheat	−497.2	−62.2
catch crop $^{\#}$	no	no	phacelia $^{\#}$		catch crop $^{\#}$	phacelia $^{\#}$	
spring wheat	−514.8	−64.4	−459.1	−57.4	spring wheat	−508.9	−63.6
underseed	no	no	clover grass		underseed	clover grass	
crop rotation	−258.1	−32.3	−243.0	37.8	crop rotation	−35.2	−4.4

* Fertilization: 35 Mg ha^{-1} with farmyard manure $^{\#}$ Catch crop effect is considered in the balance value of the main crop.

Table 7. Scenario for the optimized crop rotation for treatment by 06M + BD + K. Coupled C- and N-based soil organic matter balance with the HU-MOD across one crop rotation (2012–2017) in kg ha^{-1} year^{-1}. For the scenario, the CN reference was standardized, therefore, the budgets of the original scenario differed from those in Table 5.

	06M + BD + K Original		
Crop Rotation	Fertilization	SOM-C kg ha^{-1} year^{-1}	SOM-N kg ha^{-1} year^{-1}
clover grass		411.5	51.4
clover grass		450.3	56.3
winter wheat	compost 30 Mg ha^{-1}	106.9	13.4
fodder maize	compost 30 Mg ha^{-1}	−890.4	−111.3
red beet	FYM * 35 Mg ha^{-1} + compost 30 Mg ha^{-1}	846	105.7
spring wheat	compost 30 Mg ha^{-1}	242.5	30.3
crop rotation		194.5	24.3
06M + BD + K optimized (compost reduced by 50%)			
clover grass		411.5	51.4
clover grass		450.3	56.3
oats	compost 30 Mg ha^{-1}	207.2	25.9
catch crop		oil radish $^{\#}$	
red beet	FYM * 35 Mg ha^{-1}	499.6	62.4
winter wheat		−566.8	−70.9
catch crop		Phacelia $^{\#}$	
spring wheat	compost 30 Mg ha^{-1}	296.8	37.1
crop rotation		216.4	27.1

* FYM = farmyard manure; $^{\#}$ Catch crop effect is considered in the balance value of the main crop.

4. Discussion

4.1. Soil Organic Matter and Nutrients

The development of soil organic matter is a rather slow process, and it will likely take more than a 6-year crop rotation or even an 8-year observation period, until significant treatment effects emerge [38]. However, we can already observe some differentiation in this initial phase of the field experiment. As could be expected, both C and N values were higher in the compost treatments than in the other two treatments. Further, it appears that decreasing C and N stocks in the treatments without compost application could indicate an insufficient supply of organic matter to soils. However, since the turnover of organic matter in soils is not only dependent on actual management, but also on site conditions and management history [36,39,40], it is not possible to determine whether farmyard manure application corresponding to a stocking rate of 0.6 LU cattle ha^{-1} is insufficient in general, or only worked under the specific conditions of this experiment.

Unfortunately, there are almost no studies on the stocking rate and the corresponding available manure effects on soil fertility in the scientific literature today. In the well-known DOK experiments in Switzerland, farmyard manure was applied at rates corresponding to 0.7 and 1.4 LU cattle per ha, but soil carbon stocks decreased under all treatments in the experiment, except for a biodynamic treatment with composted manure application, corresponding to 1.4 LU ha^{-1} [41]. This was most likely an effect of the site history. Leithold et al. [3] assumed that 1 LU ha^{-1} would be an adequate stocking rate to maintain soil fertility in productive organic farming systems. Nevertheless, Schulz et al. [4] observed even increasing SOM levels in the Organic Arable Farming Experiment Gladbacherhof (Germany) at the FYM application, corresponding to 1 LU cattle per ha.

Soil organic matter balances provide some explanation for the observed trends of soil C and N, despite the uncertainties in the parametrization (see below). According to the model calculations, organic matter supply is not sufficient in the treatments without compost application to compensate for turnover losses, despite the low yield level. It was considered that the N uptake of crops is taken as a proxy for soil organic matter mineralization in the model [34]. The yield level of non-legume crops is, therefore, positively correlated with the demand for organic matter to compensate for SOM mineralization [42]. At present, only treatments with compost applications have the potential to build up SOM. However, the demand for organic matter would increase if the yield levels shall be improved in the experiment.

The development of K, S, and P stocks reflects the different input rates in the treatments. Therefore, treatments with compost application (06M + BD, 06M + BD + K) have higher p-values than the two other treatments, and treatments with potassium sulfate application (06M + K, 06M + BD + K) have higher K and S values.

4.2. Crop Yields

The fodder legumes obviously benefitted from potassium sulfate application. From our results we might not conclude whether this was a K- or S-effect. Both elements play a vital role in biological nitrogen fixation, and the effect of variable availability is similar [17]. Usually, K is not a limiting factor in arable soils in Germany. At the site of the field experiment, however, K supply might be limited by K fixation.

Row crops (maize and red beet) both benefitted from compost application, while K/S fertilization had a smaller effect. Here, it must be considered that all treatments received farmyard manure. According to Blake et al. [43], farmyard manure application is more effective in supplying crops with K than mineral fertilizers. Lehtinen et al. [10] found that K input with manure-based compost was higher than with plant-based compost. In fact, we found that K use efficiency was highest in the 06M treatment and lowest in the 06M + BD + K.

The reaction of the cereals to fertilization was not consistent. Higher winter wheat yields in the treatments with K fertilization are likely an effect of the preceding fodder legumes. Spring wheat at the end of the crop rotation on the other hand obviously profited from compost fertilization.

Altogether, yield levels were comparably low in the experiment. Nutrient balances revealed that the actual nutrient supply did not offer much potential for yield increases, if at all. Increasing yields would require additional efforts in soil fertility management, like the use of green manure and catch crops to improve N supply [44], and additional fertilization.

4.3. Nutrient and Soil Organic Matter Balances

Nutrient balances were negative for both S, P, and K in the control treatment (06M). This corresponded to the results of Berry et al. [45], which indicate negative K and P balances in organic farming systems without external inputs. In our field experiment, the application of compost compensated for nutrient exports (and losses), even in the treatment without additional Potassium sulfate fertilization. The utilization of additional internal organic matter resources, therefore, is a good opportunity to improve nutrient balances on arable land, but of course this measure did not close the nutrient cycle on the farm level. Instead, nutrients were transferred and re-distributed within the farm. Reimer et al. [46] addressed this situation in their meta-analysis of nutrient budgets in organic farms in Germany. The authors emphasized that positive nutrient balances at the farm gate could not be achieved without nutrient imports. In principle, biowaste compost and sewage sludge would be the appropriate sources to close nutrient cycles. However, both sources featured the risk to import mineral and organic pollutants into organic farming systems [47]. Therefore, the utilization of internal resources must be considered a viable interim option.

Soil organic matter balances were calculated with the HU-MOD model [34,35]. The tool was originally developed for decision support in farming practice, but unlike most other so-called 'Humus balance methods' it could also be used for analytical purposes [36].

The advantage of the HU-MOD model was that the utilization of N in plant biomass as a proxy for the mineralization of soil organic matter allowed us to by-pass the need for information on site factors, as it was assumed that their effect on soil organic matter mineralization became visible in the N fluxes that were considered in the model. However, the procedure made the model susceptible to errors resulting from erroneous estimates of N pools. Most importantly, biological nitrogen fixation is known to be highly variable [48], even though [49] found that there is a statistically significant average rate of approximately 0.7 kg BNF-N per kg plant shoot N. Nevertheless, the error of this average was high enough to severely impact site-specific N balance calculations. As we did not measure BNF, we could not account for any differentiation between the treatments in N yield from this process.

Regarding the congruence between the observed and predicted trends of soil C and N stocks, it must be considered that the model output does refers to the total rooted soil layer. A comparison with topsoil C and N trends, therefore, comprises the risk that C and N changes in deeper soil layers are not captured. Soil organic matter in the subsoil is usually more stable than topsoil SOM [50], and turnover mainly takes place in the topsoil. However, it was argued that organic matter turnover in subsoil is relevant for the calculation of actual C balances [51]. Therefore, it cannot be concluded whether the deviations between measured topsoil organic matter changes and the calculated changes according to the HU-MOD model indicate parametrization errors, or are caused by the limited database on C and N changes in the soil.

4.4. Practical Implications

4.4.1. Farm Compost to Open Additional Organic Matter Sources

In our experiment, we used on-farm composting as an option to increase organic matter supply to soils based on own resources of a farm. This compost provided an additional input of several nutrients. Besides N, the compost contained considerable amounts of P, K, and even S and could be considered an effective 'full-fertilizer' with a low leaching potential [52]. Compost application therefore could effectively increase crop yields [11,13,53].

As outlined by D'Hose et al. [53], the positive effect of farm compost on crop yields should not only be ascribed to the additional nutrient input, but also to the improvement of growing conditions for the crops. Compost application improves soil physical properties [52,54], and has a beneficial effect of compost on pathogen regulation in soils and plant health [54–56]. Further, compost might even facilitate the formation of arbuscular mycorrhizas [57].

It is widely acknowledged that compost builds up soil organic matter [8,10,52,58]. This improves microbial activity and related ecosystem services [59]. For example, organic matter build-up improves the accessibility of micronutrients to plants [14]. In general, there is a positive relationship between soil organic matter and crop yields [15,42].

Composting was also identified as a viable option to reduce ecological trade-offs between soil fertility management and environment and climate protection [60,61].

4.4.2. Optimization

Modeling possible adaptations of crop rotation or fertilization in selected treatments indicated a low potential of catch crops to improve the SOM balance. This is supported by findings of White et al. [58] as well as Tautges et al. [12]. However, it contradicts the results of Poeplau and Don [62] who concluded that the introduction of catch crops into crop rotations would have a considerable effect on carbon sequestration in Germany. The reason for the different observations could be a stoichiometric effect, were N availability limits C retention from catch crops in the soil [63]. As the C:N ratio of organic material is narrowed down in the turnover process, excess C is lost by respiration [64]. At the same time, it should be considered that increasing N supply causes a priming effect [65], which pushes organic matter turnover. The impact of catch crops on soil organic matter is therefore probably dependent on both C and N amounts and availability, alongside with biological and physical factors. As we included non-legume catch crops in the model that did not receive any fertilization, the model calculated only slightly positive balances based on the fertilization effect of the catch crops on the succeeding crops. However, it should be considered that soil N taken up by the catch crops might have leached in a corresponding bare fallow period. On the other hand, N leaching usually is very low under the N-limited conditions of organic farming [66], especially on heavy soils.

In contrast, the substitution of fodder maize by oats proved a very effective measure in the modeling study, as this adaptation significantly decreased N export and the related demand for organic matter. Field experiments comparing maize and cereal cultivation effects on soil organic matter are rare. Nevertheless, in a study from Poland, Rychcik et al. [67] found that maize and grain legumes had lower soil carbon values than cereals. In a recent paper, Benbi et al. [68] showed that soil C respiration was three-fold higher under maize, as compared to wheat. Our results therefore are plausible.

Of course, changes in the crop rotation need to be discussed against the background of the requirements in the farming system. A substitution of fodder maize by cereals might not always fit to the specific situation. In such cases, intercropping could provide an option to improve the soil organic matter balance of maize [69].

5. Conclusions

Sustainable nutrient supply might be threatened in organic farming systems with a stocking rate of 0.6 LU cattle per hectare, if fertilization only relies on the available manure. Additional compost application provides a solution, as compost provides a direct additional input of nutrients, and contributes to the nutrition of legumes, which in turn enhances biological N fixation. Additional supply of essential nutrients (K, S) does further improve BNF. This compost can be made from internal resources on the farm (e.g., hedge-cutting), to be independent from external inputs and to avoid the import of pollutants.

However, it must be considered that own farm compost makes new nutrient, and the organic matter sources available on the farm and is a viable interim solution, but does not solve the problem of open nutrient cycles at the farm gate level.

Author Contributions: Conceptualization, M.O., C.B., and B.S.; methodology, B.S. and M.O.; validation, H.S. (Hartmut Spieß) and C.B.; formal analysis, M.O.; investigation, C.M., H.S. (Harald Schaaf), and D.B.; resources, H.S. (Harald Schaaf), D.B. and H.H.; data curation, C.M., B.S., and M.O.; writing—original draft preparation, M.O. and B.S.; writing—review and editing, C.B. and H.S. (Hartmut Spieß); visualization, M.O.; supervision, C.B.; project administration, C.B.; funding acquisition, C.B. All authors have read and agreed to the published version of the manuscript.

Funding: This research was funded by Software AG-Stiftung, Rudolf-Steiner-Fonds and Zukunftsstiftung Landwirtschaft. The APC was funded by Software AG-Stiftung.

Institutional Review Board Statement: Not applicable.

Informed Consent Statement: Not applicable.

Data Availability Statement: All data are available at Forschungsring and Forschung&Züchtung Dottenfelderhof.

Acknowledgments: The authors acknowledge the support of member of the advisory board of the BoDyn Long-Term Field Experiment (Miriam Athmann, Johan Bachinger, Andreas Gattinger, Jürgen Fritz, Ulrich Köpke, Harald Schmidt, Klaus Wais) and of Cornelius Sträßer, the responsible project supervisor of Software AG-Stiftung.

Conflicts of Interest: The authors declare no conflict of interest.

References

1. Fageria, N.K. Role of Soil Organic Matter in Maintaining Sustainability of Cropping Systems. *Commun. Soil Sci. Plant Anal.* **2012**, *43*, 2063–2113. [CrossRef]
2. Gattinger, A.; Muller, A.; Haeni, M.; Skinner, C.; Fliessbach, A.; Buchmann, N.; Mäder, P.; Stolze, M.; Smith, P.; Scialabba, N.E.-H.; et al. Enhanced top soil carbon stocks under organic farming. *Proc. Natl. Acad. Sci. USA* **2012**, *109*, 18226–18231. [CrossRef] [PubMed]
3. Leithold, G.; Hülsbergen, K.-J.; Brock, C. Organic matter returns to soils must be higher under organic compared to conventional farming. *J. Plant Nutr. Soil Sci.* **2014**, *178*, 4–12. [CrossRef]
4. Schulz, F.; Brock, C.; Schmidt, H.; Franz, K.-P.; Leithold, G. Development of soil organic matter stocks under different farm types and tillage systems in the Organic Arable Farming Experiment Gladbacherhof. *Arch. Agron. Soil Sci.* **2013**, *60*, 313–326. [CrossRef]
5. Brock, C.; Oberholzer, H.-R.; Schwarz, J.; Fließbach, A.; Hülsbergen, K.-J.; Koch, W.; Pallutt, B.; Reinicke, F.; Leithold, G.; Fliessbach, A. Soil organic matter balances in organic versus conventional farming—modelling in field experiments and regional upscaling for cropland in Germany. *Org. Agric.* **2012**, *2*, 185–195. [CrossRef]
6. Six, J.; Conant, R.T.; Paul, E.A.; Paustian, K. Stabilization mechanisms of soil organic matter: Implications for C-saturation of soils. *Plant Soil* **2002**, *241*, 155–176. [CrossRef]
7. Amanullah; Khalid, S.; Imran; Khan, H.A.; Arif, M.; Altawaha, A.R.; Adnan, M.; Fahad, S.; Parmar, B. Organic Matter Management in Cereals Based System: Symbiosis for Improving Crop Productivity and Soil Health. *Sustain. Agric. Rev.* **2019**, *29*, 67–92. [CrossRef]
8. Martínez-Blanco, J.; Lazcano, C.; Christensen, T.H.; Muñoz, P.; Rieradevall, J.; Møller, J.; Antón, A.; Boldrin, A. Compost benefits for agriculture evaluated by life cycle assessment. A review. *Agron. Sustain. Dev.* **2013**, *33*, 721–732. [CrossRef]
9. Azim, K.; Soudi, B.; Boukhari, S.; Perissol, C.; Roussos, S.; Alami, I.T. Composting parameters and compost quality: A literature review. *Org. Agric.* **2018**, *8*, 141–158. [CrossRef]
10. Lehtinen, T.; Dersch, G.; Söllinger, J.; Baumgarten, A.; Schlatter, N.; Aichberger, K.; Spiegel, H. Long-term amendment of four different compost types on a loamy silt Cambisol: Impact on soil organic matter, nutrients and yields. *Arch. Agron. Soil Sci.* **2016**, *63*, 663–673. [CrossRef]
11. Kurzemann, F.; Plieger, U.; Probst, M.; Spiegel, H.; Sandén, T.; Ros, M.; Insam, H. Long-Term Fertilization Affects Soil Microbiota, Improves Yield and Benefits Soil. *Agronomy* **2020**, *10*, 1664. [CrossRef]
12. Tautges, N.E.; Chiartas, J.L.; Gaudin, A.C.M.; O'Geen, A.T.; Herrera, I.; Scow, K.M. Deep soil inventories reveal that impacts of cover crops and compost on soil carbon sequestration differ in surface and subsurface soils. *Glob. Chang. Biol.* **2019**, *25*, 3753–3766. [CrossRef] [PubMed]
13. Wortman, S.E.; Holmes, A.A.; Miernicki, E.; Knoche, K.; Pittelkow, C.M. First-Season Crop Yield Response to Organic Soil Amendments: A Meta-Analysis. *Agron. J.* **2017**, *109*, 1210–1217. [CrossRef]

14. Dhaliwal, S.; Naresh, R.; Mandal, A.; Singh, R.; Dhaliwal, M. Dynamics and transformations of micronutrients in agricultural soils as influenced by organic matter build-up: A review. *Environ. Sustain. Indic.* **2019**, *1–2*, 100007. [CrossRef]

15. Oldfield, E.E.; Bradford, M.A.; Wood, S.A. Global meta-analysis of the relationship between soil organic matter and crop yields. *Soil* **2019**, *5*, 15–32. [CrossRef]

16. Vance, C.P.; Graham, P.H.; Allan, D.L. Biological Nitrogen Fixation: Phosphorus–A Critical Future Need? In *Proceedings of the Nitrogen Fixation: From Molecules to Crop Productivity*; Springer: Berlin, Germany, 2005; Volume 12, pp. 509–514.

17. Divito, G.A.; Sadras, V.O. How do phosphorus, potassium and sulphur affect plant growth and biological nitrogen fixation in crop and pasture legumes? A meta-analysis. *Field Crops Res.* **2014**, *156*, 161–171. [CrossRef]

18. Israel, D.W. Investigation of the Role of Phosphorus in Symbiotic Dinitrogen Fixation. *Plant Physiol.* **1987**, *84*, 835–840. [CrossRef] [PubMed]

19. Sulieman, S.; Van Ha, C.; Schulze, J.; Tran, L.-S.P. Growth and nodulation of symbiotic Medicago truncatula at different levels of phosphorus availability. *J. Exp. Bot.* **2013**, *64*, 2701–2712. [CrossRef]

20. Almeida, J.; Hartwig, U.A.; Frehner, M.; Nösberger, J.; Lüscher, A. Evidence that P deficiency induces N feedback regulation of symbiotic N_2 fixation in white clover (*Trifolium repens* L.). *J. Exp. Bot.* **2000**, *51*, 1289–1297. [CrossRef]

21. Høgh-Jensen, H. The effect of potassium deficiency on growth and N_2-fixation in Trifolium repens. *Physiol. Plant.* **2003**, *119*, 440–449. [CrossRef]

22. Lea, P.J.; Sodek, L.; Parry, M.A.J.; Shewry, P.R.; Halford, N.G. Asparagine in plants. *Ann. Appl. Biol.* **2007**, *150*, 1–26. [CrossRef]

23. FAO. World reference base for soil resources 2014 International soil classification system for naming soils and creating legends for soil maps. *World Soil Resour. Rep.* **2015**, *106*, 162–163.

24. Steiner, R. *Geisteswissenschaftliche Grundlagen zum Gedeihen der Landwirtschaft: Landwirtschaftlicher Kurs. Koberwitz bei Breslau 1924, und ein Vortrag, Dornach 1924*; Schwabe: Karlsruhe, Germany, 2015; ISBN 3727485124.

25. van Eimern, J.; Häckel, H. *Wetter und Klimakunde. Für Landwirte, Gärtner, Winzer und Landschaftspfleger: Ein Lehrbuch der Agrarmeteorologie*; Eugen Ulmer Verlag: Stuttgart, Germany, 1979; pp. 50–51.

26. ISO 13878. *Soil Quality Determination of Total Nitrogen Content by Dry Combustion ("Elemental Analysis")*; International Organization for Standardization: Geneva, Switzerland, 1998.

27. DIN EN 16168. *Sludge, Treated Biowaste and Soil–Determination of Total Nitrogen–Dry Combustion Method*; Deutsches Institut für Normung e.V.: Berlin, Germany, 2012. [CrossRef]

28. ISO 11885. *Water Quality—Determination of Selected Elements by Inductively Coupled Plasma Optical Emission Spectrometry*; International Organization for Standardization: Geneva, Switzerland, 2009. [CrossRef]

29. Hoffmann, G. *Methodenbuch Band 1, Die Untersuchung von Böden*; VDLUFA: Darmstadt, Germany, 1991.

30. DIN 19682-10. *Methods of Soil Investigations for Agricultural Water Engineering–Field Tests–Part 10: Description and Evaluation of Soil Structure*; Deutsches Institut für Normung e.V.: Berlin, Germany, 2007. [CrossRef]

31. Naumann, C.; Bassler, R.; Seibold, R.; Barth, K. *VDLUFA-Methodenbuch Band III, Die Chemische Untersuchung von Futtermitteln*; VDLUFA: Darmstadt, Germany, 1997.

32. ISO 12099. *Animal Feeding Stuffs, Cereals and Milled Cereal Products*; International Organization for Standardization: Geneva, Switzerland, 2017. [CrossRef]

33. Stein-Bachinger, K.; Bachinger, J. *Nährstoffmanagement im ökologischen Landbau: Ein Handbuch für Beratung und Praxis; Berechnungsgrundlagen, Faustzahlen, Schätzverfahren zur Erstellung von Nährstoffbilanzen; Handlungsempfehlungen zum Effizienten Umgang mit Innerbetrieblichen Nährstoffressourcen, Insbesondere Stickstoff*; Landwirtschaftsverlag: Münster, Germany, 2004; ISBN 3784321682.

34. Brock, C.; Hoyer, U.; Leithold, G.; Hülsbergen, K.-J. The humus balance model (HU-MOD): A simple tool for the assessment of management change impact on soil organic matter levels in arable soils. *Nutr. Cycl. Agroecosystems* **2012**, *92*, 239–254. [CrossRef]

35. Knebl, L.; Leithold, G.; Brock, C. Improving minimum detectable differences in the assessment of soil organic matter change in short-term field experiments. *J. Plant Nutr. Soil Sci.* **2015**, *178*, 35–42. [CrossRef]

36. Brock, C.; Franko, U.; Oberholzer, H.-R.; Kuka, K.; Leithold, G.; Kolbe, H.; Reinhold, J. Humus balancing in Central Europe-concepts, state of the art, and further challenges. *J. Plant Nutr. Soil Sci.* **2013**, *176*, 3–11. [CrossRef]

37. Knebl, L.; Leithold, G.; Schulz, F.; Brock, C. The role of soil depth in the evaluation of management-induced effects on soil organic matter. *Eur. J. Soil Sci.* **2017**, *68*, 979–987. [CrossRef]

38. Smith, P. How long before a change in soil organic carbon can be detected? *Glob. Chang. Biol.* **2004**, *10*, 1878–1883. [CrossRef]

39. Stockmann, U.; Adams, M.; Crawford, J.W.; Field, D.J.; Henakaarchchi, N.; Jenkins, M.; Minasny, B.; McBratney, A.B.; Courcelles, V.D.R.D.; Singh, K.; et al. The knowns, known unknowns and unknowns of sequestration of soil organic carbon. *Agric. Ecosyst. Environ.* **2013**, *164*, 80–99. [CrossRef]

40. Crews, T.E.; Rumsey, B.E. What Agriculture Can Learn from Native Ecosystems in Building Soil Organic Matter: A Review. *Sustainability* **2017**, *9*, 578. [CrossRef]

41. Fließbach, A.; Oberholzer, H.-R.; Gunst, L.; Mäder, P. Soil organic matter and biological soil quality indicators after 21 years of organic and conventional farming. *Agric. Ecosyst. Environ.* **2007**, *118*, 273–284. [CrossRef]

42. Brock, C.; Fließbach, A.; Oberholzer, H.-R.; Schulz, F.; Wiesinger, K.; Reinicke, F.; Koch, W.; Pallutt, B.; Dittman, B.; Zimmer, J.; et al. Relation between soil organic matter and yield levels of nonlegume crops in organic and conventional farming systems. *J. Plant Nutr. Soil Sci.* **2011**, *174*, 568–575. [CrossRef]

43. Blake, L.; Mercik, S.; Koerschens, M.; Goulding, K.; Stempen, S.; Weigel, A.; Poulton, P.; Powlson, D. Potassium content in soil, uptake in plants and the potassium balance in three European long-term field experiments. *Plant Soil* **1999**, *216*, 1–14. [CrossRef]

44. Thorup-Kristensen, K.; Dresbøll, D.B.; Kristensen, H.L. Crop yield, root growth, and nutrient dynamics in a conventional and three organic cropping systems with different levels of external inputs and N re-cycling through fertility building crops. *Eur. J. Agron.* **2012**, *37*, 66–82. [CrossRef]

45. Berry, P.M.; Sylvester-Bradley, R.; Philipps, L.; Hatch, D.J.; Cuttle, S.P.; Rayns, F.W.; Gosling, P. Is the productivity of organic farms restricted by the supply of available nitrogen? *Soil Use Manag.* **2006**, *18*, 248–255. [CrossRef]

46. Reimer, M.; Möller, K.; Hartmann, T.E. Meta-analysis of nutrient budgets in organic farms across Europe. *Org. Agric.* **2020**, *10*, 65–77. [CrossRef]

47. Alvarenga, P.; Mourinha, C.; Farto, M.; Santos, T.; Palma, P.; Sengo, J.; Morais, M.-C.; Cunha-Queda, C. Sewage sludge, compost and other representative organic wastes as agricultural soil amendments: Benefits versus limiting factors. *Waste Manag.* **2015**, *40*, 44–52. [CrossRef]

48. Liu, Y.; Wu, L.; Baddeley, J.A.; Watson, C.A. Models of biological nitrogen fixation of legumes. A review. *Agron. Sustain. Dev.* **2011**, *31*, 155–172. [CrossRef]

49. Anglade, J.; Billen, G.; Garnier, J. Relationships for estimating N_2 fixation in legumes: Incidence for N balance of legume-based cropping systems in Europe. *Ecosphere* **2015**, *6*, art37. [CrossRef]

50. Rumpel, C.; Kögel-Knabner, I. Deep soil organic matter—A key but poorly understood component of terrestrial C cycle. *Plant Soil* **2011**, *338*, 143–158. [CrossRef]

51. Salome, C.; Nunan, N.; Pouteau, V.; Lerch, T.Z.; Chenu, C. Carbon dynamics in topsoil and in subsoil may be controlled by different regulatory mechanisms. *Glob. Chang. Biol.* **2010**, *16*, 416–426. [CrossRef]

52. Siedt, M.; Schäffer, A.; Smith, K.E.; Nabel, M.; Roß-Nickoll, M.; van Dongen, J.T. Comparing straw, compost, and biochar regarding their suitability as agricultural soil amendments to affect soil structure, nutrient leaching, microbial communities, and the fate of pesticides. *Sci. Total Environ.* **2021**, *751*, 141607. [CrossRef]

53. D'Hose, T.; Cougnon, M.; De Vliegher, A.; Willekens, K.; Van Bockstaele, E.; Reheul, D. Farm Compost Application: Effects on Crop Performance. *Compos. Sci. Util.* **2012**, *20*, 49–56. [CrossRef]

54. Termorshuizen, A.; Moolenaar, S.; Veeken, A.; Blok, W. The value of compost. *Rev. Environ. Sci. Biotechnol.* **2004**, *3*, 343–347. [CrossRef]

55. De Corato, U. Agricultural waste recycling in horticultural intensive farming systems by on-farm composting and compost-based tea application improves soil quality and plant health: A review under the perspective of a circular economy. *Sci. Total Environ.* **2020**, *738*, 139840. [CrossRef] [PubMed]

56. Milinković, M.; Lalević, B.; Jovičić-Petrović, J.; Golubović-Ćurguz, V.; Kljujev, I.; Raičević, V. Biopotential of compost and compost products derived from horticultural waste—Effect on plant growth and plant pathogens' suppression. *Process. Saf. Environ. Prot.* **2019**, *121*, 299–306. [CrossRef]

57. Cavagnaro, T.R. *Biologically Regulated Nutrient Supply Systems*; Elsevier: Amsterdam, The Netherlands, 2015; pp. 293–321.

58. White, K.E.; Brennan, E.B.; Cavigelli, M.A.; Smith, R.F. Winter cover crops increase readily decomposable soil carbon, but compost drives total soil carbon during eight years of intensive, organic vegetable production in California. *PLoS ONE* **2020**, *15*, e0228677. [CrossRef]

59. Abbott, L.K.; Manning, D.A.C. Soil Health and Related Ecosystem Services in Organic Agriculture. *Sustain. Agric. Res.* **2015**, *4*, 116. [CrossRef]

60. Pergola, M.; Persiani, A.; Pastore, V.; Palese, A.M.; D'Adamo, C.; De Falco, E.; Celano, G. Sustainability Assessment of the Green Compost Production Chain from Agricultural Waste: A Case Study in Southern Italy. *Agronomy* **2020**, *10*, 230. [CrossRef]

61. Bos, J.F.; Berge, H.F.T.; Verhagen, J.; Van Ittersum, M.K. Trade-offs in soil fertility management on arable farms. *Agric. Syst.* **2017**, *157*, 292–302. [CrossRef]

62. Poeplau, C.; Don, A. Carbon sequestration in agricultural soils via cultivation of cover crops—A meta-analysis. *Agric. Ecosyst. Environ.* **2015**, *200*, 33–41. [CrossRef]

63. Dannehl, T.; Leithold, G.; Brock, C. The effect of C:N ratios on the fate of carbon from straw and green manure in soil. *Eur. J. Soil Sci.* **2017**, *68*, 988–998. [CrossRef]

64. Schimel, J. The implications of exoenzyme activity on microbial carbon and nitrogen limitation in soil: A theoretical model. *Soil Biol. Biochem.* **2003**, *35*, 549–563. [CrossRef]

65. Kuzyakov, Y.; Friedel, J.; Stahr, K. Review of mechanisms and quantification of priming effects. *Soil Biol. Biochem.* **2000**, *32*, 1485–1498. [CrossRef]

66. Chmelíková, L.; Schmid, H.; Anke, S.; Hülsbergen, K.-J. Nitrogen-use efficiency of organic and conventional arable and dairy farming systems in Germany. *Nutr. Cycl. Agroecosystems* **2021**, *119*, 1–18. [CrossRef]

67. Rychcik, B.; Adamiak, J.; Wójciak, H. Dynamics of the soil organic matter in crop rotation and long-term monoculture. *Plant Soil Environ.* **2006**, *52*, 15–20.

68. Benbi, D.K.; Toor, A.; Brar, K.; Dhall, C. Soil respiration in relation to cropping sequence, nutrient management and environmental variables. *Arch. Agron. Soil Sci.* **2020**, *66*, 1873–1887. [CrossRef]

69. Cong, W.-F.; Hoffland, E.; Li, L.; Six, J.; Sun, J.-H.; Bao, X.-G.; Zhang, F.-S.; Van Der Werf, W. Intercropping enhances soil carbon and nitrogen. *Glob. Chang. Biol.* **2015**, *21*, 1715–1726. [CrossRef]

Article

Effect of Drip Fertigation with Nitrogen on Yield and Nutritive Value of Melon Cultivated on a Very Light Soil

Roman Rolbiecki [1], Stanisław Rolbiecki [1], Anna Figas [2], Barbara Jagosz [3], Dorota Wichrowska [4], Wiesław Ptach [5], Piotr Prus [6,*], Hicran A. Sadan [1], Pal-Fam Ferenc [7], Piotr Stachowski [8] and Daniel Liberacki [8]

[1] Department of Agrometeorology, Plant Irrigation and Horticulture, Faculty of Agriculture and Biotechnology, UTP University of Science and Technology in Bydgoszcz, 85-029 Bydgoszcz, Poland; rolbr@utp.edu.pl (R.R.); rolbs@utp.edu.pl (S.R.); hicran_sadan_76@hotmail.com (H.A.S.)

[2] Department of Agricultural Biotechnology, Faculty of Agriculture and Biotechnology, UTP University of Science and Technology, 85-029 Bydgoszcz, Poland; figasanna@utp.edu.pl

[3] Department of Plant Biology and Biotechnology, Faculty of Biotechnology and Horticulture, University of Agriculture in Krakow, 31-120 Krakow, Poland; Barbara.Jagosz@urk.edu.pl

[4] Department of Microbiology and Food Technology, Faculty of Agriculture and Biotechnology, UTP University of Science and Technology, 85-029 Bydgoszcz, Poland; wichrowska@utp.edu.pl

[5] Department of Remote Sensing and Environmental Research, Institute of Environmental Engineering, Warsaw University of Life Sciences, 02-776 Warszawa, Poland; wieslaw_ptach@sggw.edu.pl

[6] Laboratory of Economics and Agribusiness Advisory, Faculty of Agriculture and Biotechnology, UTP University of Science and Technology in Bydgoszcz, 85-029 Bydgoszcz, Poland

[7] Institute of Plant Production, Kaposvár Campus, Hungarian University of Agriculture and Life Sciences (MATE), H-7400 Kaposvár, Hungary; Pal-Fam.Ferenc.Istvan@szie.hu

[8] Department of Land Improvement, Environmental Development and Spatial Management, Faculty of Environmental Engineering and Mechanical Engineering, Poznan University of Life Sciences, 60-649 Poznań, Poland; piotr.stachowski@up.poznan.pl (P.S.); daniel.liberacki@up.poznan.pl (D.L.)

* Correspondence: piotr.prus@utp.edu.pl; Tel.: +48-52-340-8084

Citation: Rolbiecki, R.; Rolbiecki, S.; Figas, A.; Jagosz, B.; Wichrowska, D.; Ptach, W.; Prus, P.; Sadan, H.A.; Ferenc, P.-F.; Stachowski, P.; et al. Effect of Drip Fertigation with Nitrogen on Yield and Nutritive Value of Melon Cultivated on a Very Light Soil. *Agronomy* **2021**, *11*, 934. https://doi.org/10.3390/agronomy11050934

Academic Editors: Nikolaos Monokrousos and Efimia M. Papatheodorou

Received: 26 March 2021
Accepted: 5 May 2021
Published: 9 May 2021

Publisher's Note: MDPI stays neutral with regard to jurisdictional claims in published maps and institutional affiliations.

Abstract: Most species of Cucurbitaceae respond favorably to irrigation, especially when combined with fertilizers. The effect of drip irrigation combined with nitrogen fertigation in melon grown on a very light soil in Central Poland, during 2013–2015, was evaluated. The field experimental design was a split-plot with four replications. Two factors were studied: (1) irrigation treatments applied in two combinations—drip irrigation + broadcast nitrogen fertilization (control), and drip irrigation + fertigation with nitrogen; (2) two cultivars—Melba and Seledyn. The total marketable yield of fruits, weight of a single fruit, and the concentration of dry matter, total sugars, monosaccharides, ascorbic acid, total carotenoids, and polyphenols were evaluated. Tested factors presented a significant effect both on the yield and nutritive value characteristics. Drip irrigation combined with nitrogen fertigation, comparing to the control, notably improved yields and nutritional value of fruits. Seledyn produced better yields than Melba. This study shows that on very light soil, with low water and nutrient retention capacity, melon should be drip-irrigated and nitrogen-fertigated to obtain the best cultivation results.

Keywords: *Cucumis melo* L.; chemical composition; cultivar; drip irrigation; fruit quality

1. Introduction

Melon (*Cucumis melo* L.) belongs taxonomically to the Cucurbitaceae family, which also includes vegetables, such as cucumber, pumpkin, squash, watermelon, and gourds. In many countries around the world, melon fruit is of considerable economic importance. World production of this species in 2018 was estimated at 40 million tons per year. The main melon-producing country is China (12.7 million tons per year), followed by Turkey, Iran, and India (1.8 to 1.2 million tons per year) [1].

There are many cultivars of melon, which differ mainly in shape, color, and taste [2]. Melon fruits are valuable in terms of nutritional and bioactive properties. This species is a very good source of carotenoids (α-, β-carotene, and β-cryptoxanthin), folic acid, pectins, as well as many vitamins (including B group) and minerals (mostly potassium, iron, and magnesium), polyphenols, such as flavonoids and phenolic acids, and fatty acids (including oleic, linoleic, and palmitoleic acids). Melon is a fruit appreciated not only for its taste and dietary qualities, but also its healing properties, thanks to which it is also used in the cosmetics industry [3–6].

In Poland, due to unfavorable climatic conditions, the melon is grown as a non-commercial species. However, the interest in melon cultivation is clearly growing every year. Currently, the Polish National List of Vegetable Plant Varieties includes seven cultivars that are suitable for cultivation in Poland, and their number is systematically growing [7]. This species is photophilous and thermophilic, with a very high water requirement. Melon plants are very sensitive to spring and autumn frosts, which negatively affect growth and development, and thus also the fruit yield [8]. The highest sensitivity of melon plants to water deficit is observed during the fruit setting period [9].

Due to the rising interest in melon growing in Poland, it is necessary to broaden the knowledge about the methods of its cultivation in temperate climatic conditions. Field production of melon largely depends on the thermal conditions and precipitation during the growing season. An important factor in obtaining high- and good-quality crops is ensuring optimal soil moisture during the vegetation period of this species. In Poland, the water requirements of plants from the Cucurbitaceae family are estimated at around 400 mm during the growing season. The main reason for the high water needs of plants belonging to this family is their high fertility and the production of much aboveground mass with a high coefficient of transpiration (as the ratio of the amount of water excreted to the production of dry matter) [10]. It is generally accepted that irrigation significantly affects both the melon yield and the components of the melon yield grown under semi-arid climatic conditions [9,11–13]. Many studies have shown that the field cultivation of melon should be carried out using irrigation treatments [11–16]. It was found that production factors such as water and nutrients (nitrogen, phosphorus, potassium) most often limit the possibility of obtaining a higher yield of melon fruit [17,18]. Drip irrigation combined with fertigation is a good way to increase the efficiency of water use and yield of Cucurbitaceae. It was also found that drip irrigation performed during the cultivation of Cucurbitaceae and other vegetables sensitive to climatic conditions during the growing season clearly increases their nutrient concentration [19,20]. Drip fertigation ensures precise administration of appropriate amounts of nutrients directly to the root zone. Accurate and uniform application of macro- and micronutrients adequately meets the needs of crops during the growing season [21–23].

The objective of this study was to evaluate the effect of drip irrigation combined with nitrogen fertigation on the melon fruit yield and nutritive value characteristics. As a control, drip irrigation combined with broadcast fertilization was used. The total marketable yield of fruits, weight of a single fruit, and the concentration of dry matter, total sugars, monosaccharides, ascorbic acid, total carotenoids, and polyphenols of two melon cultivars (Melba and Seledyn) were evaluated. The experiment was carried out on very light soil in a region of high precipitation deficit; hence, the advisability of irrigation treatments in this area is justified.

2. Materials and Methods

2.1. Field Experiment Description

The field experiment involving the drip fertigation of two melon (*Cucumis melo* L.) cultivars, namely Melba and Seledyn, was conducted in Kruszyn Krajeński near Bydgoszcz 53°04′53″ N, 17°51′52″ E (Central Poland). The area has precipitation deficits, an extremely unfavorable water balance, and high frequency of long periods without rainfall [19,22–29]. The study was carried out in the years 2013–2015. The plants were grown using standard crop management practices recommended for melon cultivation in Poland. The study was carried out on very light soil with a weak and very weak rye–soil complex. Based on the percentage content of individual granulometric fractions, this soil was classified as sand [30]. The soil of the experimental field contained such fractions as: sand—86.97% (from 2.0 mm to 0.05 mm), silt—12.28% (from 0.05 mm to 0.002 mm), and clay 0.75% (<0.002 mm). The average content of total organic carbon and concentration total nitrogen in the soil was 9.6 g kg^{-1} and 0.9 g kg^{-1}, respectively. The experimental soil was characterized by a low capacity for water retention. The water reserve to 0.6 m depth of soil at field capacity was 72.7 mm, at wilting point 29.1 mm, and the available water 43.6 mm.

The experiment was conducted as a split-plot design with four replications. Two factors were used in the study. The first factor was the drip fertigation with nitrogen applied in two combinations: (1) drip irrigation + broadcast nitrogen fertilization (control); (2) drip irrigation + fertigation with nitrogen. The second factor was two melon cultivars: Melba and Seledyn.

Melon seedlings were transplanted at 0.6 m within rows and 1.6 m between rows. The area of each harvest plot was 12 m^2 and included 15 melon plants, and the whole experimental plot size was 274 m^2. Before planting the seedlings, cultivating and harrowing were performed. The fertilization consisted of 120:100:150 kg ha^{-1} of nitrogen: phosphorus: potassium. The fertilization of phosphorus and potassium was carried out every year in early spring. The doses of potassium (potash salt) and phosphorus (superphoshate) fertilization depended on the abundance of these nutrients in the soil, based on the soil analysis carried out each year. Nitrogen fertilization (ammonium nitrate) was applied in three doses of 40 kg N ha^{-1} during the growing season for both variants of fertilization. Fertigation was carried out using a proportional fertilizer dispenser. Drip irrigation and drip fertigation were carried out using the "T-Tape" drip line with a distance of 20 cm between the emitters. The efficiency of a single emitter was 1 l h^{-1}. The distance between the drip lines was 1.6 m. Water from the subsurface well was used for irrigation. The quality and physical and chemical properties of the irrigation water used complied with the quality standards for irrigation water. Drip irrigation was started when the water potential in the soil was close to –40 kPa and finished when the water potential in the soil was close to –10 kPa. The end of irrigation treatments was determined on the basis of soil water potential at field water capacity, measured with a tensiometer. The tensiometers have been installed at every variant of the experiment at the depth of 25 cm. The dates of planting during the particular growing seasons were in the second week of June. Harvesting took place at the physiological stage of fruit ripeness (from the beginning of 3rd week of August till 1st week of September). Ripe fruits were picked progressively as they matured. In the experiment, the total marketable yield of melon fruits (t ha^{-1}) and weight of single melon fruit (kg) were assessed.

2.2. Nutritive Value Assessment

To carry out a nutritional assessment the fresh melon fruits, one fruit from all plants in one plot was cut into a 5 cm wedge and then cut into 1-cm-thick slices. The frozen material was lyophilized (model Alpha 1–4 LDplus, Donserv, Warszawa, Poland), in order to achieve a permanent weight, and then it was ground to a fine powder (the particles were 0.3–0.5 mm in size) and was milled using the ultracentrifuge (Model FW177, Chemland, Stargard, Poland). The ground samples were stored in the dark, in bags, which were placed in desiccators for further analysis.

The total dry matter content of 'Melba' and 'Seledyn' melon fruits was determined using the drying technique according to the methodology of the Association of Official Analytical Chemists [31].

Carbohydrate analyses were performed according to Talburt and Smith's [32] procedures. For reducing sugar concentration assessment, one gram of freeze-dried material sample was placed in a 250 mL bottle; 150 mL of distilled water was then added and it was shaken vigorously. One milliliter of the filtrate was mixed with 3 mL of DNP reagent in a test tube and then heated in a water bath at 95 °C for 6 min. Absorbance of the mixture was measured using a spectrophotometer at a wavelength of 600 nm. The reducing sugar concentration was then estimated using the standard curve of glucose. The total soluble carbohydrate was determined after hydrolysis of sugars. After filtration, 40 mL of the filtrate was taken, and 2 drops of concentrated HCl were added. The samples were warmed in a water bath for 30 min. After cooling, the mixture was neutralized using concentrated NaOH until pH 8.0 was reached. Next, 1 mL of the filtrate was mixed with 3 mL of DNP reagent and the procedure for determining the concentration of reducing sugars was followed. The results were converted to fresh weight taking into account the percentage of dry weight in the fresh matter.

Ascorbic acid reducing sugar concentration was assessed according to Kapur et al. [33]. Ten grams of fresh melon sample was homogenized with 25 mL of 2% oxalic acid solution and quantitatively transferred into a 50 mL volumetric flask and shaken gently to homogenize the solution. Then, it was diluted up to the mark with oxalic acid solution. The obtained solution was then filtered and centrifuged at 4000 rpm for 15 min, after which the supernatant solution was used for spectrophotometric determination (UV–1800, UV Spectrophotometer System, Shimadzu, Kyoto, Japan) of ascorbic acid concentration. Ascorbic acid is oxidized to dehydroascorbic acid by adding bromine water. After this, L—dehydroascorbic acid reacts with 2,4—DNPH and produces an osazone, which, treated with 85% H_2SO_4, forms a red-colored solution. A typical calibration plot was made and used to determine the concentration of ascorbic acid in the investigated samples.

Total carotenoids in melon samples were extracted by procedures described by Herrero-Martinez et al. [34]. Ten grams of lyophilized melon was blended with 100 mL saturated anhydrous sodium carbonate and mixed with a mechanical blender. Ten grams of the mixture was transferred into a centrifuge tube, 20 mL tetrahydrofuran was added, and it was mixed for 2 min under cold water. The mixture was centrifuged at 5000 rpm for 5 min and the supernatant was collected. Extraction was performed by adding 15 mL dichloromethane and 15 mL of 10% w/v NaCl into the supernatant and shaking it for 2 min. The extraction was repeated twice; the organic layer was collected and evaporated under nitrogen steam. The residue was kept at –20 °C, reconstituted with 5 mL dichloromethane, and diluted (1/40-fold) with dichloromethane prior UV measurements (Shimadzu UV-1800, UV–Vis spectral photometer system, Japan). Detection was performed at 450 nm according to the procedure reported in the Polish Standard [35]. Standard β-carotene for identification was prepared in dichloromethane to obtain 4 μg mL^{-1}.

Total phenolic reducing sugar concentration was determined using the Folin–Ciocalteu reagent (Sigma-Aldrich, Darmstadt, Germany) according to the method of Singleton and Orthofer [36]. A volume of 0.5 mL of Folin–Ciocalteu reagent previously diluted with distilled water (1:10) was mixed with 0.1 mL of each sample. The solution was allowed to stand for 5 min at 25 °C before adding 1.7 mL of sodium carbonate solution (20%). Then, 10 mL of distilled water was added to the mixture, and the absorbance was measured at $\lambda = 735$ nm after 20 min of incubation with agitation at room temperature. Results were expressed in mg of gallic acid equivalents (GAE) per kg of fresh sample.

2.3. Statistical Analysis

All the experimental data were tested for differences by two-way ANOVA using of Statistica® 13.1 package. The significance of differences (LSD—lowest significant difference) was evaluated using the Tukey multiple confidence intervals for the significance level of $p = 0.05$.

2.4. Weather Conditions

The average air temperature in Kruszyn Krajeński in the vegetation period, i.e., from 1 April to 30 September in the years 2013–2015, was 14.9 °C and was 0.3 °C higher than the mean for the long-term period 1986–2015 (Table 1). The warmest month of the growing season in 2013–2015 was July, with a mean temperature of 19.6 °C (0.8 °C above the mean for long-term period). In 2014, the highest average air temperature (15.4 °C) was recorded, which was 0.8 °C higher compared to the mean for the long-term period.

Table 1. Average air temperature (°C) data during the vegetation period of Melba and Seledyn melon cultivars in the years 2013–2015.

Study Years	Months of Vegetation Period						Mean
	IV	V	VI	VII	VIII	IX	
2013	7.0	14.2	17.4	18.9	18.1	10.7	14.4
2014	9.9	13.3	16.0	21.5	17.2	14.4	15.4
2015	7.5	12.4	15.7	18.5	20.9	13.8	14.8
Mean for 2013–2015	8.1	13.3	16.4	19.6	18.7	10.0	14.9
Mean for long-term period 1986–2015	8.1	13.3	16.3	18.8	18.0	13.1	14.6

The mean sum of precipitation in Kruszyn Krajeński in the period from 1 April to 30 September, for the years 2013–2015, amounted to 279.2 mm and was 31.4 mm lower than the mean for the long-term period 1986–2015 (Table 2). The highest precipitation during the vegetation period occurred in 2013 and amounted to 354.3 mm (43.7 mm above the mean for the long-term period). In the 2015 growing season, the lowest total precipitation was recorded, amounting to 193.3 mm, and was 117.3 mm below the mean for the long-term period. The mean precipitation in April, June, July, and August in 2013–2015 was lower than the mean for the long-term period. The highest monthly precipitation (91.7 mm in May and 79.0 mm in July) was noted in the growing season of 2013.

Table 2. Precipitation (mm) data during the vegetation period of Melba and Seledyn melon cultivars in the years 2013–2015.

Study Years	Months of Vegetation Period						Sum
	IV	V	VI	VII	VIII	IX	
2013	13.6	91.7	49.3	79.0	56.6	64.1	354.3
2014	40.7	65.7	44.9	55.4	57.3	25.9	289.9
2015	15.6	21.6	33.0	50.4	20.3	52.4	193.3
Mean for 2013–2015	23.3	59.7	42.4	61.6	44.7	47.5	279.2
Mean for long-term period 1986–2015	26.9	50.2	54.9	71.4	59.7	47.5	310.6

2.5. Irrigation Water Rates

The seasonal irrigation water rates used in the growing of Melba and Seledyn melon cultivars were inversely proportional to rainfall amount during the irrigation period. Relationship between precipitation (mm) and seasonal irrigation water rates (mm) of Melba and Seledyn melon cultivars in June–August in the years 2013–2015 is shown in Figure 1. The melon irrigation period, mean for 2013–2015, began on 13 June and ended on 4 August and lasted for an average of 53 days. The shortest irrigation period, only 11 days, was carried out in 2013. On average, in 2013–2015, during the irrigation period, 14 single waterings took place. The average seasonal dose, in the years 2013–2015, was 142.2 mm and ranged from 104.5 mm in 2013 to 169.0 mm in 2014. Both experimental treatments received the same amount of irrigation water.

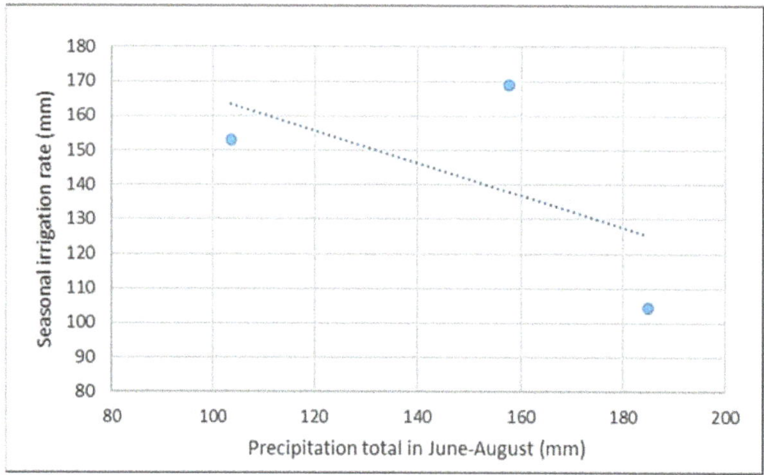

Figure 1. Relationship between precipitation (mm) and seasonal irrigation water rates (mm) of Melba and Seledyn melon cultivars in June–August in the years 2013–2015.

3. Results and Discussion

In the control field (the drip irrigation combined with broadcast nitrogen fertilization), the mean total marketable melon yield in the years 2013–2015 was 26.38 t ha^{-1} and 32.69 t ha^{-1} for Melba and Seledyn cultivars, respectively (Table 3). On average, for the two studied melon cultivars, the highest marketable fruit yield was recorded in 2014. With the drip irrigation and broadcast nitrogen fertilization, the marketable yield was 34.95 t ha^{-1}, and in the field with the drip irrigation and fertigation with nitrogen, the yield of fruits was 39.85 t ha^{-1}. The lowest values of this parameter were recorded in 2015, when the marketable yield was 19.96 t ha^{-1} and 22.68 t ha^{-1}, respectively, for the control and for drip irrigation combined with nitrogen fertigation. Compared to the control, the liquid fertigation significantly increased the fruit yield of Melba and Seledyn cultivars by 12.5% and 13.6%, respectively. There was no significant interaction between irrigation treatment and cultivars.

Table 3. Influence of drip fertigation on the total marketable yield of fruits (t ha^{-1}) of Melba and Seledyn melon cultivars in the years 2013–2015.

Irrigation Treatment	Cultivar	Years of Study			Mean for 2013–2015
		2013	2014	2015	
Drip irrigation + broadcast nitrogen fertilization (control)	Melba	32.47	31.62	15.04	26.38
	Seledyn	34.92	38.27	24.89	32.69
	Mean	33.70	34.95	19.96	29.54
Drip irrigation + nitrogen fertigation	Melba	35.37	35.66	18.01	29.68
	Seledyn	39.95	44.04	27.36	37.12
	Mean	37.66	39.85	22.68	33.40
LSD$_{0.05}$ for irrigation treatment [1]		2.182	4.624	1.828	1.902
LSD$_{0.05}$ for cultivar		1.913	2.380	2.053	2.198
Interaction		n.s.	n.s	n.s.	n.s.

[1] LSD = the lowest significant difference (Tukey's confidence half-interval) at $p < 0.05$; n.s.—not significant at $p < 0.05$.

Some studies have previously confirmed the beneficial effect of irrigation combined with fertilization on the development of plants of the Cucurbitaceae family during cultivation on light soils. The marketable yield of watermelon grown on light soil in Central Poland under the influence of irrigation combined with nitrogen fertilization increased by an average of 21% [23]. In addition, in the cultivation of watermelon performed in semi-arid regions of Brazil, on the sandy soil with low retention capacity of water and low nutrient levels, the authors reported a significant effect of irrigation and nitrogen fertigation on the plant growth, increasing the yield by 64% [37]. In the study presented in this paper, the marketable yield of melon fruits was comparable to the yields obtained in research performed in other soil and climate conditions in other regions of the world. In a study carried out in Turkey, the yield of melon ranged from 18.0 t ha^{-1} to 32.4 t ha^{-1} depending on the irrigation method [38], and from 8.2 t ha^{-1} to 43.8 t ha^{-1} depending on the year of research and the type of irrigation system [39]. In an experiment carried out under field conditions with furrow irrigation in Northwest China, near to the Tengger dessert, the yield of melon ranged from 19.6 t ha^{-1} to 27.8 t ha^{-1} [40]. In research performed in Northern Jordan, melon fruit yields ranged from 15.6 t ha^{-1} to 23.5 t ha^{-1} depending on irrigation quantity [41]. In other studies carried out in Turkey, different irrigation systems and nitrogen levels affected the fruit yield of melon [42]. In the above research, by analyzing different levels of melon irrigation and fertilization, positive effects of combining nitrogen fertilization at a dose 60 kg N ha^{-1} and drip irrigation were observed. As a result of these experiments, the fruit yield of melon was 59.77 t ha^{-1}.

In the present study, a significant influence of the cultivar on the marketable yield of melon fruit was also noticed. The marketable yield of fruits of the Seledyn cultivar was higher by 24.5% compared to the Melba cultivar. Significant relationships between the yield characteristics and the cultivar of watermelon have already been observed in previous studies, the purpose of which was to compare the effects of irrigation and fertilization on the fruit yield [23,43,44].

The increase in the marketable yield of melon fruits results primarily from a significant increase in the single fruit weight. The lowest total marketable yield of fruits and weight of a single fruit was obtained in 2015 (Tables 3 and 4). According to meteorological data, 2015 was very dry. The total rainfall in the period from April to September was only 193.3 mm (62% of the mean for the long-term period 1986–2015). The average air temperatures during the growing season in April, May, June, and July were lower than the mean for the long-term period, 1986–2015, by 0.6 °C, 0.9 °C, 0.6 ° C, and 0.3 °C, respectively (Table 1). Water deficits negatively affect the development of the melon, as it is a photophilous and thermophilic species, with a very high water requirement [8,9]. The melon plants of the Seledyn cultivar produced fruits of significantly greater weight than the plants of the Melba cultivar (Table 4). The use of drip fertigation with nitrogen significantly increased the average melon fruit weight of both the Melba cultivar by 0.12 kg (average fruit weight 0.76 kg) and the Seledyn cultivar by 0.19 kg (average fruit weight 1.15 kg). There was no significant interaction between irrigation treatment and cultivar. Melba is an early cultivar with an average weight of one fruit ranging from 0.5 kg to 0.7 kg [8]. In turn, the Seledyn cultivar is one of the very early ones with fruit larger than Melba, weighing up to 1.4 kg. For comparison, in the study carried out in Turkey, depending on the irrigation method, the weight of melon fruit ranged from 0.8 kg to 1.2 kg [38].

The nutritive values of melons are presented in the Table 5. The content of dry matter, total sugars, monosaccharides, ascorbic acid, total carotenoids, and polyphenols depended on both studied factors: drip irrigation and the cultivar. Drip irrigation combined with nitrogen fertigation significantly increased the concentration of studied components in relation to the control: for dry matter, by 1.7 points on average, total sugars by 14.1 points, monosaccharides by 17.3 points, ascorbic acid by 10.4 points, total carotenoids by 4.5 points, and total polyphenols by 10.8 points. In the study published by Ouzounidou et al. [45], melon fruit concentrated up to 5.1 g 100 g^{-1} fresh weight of monosaccharides and from 0.8 g 100 g^{-1} fresh weight to 4.0 g 100 g^{-1} fresh weight of saccharose (total sugars). In the present experiment, similar results were obtained if we converted our figures into g 100 g^{-1} fresh weight; the levels of total carbohydrates and monosaccharides in the edible parts of melon were significantly affected by the cultivar and irrigation treatments. Seledyn contained significantly more dry matter, total sugars, and monosaccharides than Melba. In the studies presented by Wichrowska et al. [46], irrigation also had a positive effect on the concentration of reducing sugars and vitamin C in Cucurbitaceae, as in the present study. Melba contained significantly more ascorbic acid, total carotenoids, and polyphenols than Seledyn. Ouzounidou et al. [45] reported that the L-ascorbic acid concentration of melon fruit ranged from 13 mg 100 g^{-1} fresh weight to 28 mg 100 g^{-1} fresh weight. Substantially lower concentrations of this acid, in the range of 8 mg 100 g^{-1} fresh weight to 13 mg 100 g^{-1} fresh weight, were noted by Lin et al. [47]. The results of ascorbic acid concentration in the presented studies ranged between 14.5 mg 100 g^{-1} fresh weight and 20.3 mg 100 g^{-1} fresh weight and depended also on the irrigation treatment. Moreover, irrigation with fertilization increased the concentration of nutrients also in *Cucurbita maxima* Duch. [22].

Table 4. Influence of drip fertigation on the single fruit weight (kg) of Melba and Seledyn melon cultivars in the years 2013–2015.

Irrigation Treatment	Cultivar	Year of Study			Mean for 2013–2015
		2013	2014	2015	
Drip irrigation + broadcast nitrogen fertilization (control)	Melba	0.55	0.87	0.50	0.64
	Seledyn	0.93	1.26	0.69	0.96
	Mean	0.74	1.06	0.59	0.80
Drip irrigation + nitrogen fertigation	Melba	0.79	0.94	0.56	0.76
	Seledyn	1.17	1.39	0.89	1.15
	Mean	0.98	1.17	0.73	0.96
$LSD_{0.05}$ for irrigation treatment [1]		0.175	0.100	0.109	0.133
$LSD_{0.05}$ for cultivar		0.199	0.168	0.096	0.134
Interaction		n.s.	n.s	n.s.	n.s.

[1] LSD = the lowest significant difference (Tukey's confidence half-interval) at $p < 0.05$; n.s.—not significant at $p < 0.05$.

Table 5. Influence of drip fertigation on the selected components of nutritive value of Melba and Seledyn melon cultivars (mean for the years 2013–2015).

Irrigation Treatment (I)	Cultivar (I)	Dry Matter (%)	Total Sugar (g kg^{-1} FM)	Monosaccharides (g kg^{-1} FM)	Ascorbic Acid (mg kg^{-1} FM)	Total Carotenoids (mg kg^{-1} FM)	Total Polyphenols (mg GAE kg^{-1} FM)
Drip irrigation + broadcast nitrogen fertilization (control)	Melba	7.75	38.47	25.50	182.47	248.13	58.50
	Seledyn	8.54	73.53	57.53	144.67	88.50	52.23
	Mean	8.15	56.00	41.52	163.57	168.32	55.37
Drip irrigation + nitrogen fertigation	Melba	7.95	45.80	36.47	202.80	256.37	62.80
	Seledyn	8.63	84.53	63.97	162.27	96.27	61.27
	Mean	8.29	65.17	50.22	182.54	176.32	62.04
Mean	Melba	7.85	42.14	30.99	192.64	252.25	60.65
	Seledyn	8.59	79.03	60.75	153.47	92.39	56.75
Mean		8.22	60.58	45.87	173.05	172.32	58.70
$LSD_{0.05}$ for irrigation treatment (I) [1]		0.076	2.050	0.745	3.867	3.462	1.926
$LSD_{0.05}$ for cultivar (II)		0.321	0.967	0.953	2.501	2.028	0.895
$LSD_{0.05}$ for interaction (I/II)		n.s.	2.226	1.187	n.s.	n.s.	2.086

[1] LSD = the lowest significant difference (Tukey's confidence half-interval) at $p < 0.05$; n.s.—not significant at $p < 0.05$.

4. Conclusions

The results of this study indicate that on loose sandy soil with low water capacity and nutrients, melon plants should be drip-fertigated with nitrogen, in order to obtain the best effects. As compared to the control (drip irrigation combined with nitrogen fertilization), the fertigation supplying nitrogen to the plants, used during the cultivation of two melon cultivars, Melba and Seledyn, on a loose sandy soil in Central Poland, significantly increased the total marketable fruit yield by 12.5% for Melba and by 13.6% for Seledyn. Compared to the control, drip fertigation also increased the weight of single fruits by 0.12 kg for Melba and by 0.19 kg for Seledyn. Drip irrigation combined with nitrogen fertigation significantly and positively influenced nutritive value, affecting the increase in dry matter, total sugars, monosaccharides, ascorbic acid, total carotenoids, and polyphenols. Melba contained significantly more ascorbic acid, total carotenoids, and polyphenols than Seledyn, while Seledyn contained significantly more dry matter, total sugars, and monosaccharides.

Our studies fill a gap in the existing scientific literature and show that on a very light soil in a region with very low precipitation within Central Europe (Central Poland), the use of drip irrigation and nitrogen fertigation is effective for the cultivation of melon.

Author Contributions: Conceptualization, R.R., S.R. and A.F.; methodology, R.R., S.R. and A.F.; software, R.R., S.R. and A.F.; validation, R.R., P.-F.F. and P.S.; formal analysis, R.R., P.-F.F. and P.S.; investigation, R.R., S.R., A.F. and D.W.; resources, R.R. and S.R.; data curation, R.R., S.R., A.F. and D.W.; writing—original draft preparation, R.R., S.R., A.F., B.J., D.W., W.P., P.P., H.A.S., P.S. and D.L.; writing—review and editing, R.R., S.R., A.F., B.J., D.W., W.P., P.P. and P.S.; visualization, R.R., S.R., B.J., P.P., H.A.S., P.-F.F. and D.L.; supervision, R.R., S.R., B.J., P.P. and P.-F.F.; project administration, R.R.; funding acquisition, P.P. All authors have read and agreed to the published version of the manuscript.

Funding: This research received no external funding.

Institutional Review Board Statement: Not applicable.

Informed Consent Statement: Not applicable.

Data Availability Statement: Not applicable.

Conflicts of Interest: The authors declare no conflict of interest.

References

1. FAO (Food and Agriculture Organization of the United Nations). Faostat Database Results. 2018. Available online: http://www.fao.org/home/en/ (accessed on 2 October 2020).
2. Pitrat, M. Melon. In *Vegetables I. Handbook of Plant Breeding*; Prohens, J., Nuez, F., Eds.; Springer: New York, NY, USA, 2008; Volume 1.
3. Parle, M.; Singh, K. Musk melon is Eat-Must melon. *Int. Res. J. Pharm. Med. Sci.* **2011**, *2*, 52–57.
4. Vishwakarma, V.K.; Gupta, J.K.; Upadhyay, P.K. Pharmacological importance of *Cucumis melo* L.: An overview. *Asian J. Pharm. Clin. Res.* **2017**, *10*, 8–12. [CrossRef]
5. Gómez-García, R.; Campos, D.A.; Oliveira, A.; Aguilar, C.N.; Madureira, A.R.; Pintado, M. A chemical valorisation of melon peels towards functional food ingredients: Bioactives profile and antioxidant properties. *Food Chem.* **2021**, *335*, 127579. [CrossRef] [PubMed]
6. Mallek-Ayadi, S.; Bahloul, N.; Kechaou, N. Characterization, phenolic compounds and functional properties of *Cucumis melo* L. peels. *Food Chem.* **2017**, *221*, 1691–1697. [CrossRef]
7. RCCT. *Polish National List of Vegetable Plant Varieties*; Research Centre for Cultivar Testing: Słupia Wielka, Poland, 2020; p. 22.
8. Gajc-Wolska, J.; Przybył, J. *Warzywa Dyniowate [Pumpkin Vegetables]*; Działkowiec: Warszawa, Poland, 2005.
9. Fabeiro, C.; Martin de Santa Olalla, F.; De Juan, J.A. Production of muskmelon (*Cucumis melo* L.) under controlled deficit irrigation in a semi-arid climate. *Agric. Water Manag.* **2002**, *54*, 93–105. [CrossRef]
10. Kaniszewski, S. *Nawadnianie Warzyw Polowych [Irrigation of Field Vegetables]*; Plantpress: Kraków, Poland, 2005; pp. 1–85.
11. Hartz, T.K. Effects of drip irrigation scheduling on muskmelon yield and quality. *Sci. Hortic.* **1997**, *69*, 117–122. [CrossRef]
12. Barboza, D.S.; Ferreira, J.A.; Rammana, T.V.; Rodríguez, V.P. Crop water stress index and water use efficiency for melon (*Cucumid melo* L.) on different irrigation regimes. *Agric. J.* **2007**, *2*, 31–37.
13. Yildirim, O.; Halloran, N.; Cavusoglu, S.; Sengul, N. Effects of different irrigation programs on the growth, yield, and fruit quality of drip-irrigated melon. *Turk. J. Agric. For.* **2009**, *33*, 243–255.
14. Leskovar, D.I.; Ward, J.C.; Sprague, R.W.; Meiri, A. Yield, quality and water use efficiency of muskmelon affected by irrigation and transplanting versus direct seeding. *HortScience* **2001**, *36*, 286–291. [CrossRef]
15. Sengul, N.; Yildirim, O.; Halloran, N.; Cavusoglu, S.; Dogan, E. Yield and fruit quality response of drip-irrigated melon to the duration of irrigation season. *Toprak Su. Dergisi.* **2014**, *3*, 90–101. [CrossRef]
16. Coolong, T. Evaluation of shallow subsurface drip irrigation for the production of acorn squash. *HortTechnology* **2016**, *26*, 436–443. [CrossRef]
17. Ayoola, O.T.; Adeniyan, O.N. Influence of poultry manure and NPK fertilizer on yield and yield components of crops under different cropping systems in south west Nigeria. *Afr. J. Biotechnol.* **2006**, *5*, 1386–1392.
18. Paula, J.A.A.; Medeiros, J.F.; Miranda, N.O.; Oliveira, F.A.; Lima, C.J.G.S. Metodologia para determinação das necessidades nutricionais de melão e melancia. *Rev. Bras. Eng. Agríc. Ambient.* **2011**, *15*, 911–916. [CrossRef]

19. Wichrowska, D.; Rolbiecki, R.; Rolbiecki, S.; Figas, A.; Jagosz, B.; Ptach, W. Influence of drip irrigation on nutritive value of winter squash 'Rouge vif d'Etampes' after harvest and storage. *Infrastruct. Ecol. Rural Areas* **2017**, *III/2*, 1167–1175.

20. Wichrowska, D.; Rolbiecki, R.; Rolbiecki, S.; Jagosz, B.; Ptach, W.; Kazula, M.; Figas, A. Concentrations of some chemical components in white asparagus spears depending on the cultivar and post-harvest irrigation treatments. *Folia Hortic.* **2018**, *30*, 147–154. [CrossRef]

21. Battilani, A.; Solimando, D. Yield, quality and nitrogen use efficiency of fertigated watermelon. *Acta Hortic.* **2006**, *700*, 85–90. [CrossRef]

22. Rolbiecki, R.; Rolbiecki, S.; Figas, A.; Wichrowska, D.; Jagosz, B.; Ptach, W. The efficiency of drip fertigation in cultivation of winter squash 'Gomez' on the very light soil. *Infrastruct. Ecol. Rural Areas* **2017**, *III/2*, 1201–1211.

23. Rolbiecki, R.; Rolbiecki, S.; Piszczek, P.; Figas, A.; Jagosz, B.; Ptach, W.; Prus, P.; Kazula, M.J. Impact of nitrogen fertigation on watermelon yield grown on the very light soil in Poland. *Agronomy* **2020**, *10*, 213. [CrossRef]

24. Żakowicz, S.; Hewelke, P. Analiza susz atmosferycznych i glebowych jako kryterium potrzeby nawodnień w danym regionie kraju [Analysis of atmospheric and soil droughts as a criterion for the need for irrigation in a given region of the country]. *Zesz. Probl. Postęp. Nauk Rol.* **1990**, *387*, 193–198.

25. Rzekanowski, C. Perspektywy rozwoju nawodnień w Polsce [Prospects for the development of irrigation in Poland]. *Wiadomości Melioracyjne i Łąkarskie* **2010**, *2*, 55–58.

26. Łabędzki, L.; Kanecka-Geszke, E.; Bąk, B.; Słowińska, S. Estimation of reference evapotranspiration using the FAO Penman-Monteith method for climatic conditions of Poland. In *Evapotranspiration*; Łabędzki, L., Ed.; InTech: Rijeka, Croatia, 2011; pp. 1–446.

27. Żarski, J. Tendencje zmian klimatycznych wskaźników potrzeb nawadniania roślin w rejonie Bydgoszczy [Trends in changes of climatic indices for irrigation needs of plants in the region of Bydgoszcz]. *Infrastruct. Ecol. Rural Areas* **2011**, *5*, 29–37.

28. Rolbiecki, R.; Rolbiecki, S. Wpływ nawadniania kroplowego na plonowanie dyni olbrzymiej odmiany 'Rouge vif d'Etampes' uprawianej na glebie bardzo lekkiej [Influence of drip irrigation on yields of winter squash cv. 'Rouge vif d'Etampes' cultivated on the very light soil]. *Infrastruct. Ecol. Rural Areas* **2012**, *2*, 191–197.

29. Rolbiecki, R.; Rolbiecki, S.; Podsiadło, C.; Wichrowska, D.; Figas, A.; Jagosz, B.; Ptach, W. Influence of drip irrigation on the yielding of summer squash 'White Bush' under rainfall-thermal conditions of Bydgoszcz and Stargard. *Infrastruct. Ecol. Rural Areas* **2017**, *III/2*, 1229–1240.

30. PSSS (Polish Society of Soil Science). Klasyfikacja uziarnienia gleb i utworów mineralnych [Classification of the grain size of soils and mineral deposits]. *Roczniki Gleboznawcze* **2009**, *60*, 5–16.

31. AOAC (Association of Official Analytical Chemists). *Official Methods of Analysis*; AOAC International: Arlington, VA, USA, 2002.

32. Talburt, W.F.; Smith, O. *Potato Processing*; Van Nostrand Reinhold Co.: New York, NY, USA, 1987; pp. 371–474.

33. Kapur, A.; Hasković, A.; Čopra-Janićijević, A.; Klepo, L.; Topčagić, A.; Tahirović, I.; Sofić, E. Spectrophotometric analysis of total ascorbic acid content in various fruits and vegetables. *Bull. Chem. Technol. Bosnia Herzeg.* **2012**, *38*, 39–42.

34. Herrero-Martínez, J.M.; Eeltink, S.; Schoenmakers, P.J.; Kok, W.T.; Ramis-Ramos, G. Determination of major carotenoids in vegetables by capillary electrochromatography. *J. Sep. Sci.* **2006**, *29*, 660–665. [CrossRef]

35. Polish Standard. *Fruit and Vegetable Juices–Journal of Separation Science Determination of Total Carotenoid Content and Individual Carotenoid Fractions (PN–EN 12136)*; Polish Committee for Standardization: Warszawa, Poland, 2000.

36. Singleton, V.L.; Orthofer, R. Analysis of total phenols and other oxidation substrates and antioxidants by means of Folin-Ciocalteu reagent. In *Methods in Enzymology*; Abelson, J.N., Simon, M.I., Sies, H., Eds.; Academic Press: Burlington, MA, USA, 1999; pp. 152–178.

37. Fernandes, C.N.V.; Azevedo, B.M.D.; Neto, J.R.N.; Viana, T.V.D.A.; Sousa, G.G.D. Irrigation and fertigation frequencies with nitrogen in the watermelon culture. *Bragantia* **2014**, *73*, 106–112. [CrossRef]

38. Sensoy, S.; Ertek, A.; Gedik, I. Irrigation frequency and amount affect yield and quality of field-grown melon (*Cucumis melo* L.). *Agric. Water Manag.* **2007**, *88*, 269–274. [CrossRef]

39. Dogan, E.; Kirnak, H.; Berakatoglu, K.; Bilgel, L.; Surucu, A. Water stress imposed on muskmelon with subsurface and surface drip irrigation systems under semi-arid climatic conditions. *Irrig. Sci.* **2008**, *20*, 131–138. [CrossRef]

40. Huang, C.-H.; Xue, X.; You , Q.-G.; Zong, L.; Luo, J. Effects of irrigation frequency on yield and quality of melon under fieldconditions in Minqin oasis. *Int. Res. J. Public Environ. Health* **2016**, *3*, 293–299.

41. Al-Mefleh, N.K.; Samarah, N.; Zaitoun, S.; Al-Ghzawi, A.A.M. Effect of irrigation levels on fruit characteristics, total fruit yield and water use efficiency of melon under drip irrigation system. *J. Food Agric. Environ.* **2012**, *10*, 540–545.

42. Simsek, M.; Comlekcioglu, N. Effects of different irrigation regimes and nitrogen levels on yield and quality of melon (*Cucumis melo* L.). *Afr. J. Biotechnol.* **2011**, *10*, 10009–10018.

43. Wakindiki, I.I.C.; Kirambia, R.K. Supplemental irrigation effects on yield of two watermelon (*Citrulus lanatus*) cultivars under semiarid climate in Kenya. *Afr. J. Agric. Res.* **2011**, *6*, 4862–4870.

44. da Costa, A.R.; de Medeiros, J.F.; Porto Filho, F.D.Q.; da Silva, J.S.; Costa, F.G.; de Freitas, D.C. Production and quality of watermelon cultivated with water of different salinities and doses of nitrogen. *Rev. Bras. Eng. Agríc. Ambient.* **2013**, *17*, 947–954.

45. Ouzounidou, G.; Papadopoulou, P.; Giannakoula, A.; Ilias, I. Effect of plant growth regulators on growth, physiology and quality characteristics of *Cucumis melo* L. *Veg. Crops Res. Bull.* **2006**, *65*, 127–135.

46. Wichrowska, D.; Wojdyła, T.; Rolbiecki, S.; Rolbiecki, R.; Piszczek, P. The content of chosen components in fruit of polish watermelon cultivar 'Bingo' as dependent on way of the seedling production and irrigation. *Pol. J. Food Nutr. Sci.* **2007**, *57*, 147–149.

47. Lin, D.; Huang, D.; Wang, S. Effect of potassium levels on fruit quality of muskmelon in soilless medium culture. *Sci. Hortic.* **2004**, *102*, 53–60. [CrossRef]

Article

Genotype × Environment Interaction of Yield and Grain Quality Traits of Maize Hybrids in Greece

Nikolaos Katsenios [1], Panagiotis Sparangis [1], Sofia Chanioti [2], Marianna Giannoglou [2], Dimitris Leonidakis [3], Miltiadis V. Christopoulos [2], George Katsaros [2] and Aspasia Efthimiadou [1,*]

[1] Department of Soil Science of Athens, Institute of Soil and Water Resources, Hellenic Agricultural Organization—Demeter, Sofokli Venizelou 1, Lycovrissi, 14123 Attica, Greece; nkatsenios@gmail.com (N.K.); pansparangis@gmail.com (P.S.)

[2] Institute of Technology of Agricultural Products, Hellenic Agricultural Organization—Demeter, Sofokli Venizelou 1, Lykovrissi, 14123 Attica, Greece; schanioti@gmail.com (S.C.); giannoglou@chemeng.ntua.gr (M.G.); miltchrist@yahoo.gr (M.V.C.); gkats@chemeng.ntua.gr (G.K.)

[3] Farmacon G.P., K. Therimioti 25, Giannouli, 41500 Larisa, Thessaly, Greece; leo@farmacon.gr

* Correspondence: sissyefthimiadou@gmail.com

Citation: Katsenios, N.; Sparangis, P.; Chanioti, S.; Giannoglou, M.; Leonidakis, D.; Christopoulos, M.V.; Katsaros, G.; Efthimiadou, A. Genotype × Environment Interaction of Yield and Grain Quality Traits of Maize Hybrids in Greece. *Agronomy* **2021**, *11*, 357. https://doi.org/10.3390/agronomy11020357

Academic Editors: Nikolaos Monokrousos and Efimia M. Papatheodorou

Received: 23 January 2021
Accepted: 12 February 2021
Published: 17 February 2021

Publisher's Note: MDPI stays neutral with regard to jurisdictional claims in published maps and institutional affiliations.

Abstract: The interaction of genotype by the environment is very common in multi-environment trials of maize hybrids. This study evaluates the quantity and the quality of grain production and the stability of four maize genotypes in a field experiment that was conducted in five different locations for two years. In order to make a reliable evaluation of the performance of genotypes in the environments, principal components analysis (PCA) was used to investigate the correlation of the yield, soil properties and quality characteristics, while the additive main effects and multiplicative interaction (AMMI) analysis detected the narrow adaptations of genotypes at specific mega-environments. For the yield, AMMI analysis indicated that a group of five environments (ENV1, ENV8, ENV6 ENV10 and ENV9) gave higher yields than the mean value and at the same time had low first interaction principal components axis (IPC1) scores, indicating small interactions. Regarding protein and fiber contents, ENV1 and ENV2, gave the highest values and this could be attributed to the high concentration rates of nutrients like Mg, Ca and the soil texture (C). Specifically for the protein, the results of the analysis indicated that certain environment would provide more protein content, so in order to obtain higher grain protein, growers should grow in certain locations in order to improve the content of this quality characteristic, certain genotypes should be used in certain environments.

Keywords: genotype × environment interaction; maize; yield

1. Introduction

Maize (*Zea mays* L.) is a crop of major importance for the nutrition of the Earth's population. Thus, there has been an urgent need to increase its yield and its quality. There are two main factors that have approximately the same influence on yield increase; improved management practices along with plant breeding have made an impact on this cause. It is notable that their interaction is the one that made such a huge progress to this matter that neither could do alone to this extent [1]. Maize yields have been increasing over the years globally and according to recent data have been doubled from 1961 to 2002 [1–3]. Even before the use of hybrids, farmers used to breed plants that seemed to fit their needs, with good adaptation at their specific environmental conditions, while maintaining their quality and morphological traits. Wherever hybrids have been adopted there has been an increase in the maize yield. Even though there was a tendency to select the high yield hybrids, the need for overall stability and dependability favors the selection of hybrids with stress tolerance. The main focus of the new hybrids now is to aim for a high and stable yield in both favorable and unfavorable growing conditions [1]. Indeed, as Rosegrant et al. [4] mention maize production may suffer a huge input reduction since the available water resources will be limited especially in regions where irrigation is essential. So, it is implied that for the future of crop production it is a mandatory need, to include the environment as an important factor in the selection of better adaptive genotypes.

Climatic conditions along with soil characteristics are the main environmental factors that affect plant growth [5]. Atmosphere and soil water availability, soil temperature and composition can make an impact to the growth of the plants along with other factors such as soil pH and its nutrient status that can influence their development [6]. Soil plays a major role in the plant's life with emphasis on soil carbon, water quality and content. It is often related to problems that occur in plants such as nutrient deficiencies, water stress and toxicities. Its structure not only affects plant growth but also influences their ability to absorb nutrients and water [7]. Each environment has its own soil characteristics and climatic conditions that can affect the productivity of crop production. Thus, it is mandatory to take into consideration the effect of the environment while investigating the most suitable cultivation.

The genotype by environment (GE) interaction is a phenomenon recognized globally by everyone involved to the goal of crop improvement and maintenance. It refers to the various responses of genotypes across a wide range of environments [8,9]. Quantitative characteristics that are economically and agronomically important such as grain yield can be significantly affected by the GE interaction and can provide relief to the breeding actions that can be avoided, by reducing futile subsequent analyses, restricting the significance of questionable deductions and limiting the selection of superior genotypes [8,10–13]. In general, genotypes adapt differently to different environments and the evaluation of each one of them differs according to each purpose [14]. These evaluations in order to be valid are submitted to a series of multiple-environment trials (MET) in an advanced selection stage [11,15]. According to Lu'quez et al. [16], cultivars with high yield and better stability can be identified when growing cultivars in various environments. Every newly registered cultivar needs to be evaluated for adaptation for several years in many locations in order to be recommended to a specific area. To accomplish this procedure METs are conducted in many countries through varietal testing programs for more reliable results. The GE is the main interaction that needs to be evaluated. By submitting genotypes to a variety of environments, differential genotypic responses are recorded and can provide better identification of a superior and stable hybrid [17].

The environment, the genotype and the GE interaction are also responsible for variations in the quality properties of grains, including the color, the texture, the protein and the fiber content. Among the quality parameters, the protein content of the grains is highly affected by the environment [18]. The evaluation of different genotypes on quality traits associating with the improvement of the yield can also contribute to future breeding strategies.

The additive main effects and multiplicative interaction (AMMI) analysis is extensively used among agricultural researchers who want to evaluate the yield performance of different genotypes under multienvironment trials. This analysis is widely used because it contributes to better understanding of the complicated interactions between genotypes and environment, and it has high accuracy [19]. The AMMI analysis is a combination of ANOVA and PCA (principal component analysis) and a major output of the results is a biplot that presents the means of the genotypes used and their relation to the first PC [20]. This biplot is an effective tool to evaluate the GE interactions graphically [21]. The results of AMMI analysis are considered useful for the evaluation of yield stability of crops across different environmental conditions and for the determination of suitable environments for all examined genotypes [22–25].

The aim of this study was to evaluate four maize hybrids at five locations, for two years (the combination of year-location created the 10 environments), in order to investigate how maize yield and quality characteristics of the grain are affected by the GE interaction. Furthermore, principal components analysis (PCA) was used to investigate the correlation of the yield, soil properties and quality characteristics, while the AMMI analysis contributed to detect the narrow adaptations of genotypes at specific mega-environments.

2. Materials and Methods

2.1. Experimental Site and Design

The effect of GE interaction in relation to the final grain yield and the various quality characteristics were assessed. The field trials were conducted at 2019 and 2020 and a randomized complete block design was used with three replications. The experimental design had two main factors (genotypes and environments). The four maize genotypes used in the experimental fields were GEN1 (P0937, Pioneer Hi-Bred Hellas S.A.), GEN2 (DKC6050, K. and N. Efthymiadis S.A.), GEN3 (DKC6980, K. and N. Efthymiadis S.A.) and GEN4 (P2105, Pioneer Hi-Bred Hellas S.A.). P0937 hybrid belongs to the 500 FAO group with 120–125 days until maturity, it provides well-balanced plants, resistant to the northern leaf blight of maize, suitable especially for environments that traditionally produce good yield results. DKC6050 belongs to the 600 FAO group with 116–123 days until maturity. It produces average height plants with a very strong stem and root system and provides well-balanced plants. It has solid, filled until the top ears with 16-18 rows of kernels. It is considered to have great agronomic characteristics with high yield and good adaptation in many environments and with great quality grains suitable for human consumption. DKC6980 belongs to the 700 FAO group with 130–136 days until maturity. It produces high plants with a very strong stem and root system. It has big ears with 18-20 rows of kernels. It has stable yield even in high temperatures. P2105 is a hybrid that belongs to the 700 FAO group with 135–140 days until maturity, it provides well-balanced plants with a strong root system. It has a fast, lively growth and high dynamic production especially in early cultivated and well-drained fields. These four hybrids were selected because they are considered to have great characteristics in Greece's environmental variability.

The five locations were Giannouli, in the Prefecture of Larissa, Thessaly; Nea Tyroloi and Kamila, in the Prefecture of Serres, Central Macedonia; Kalamonas, in the Prefecture of Drama, Eastern Macedonia and Thrace and Koutso, in the Prefecture of Xanthi, Eastern Macedonia and Thrace, Greece. Each location–year combination created a different environment; thus 10 environments were used to evaluate the four genotypes.

The environments had different soil texture and variable microclimate conditions (Tables S1 and S2). Planting dates were between 22 and 25 March for the first year and 18 and 26 April for the second year and harvesting dates between 9 and 11 September for the first year and 19 and 21 September for the second year (Table S1). The plant density for all cultivars was 85,000 plants per hectare and the planting depth was 3.5 cm. The field plots (60 m^2) consisted of 4 rows with spacing between rows 0.75 m. The measurement of yield was made at a length of 12 m in the 2 middle rows of each plot to avoid the border effect. All five locations are traditionally used for maize production in Greece since they appear to be suitable for this crop.

All field management procedures were standard to ensure the avoidance of deficiencies and the balancing of soil nutrients in all environments. After grain harvesting, the measurement of moisture content was conducted with a portable humidity meter in order to calculate the production per hectare adjusted to 15% humidity. Afterward, grain samples were directed for the determination and measurement of the quality characteristics.

2.2. Soil Sample Analyses

At the soil samples, the elements Ca, Mg, K and Na were determined by atomic absorption spectrometry after extraction using $BaCl_2$ [26]. The measurements of Zn, Mn, Cu and Fe were conducted by atomic absorption spectrometry after extraction using DTPA [27]. The available B was determined using a spectrophotometer, using azomethine-H as the color (yellow) development reagent [28]. Total nitrogen was determined by the Kjeldahl method [29]. Organic carbon was determined with oxidization by $K_2Cr_2O_7$ [30]. Available phosphorus was determined after extraction with $NaHCO_3$ [31]. Cation exchange capacity was determined according to ISO 11260 [26]. Soil texture was determined using the method of Bouyoucos [32] and the soil taxonomy of USDA (1999). The moisture content was determined in a furnace at 105 °C for 24 h. The value of pH was measured with a pH-meter equipped with a glass electrode in the saturated paste extract.

2.3. Quality Characteristics of Harvested Corn Grains

After their harvest, the corn grains were dried in the shade following the farming practices. The moisture content across all cultivars varied from 10.85 to 16.04% with an average of 12.74%.

The color of the corn grains was measured using Minolta Colorimeter (CR-300, Minolta Company, Chuo-Ku, Osaka, Japan). The lightness or brightness of the samples was indicated by the L value where 0–100 represents darkness to lightness color. The index a indicates the redness or greenness of the corn grains, with a positive a value representing more red color. The index b value represents the degree of the yellow-blue color, with a positive b value illustrating more yellow color.

The texture analysis was carried out by HD-Plus texture analyzer (Stable Micro Systems Ltd., Godalming, UK) and the Texture Expert Exceed Software for the data analysis. The determination of the textural characteristics of corn grains was performed by a puncture probe of 5 mm diameter. Probe speeds of 1 mm/s during the test, 2 mm/s for the pretest and 10 mm/s for the post-test were used throughout the study. All the measurements were performed at 25 ± 1 °C and the hardness of the corn seeds was determined and expressed at N.

The corn grains were grinded by using a grinding mill for the determination of the moisture, the ash, the total protein and the total crude fiber content. Ash and crude fiber content of corn flours were determined according to AOAC Official Method 923.03 and 984.04 (Weende Method), respectively and recorded manually. Total protein content analysis of corn flours was conducted by applying the Kjeldahl method (IDF 2008), using a Kjeldahl rapid distillation unit (Protein Nitrogen Distiller DNP-1500-MP, Raypa Spain).

2.4. Statistical Analysis

A two-way analysis of variance (ANOVA) was used to evaluate the effect of genotype, environments and their GE interaction on the cultivation and quality characteristics of maize hybrids. The experimental data were analyzed using IBM SPSS software ver. 24 (IBM Corp., Armonk, N.Y., USA). The comparisons of means were calculated using the Duncan test at the 5% level of significance ($p < 0.05$). Multivariate analysis was conducted by means of principal component analysis (PCA) by using STATISTICA 7 (Statsoft Inc., Tulsa, OK, USA). Additive main effects and multiplicative interaction (AMMI) analysis was conducted by using AMMISOFT version 1.0 (Soil and Crop Sciences, Cornell University, Ithaca, NY, USA).

3. Results and Discussion

3.1. Grain Yield

Yield (kg/ha) of maize was statistically significantly influenced by genotype, environment and their interaction (Table S3). Genotype and environment effects on maize yield are presented in Table 1. GEN1 had mean yield that was significantly different in all the tested environments. Its highest yield value was recorded in ENV10 and ENV9 (19,288 ± 289 and 19,087 ± 471 kg/ha respectively) and the lowest in ENV2 (12,805 ± 1361 kg/ha). GEN2 presented similar results; the highest mean yield was recorded in ENV10 and ENV9 (18,244 ± 182 and 18,113 ± 86 kg/ha respectively) but the lowest mean yield in this case was recorded in ENV4 and ENV5 (13,617 ± 370 and 13,657 ± 51 kg/ha respectively). On the other hand, GEN3 presented its highest mean yield in ENV10 where it recorded 20,032 ± 179 kg/ha, and significantly lower results in ENV4 (13,220 ± 569 kg/ha). GEN4 had high yields in ENV6 (20,070 ± 346 kg/ha) and the significant lower value was presented in ENV4 (12,077 ± 166 kg/ha). As for the environments, in ENV4, ENV8 and ENV9, GEN1 presented significant higher mean yields. ENV4 had also a statistically significant performance in GEN2 along with ENV2 and ENV3. Environments ENV1, ENV6 and ENV7 showed significant high performance with GEN4. All the environments presented great results when GEN3 was used except ENV6 and ENV7, which had their lowest performance with this genotype.

Table 1. Effect of genotype and environments on yield and quality characteristics (protein, fiber, color parameters, texture and ash content) of corn grains.

Genotype	Environment	Yield (kg/ha)	Protein (%)	Fiber (%)	Lightness
GEN1	ENV1	17,643 ± 456 ^{Bb}	8.10 ± 0.02 ^{Ad}	3.60 ± 0.01 ^{Bb}	37.85 ± 0.69 ^{Ba}
	ENV2	12,805 ± 1361 ^{Eb}	8.15 ± 0.04 ^{Ab}	4.08 ± 0.06 ^{Aa}	78.88 ± 0.24 ^{A}
	ENV3	14,466 ± 74 ^{Db}	7.24 ± 0.02 ^{Dd}	2.24 ± 0.02 ^{Gd}	37.83 ± 0.25 ^{Bc}
	ENV4	13,993 ± 785 ^{Da}	7.42 ± 0.08 ^{Ca}	3.53 ± 0.02 ^{Ba}	77.17 ± 0.36 ^{A}
	ENV5	15,860 ± 710 ^{Cc}	7.91 ± 0.01 ^{Bb}	3.12 ± 0.01 ^{Db}	35.53 ± 0.57 ^{Bc}
	ENV6	18,659 ± 226 ^{ABb}	6.83 ± 0.02 ^{Ec}	3.44 ± 0.01 ^{Ca}	77.17 ± 0.25 ^{A}
	ENV7	18,456 ± 784 ^{ABab}	6.23 ± 0.01 ^{Hd}	1.37 ± 0.02 ^{Id}	36.96 ± 0.45 ^{Bb}
	ENV8	18,600 ± 306 ^{ABa}	6.65 ± 0.03 ^{Fc}	2.38 ± 0.01 ^{Fb}	78.23 ± 0.52 ^{Aa}
	ENV9	19,087 ± 471 ^{Aa}	6.38 ± 0.01 ^{Gb}	1.99 ± 0.01 ^{Hd}	36.83 ± 0.26 ^{Bc}
	ENV10	19,288 ± 289 ^{Ab}	6.81 ± 0.02 ^{Ea}	2.75 ± 0.02 ^{Eb}	75.52 ± 0.85 ^{A}
GEN2	ENV1	17,137 ± 191 ^{Bc}	8.61 ± 0.02 ^{Ac}	3.52 ± 0.01 ^{Ac}	38.05 ± 0.81 ^{Ca}
	ENV2	15,470 ± 449 ^{Ca}	8.62 ± 0.02 ^{Aa}	3.51 ± 0.01 ^{Ac}	73.05 ± 0.25 ^{A}
	ENV3	15,917 ± 338 ^{Ca}	7.47 ± 0.02 ^{Cc}	2.39 ± 0.02 ^{Gc}	41.99 ± 0.65 ^{BCa}
	ENV4	13,617 ± 370 ^{Da}	7.42 ± 0.03 ^{Ca}	2.69 ± 0.02 ^{Ed}	72.72 ± 0.88 ^{A}
	ENV5	13,657 ± 51 ^{Dd}	7.42 ± 0.03 ^{Cc}	3.02 ± 0.02 ^{Bc}	42.58 ± 0.65 ^{Ba}
	ENV6	17,317 ± 613 ^{Bb}	7.95 ± 0.03 ^{Ba}	2.81 ± 0.02 ^{Db}	71.53 ± 0.48 ^{A}
	ENV7	16,747 ± 417 ^{Bb}	6.99 ± 0.03 ^{Dc}	2.82 ± 0.02 ^{Da}	41.75 ± 0.52 ^{BCa}
	ENV8	17,410 ± 676 ^{Bb}	6.31 ± 0.02 ^{Fd}	2.60 ± 0.03 ^{Fa}	71.53 ± 0.62 ^{Ab}
	ENV9	18,113 ± 86 ^{Ab}	6.10 ± 0.01 ^{Jb}	2.86 ± 0.03 ^{Ca}	42.53 ± 0.45 ^{Ba}
	ENV10	18,244 ± 182 ^{Ac}	6.69 ± 0.02 ^{Ea}	3.52 ± 0.02 ^{Aa}	75.20 ± 0.48 ^{A}
GEN3	ENV1	18,557 ± 88 ^{Ba}	8.86 ± 0.02 ^{Ab}	3.14 ± 0.01 ^{Cd}	35.50 ± 0.25 ^{Ba}
	ENV2	16,986 ± 386 ^{Ca}	8.07 ± 0.04 ^{Cc}	3.90 ± 0.01 ^{Ab}	74.83 ± 0.93 ^{A}
	ENV3	15,670 ± 625 ^{Da}	8.66 ± 0.01 ^{Ba}	2.56 ± 0.01 ^{Gb}	36.68 ± 0.43 ^{Bd}
	ENV4	13,220 ± 569 ^{Ea}	7.04 ± 0.08 ^{FGc}	2.85 ± 0.03 ^{Eb}	72.22 ± 0.26 ^{A}
	ENV5	19,103 ± 440 ^{ABa}	7.13 ± 0.02 ^{Fd}	3.33 ± 0.02 ^{Ba}	36.49 ± 0.29 ^{Bc}
	ENV6	19,312 ± 112 ^{ABc}	7.06 ± 0.02F ^{Gb}	2.72 ± 0.02 ^{Fc}	70.15 ± 0.48 ^{A}
	ENV7	14,787 ± 1140 ^{Dc}	7.81 ± 0.03 ^{Db}	2.43 ± 0.01 ^{Ib}	34.28 ± 0.52 ^{Bb}
	ENV8	18,810 ± 608 ^{Ba}	6.89 ± 0.01 ^{Gb}	1.79 ± 0.01 ^{Jd}	70.01 ± 0.48 ^{Ab}
	ENV9	19,107 ± 179 ^{ABa}	7.56 ± 0.03 ^{Ea}	2.49 ± 0.03 ^{Hc}	39.48 ± 0.47 ^{Bb}
	ENV10	20,032 ± 179 ^{Aa}	6.86 ± 0.03 ^{Ga}	2.89 ± 0.02 ^{Db}	71.65 ± 0.45 ^{A}
GEN4	ENV1	18,430 ± 161 ^{BCa}	8.96 ± 0.02 ^{Aa}	3.90 ± 0.01 ^{Aa}	34.62 ± 0.34 ^{Eb}
	ENV2	15,887 ± 659 ^{Ea}	8.13 ± 0.02 ^{Bb}	3.12 ± 0.01 ^{Cd}	76.35 ± 0.45 ^{A}
	ENV3	14,083 ± 245 ^{Fb}	7.61 ± 0.01 ^{Db}	3.55 ± 0.02 ^{Ba}	39.62 ± 0.70 ^{Db}
	ENV4	12,077 ± 166 ^{Gb}	7.24 ± 0.03 ^{Eb}	2.73 ± 0.02 ^{Dc}	73.31 ± 0.55 ^{B}
	ENV5	17,670 ± 207 ^{CDb}	8.02 ± 0.02 ^{Ca}	2.37 ± 0.02 ^{Ed}	38.22 ± 0.45 ^{Db}
	ENV6	20,070 ± 346 ^{Aa}	6.48 ± 0.02 ^{Fd}	2.56 ± 0.03 ^{DEd}	71.72 ± 0.35 ^{BC}
	ENV7	19,040 ± 1151 ^{Ba}	8.00 ± 0.02 ^{Ca}	1.43 ± 0.01 ^{Hc}	34.79 ± 0.47 ^{Eb}
	ENV8	15,946 ± 423 ^{Ec}	7.60 ± 0.03 ^{Da}	2.11 ± 0.02 ^{Fc}	69.42 ± 0.46 ^{Cb}
	ENV9	17,477 ± 625 ^{Db}	6.13 ± 0.02 ^{Hb}	2.53 ± 0.03 ^{DEb}	38.73 ± 0.74 ^{Dbc}
	ENV10	17,887 ± 160 ^{CDc}	6.33 ± 0.02 ^{Gb}	1.64 ± 0.04 ^{Gc}	72.98 ± 0.82 ^{B}

Table 1. *Cont.*

Genotype	Environment	a	b	Texture (N)	Ash (%)
GEN1	ENV1	3.88 ± 0.29 [ABc]	34.77 ± 0.55 [Bb]	20.49 ± 0.72 [CD]	1.65 ± 0.01 [Aa]
	ENV2	4.81 ± 0.55 [AB]	49.41 ± 0.20 [A]	19.99 ± 0.30 [Dab]	1.18 ± 0.02 [Jc]
	ENV3	4.37 ± 0.15 [ABd]	35.37 ± 0.26 [Bc]	12.53 ± 0.59 [Fb]	1.41 ± 0.01 [Ba]
	ENV4	3.87 ± 0.74 [AB]	50.67 ± 0.72 [A]	23.69 ± 0.62 [BCa]	1.39 ± 0.02 [Cc]
	ENV5	4.36 ± 0.15 [ABc]	35.09 ± 0.65 [Bc]	17.97 ± 0.25 [DEb]	1.37 ± 0.02 [Cb]
	ENV6	2.70 ± 0.15 [Bb]	47.57 ± 0.56 [A]	19.44 ± 0.32 [Abc]	1.23 ± 0.04 [Fb]
	ENV7	5.13 ± 0.26 [A]	34.47 ± 0.54 [Bbc]	14.80 ± 0.40 [EFbc]	1.15 ± 0.01 [Hc]
	ENV8	5.66 ± 0.09 [A]	49.38 ± 0.29 [Aa]	26.28 ± 0.20 [Ha]	1.09 ± 0.02 [Ic]
	ENV9	5.12 ± 0.15 [Ab]	38.15 ± 0.63 [Bc]	17.05 ± 0.52 [DE]	1.31 ± 0.03 [Db]
	ENV10	2.80 ± 0.08 [B]	52.98 ± 0.27 [A]	30.09 ± 0.39 [Aa]	1.27 ± 0.03 [Ec]
GEN2	ENV1	7.47 ± 0.71 [a]	47.08 ± 0.46 [Ba]	21.63 ± 0.41 [BCD]	1.39 ± 0.01 [Dd]
	ENV2	7.17 ± 0.15	50.57 ± 0.26 [AB]	22.48 ± 0.59 [ABCa]	1.46 ± 0.03 [Ca]
	ENV3	6.45 ± 0.08 [a]	51.10 ± 0.77 [ABa]	17.96 ± 0.52 [DEa]	1.43 ± 0.02 [Ca]
	ENV4	4.90 ± 0.26	53.07 ± 0.86 [AB]	25.68 ± 0.72 [Aa]	1.74 ± 0.02 [Bb]
	ENV5	7.70 ± 0.28 [a]	55.22 ± 0.74 [ABa]	23.18 ± 0.65 [ABa]	1.21 ± 0.02 [Fc]
	ENV6	5.21 ± 0.22 [b]	49.70 ± 0.52 [AB]	24.88 ± 0.72 [ABa]	2.40 ± 0.03 [Aa]
	ENV7	7.64 ± 0.09 [a]	52.40 ± 0.48 [ABa]	13.51 ± 0.55 [Fc]	1.11 ± 0.03 [Hd]
	ENV8	5.21 ± 0.14	49.70 ± 0.52 [ABa]	18.89 ± 0.65 [CDEb]	0.73 ± 0.04 [Id]
	ENV9	7.74 ± 0.26 [a]	51.15 ± 0.74 [ABa]	15.47 ± 0.35 [EF]	1.31 ± 0.01 [Eb]
	ENV10	5.25 ± 0.21	58.90 ± 0.59 [A]	23.48 ± 0.62 [ABb]	1.17 ± 0.02 [Gd]
GEN3	ENV1	4.66 ± 0.45 [Bbc]	37.70 ± 0.41 [Cb]	22.30 ± 0.42 [A]	1.54 ± 0.01 [Ab]
	ENV2	3.61 ± 0.46 [B]	44.27 ± 0.82 [ABC]	15.68 ± 0.15 [Bc]	1.13 ± 0.02 [Ed]
	ENV3	4.38 ± 0.06 [Bc]	41.81 ± 0.46 [BCb]	15.84 ± 0.43 [Bab]	1.27 ± 0.02 [Cb]
	ENV4	6.91 ± 0.52 [A]	52.27 ± 0.74 [A]	17.40 ± 0.83 [Bb]	1.05 ± 0.02 [Fd]
	ENV5	5.15 ± 0.09 [Bd]	37.60 ± 0.14 [Cb]	22.49 ± 0.25 [Aa]	1.01 ± 0.03 [Fd]
	ENV6	4.64 ± 0.08 [Bb]	39.35 ± 0.25 [C]	21.21 ± 0.56 [Ab]	1.11 ± 0.04 [Ec]
	ENV7	5.15 ± 0.12 [Bb]	37.97 ± 0.18 [Cb]	17.06 ± 0.48 [Ba]	1.20 ± 0.03 [Db]
	ENV8	4.64 ± 0.07 [B]	39.35 ± 0.22 [Ca]	17.22 ± 0.25 [Bbc]	1.25 ± 0.02 [Ca]
	ENV9	5.05 ± 0.14 [Bb]	44.67 ± 0.26 [ABCb]	15.30 ± 0.45 [B]	1.50 ± 0.02 [Aa]
	ENV10	4.84 ± 0.15 [B]	50.06 ± 0.28 [AB]	20.59 ± 0.25 [Ac]	1.34 ± 0.02 [Bb]
GEN4	ENV1	5.26 ± 0.14 [Bb]	35.16 ± 0.94 [Cb]	21.38 ± 0.48 [A]	1.48 ± 0.01 [Bc]
	ENV2	5.48 ± 0.25 [B]	50.64 ± 0.84 [A]	18.64 ± 0.53 [Bbc]	1.33 ± 0.03 [Db]
	ENV3	5.70 ± 0.08 [Bb]	40.95 ± 0.77 [ABCb]	13.47 ± 0.52 [Db]	1.25 ± 0.01 [Eb]
	ENV4	5.97 ± 0.63 [B]	50.08 ± 0.74 [A]	18.75 ± 0.72 [Bb]	1.84 ± 0.02 [Aa]
	ENV5	5.49 ± 0.49 [Bb]	39.35 ± 0.72 [BCb]	18.34 ± 0.65 [Bb]	1.43 ± 0.01 [Ca]
	ENV6	8.80 ± 0.15 [Aa]	47.72 ± 0.74 [AB]	18.63 ± 0.37 [Ba]	1.03 ± 0.01 [Gd]
	ENV7	6.62 ± 0.12 [Ba]	31.85 ± 0.65 [Cc]	15.84 ± 0.36 [Cab]	1.32 ± 0.02 [Da]
	ENV8	5.91 ± 0.23 [B]	35.91 ± 0.67 [Db]	13.38 ± 0.47 [Dc]	1.21 ± 0.03 [Fb]
	ENV9	5.47 ± 0.27 [Bb]	40.53 ± 0.45 [ABCc]	15.46 ± 0.45 [C]	1.49 ± 0.01 [Ba]
	ENV10	6.90 ± 0.18 [B]	49.78 ± 0.68 [AB]	18.10 ± 0.52 [Bc]	1.50 ± 0.01 [Ba]

Mean value of three replicates ± standard deviation. Values with different capital letter (A, B, C, D, E, F, G, H, I, J) denotes significant difference between environments, and small letter (a, b, c, d) denotes significant difference between genotypes in each environment according to the Duncan's multiple range test at $p < 0.05$. Where there are no letters, no significant differences were recorded.

3.2. Quality Characteristics of Harvested Corn Grains

The performance of genotypes on environments for quality characteristics (color parameters, texture, ash, protein and fiber content) of the harvested corn grains is presented in Table 1. The performance (color parameters) for corn grains varied across different environments. The lightness (L) across all cultivars ranged from 34.28 to 78.23, the yellow index b ranged from 34.47 to 58.90 and the red index a ranged from 3.61 to 8.80. For L, a and b color parameters, there was a significant difference between genotype ($p < 0.001$), environments ($p < 0.001$) and for their GE interaction, except for the interaction on the red index. The highest values of L color parameter of corn grains were obtained from the environments

ENV2, ENV4, ENV6, ENV8 and ENV10 and the lowest from the environments ENV1, ENV3, ENV5 ENV7 and ENV9. According to the average of the tested environments, the highest values of L, a and b color parameters were achieved from genotype GEN2, and the lowest corresponding values were obtained from the genotypes GEN1 and GEN4 (Table 1).

The texture of the harvested corn grains was influenced by the environmental and genotype effects ($p < 0.001$) (Table 1). Texture (hardness) of corn grains varied between 12.53 and 26.28 N across the cultivars and environments. The hardness of corn grains at ENV1 (21.38 N), ENV4 (18.75 N), ENV2 (18.64 N), ENV6 (18.63 N), ENV5 (18.33 N) and ENV10 (18.10 N) was relatively higher than the one at ENV3 (13.47 N) and ENV8 (13.38 N). According to the average of the tested environments, the highest values of hardness were achieved from genotypes GEN2 (20.72 N) and GEN1 (20.23 N), and the lowest corresponding values were obtained from the genotypes GEN3 (18.50 N) and GEN4 (17.20 N) (Table 1).

The ash content of corn grains was significantly affected by genotype ($p < 0.001$), environments ($p < 0.001$) and their GE interaction ($p < 0.001$) (Table 1). The ash content across all cultivars ranged from 0.73 to 2.40%. The highest ash content was found in corn grains obtained at the environments of ENV1 (1.51%) and ENV4 (1.50%) and the lowest at ENV8 (1.07%). The ash content of corn grains from GEN2 (1.39%) and GEN4 (1.38%) was significantly higher than the one from GEN1 (1.30%) and GEN3 (1.23%).

The genotype ($p < 0.001$), the environments ($p < 0.001$) and their GE interaction ($p < 0.001$) highly influenced the protein content of corn grains (Table 1). The protein content across all genotypes varied from 6.13 to 8.96%. The highest protein content was found in corn grains obtained at the ENV1 (8.63%). The protein content of corn grains from GEN3 (7.59%) and GEN4 (7.45%) was significantly higher than the ones from the other genotypes. Protein content is a primary quality indicator for corn grains. Mut et al. [33] and Peterson et al. [34] reported that the grain protein content changed from 3 to 4% and 10.0 to 18.0%, respectively within different oat genotypes cultivated in different environments. The protein content was mainly affected by the environment rather than the genotype. This finding is in accordance with other scientific studies [18,35,36]. The protein content of corn grain is illustrated the quality of corn flour and is a desirable trait for the food industry.

The fiber content of corn grains was significantly influenced by genotype ($p < 0.01$), environments ($p < 0.001$) and their GE interaction ($p < 0.001$) (Table 1). The fiber content across all cultivars ranged from 1.37 to 4.08%. The highest fiber content was found in corn grains obtained at the environments of ENV2 (3.65%) and ENV1 (3.54%) while the lowest at ENV7 (2.01%). The fiber content of corn grains from GEN2 (2.97%) and GEN1 (2.85%) was significantly higher than others. It was observed that the effect of environment on the fiber content of corn grains was stronger as compared to the genotype. This finding is in accordance with other scientific studies [18,37].

3.3. Correlation and Evaluation of the Yield, Soil Properties and Quality Characteristics vs. the Genotype and Environment on Maize Cultivation

To investigate the correlation of the yield, soil properties and quality characteristics by using four different genotypes of maize hybrids at ten different environmental conditions, principal components analysis (PCA) was used (Figure 1). Each point on the loading plot represented the contribution of a variable (yield, soil properties: clay, silt, sand, pH, organic matter, total nitrogen, $CaCO_3$, K, Ca, Mg, P, Fe, Cu, Zn, Mn, B and quality characteristics: color, texture, moisture, ash, protein and fiber content) to the score, while each point on the score plot represented a tested sample. The first principal component (PC1) described 41.38% of the variation of extraction experiments, whereas the second principal component (PC2) 25.63% respectively, so that they contributed 67.01% of the total variation of extraction experiments.

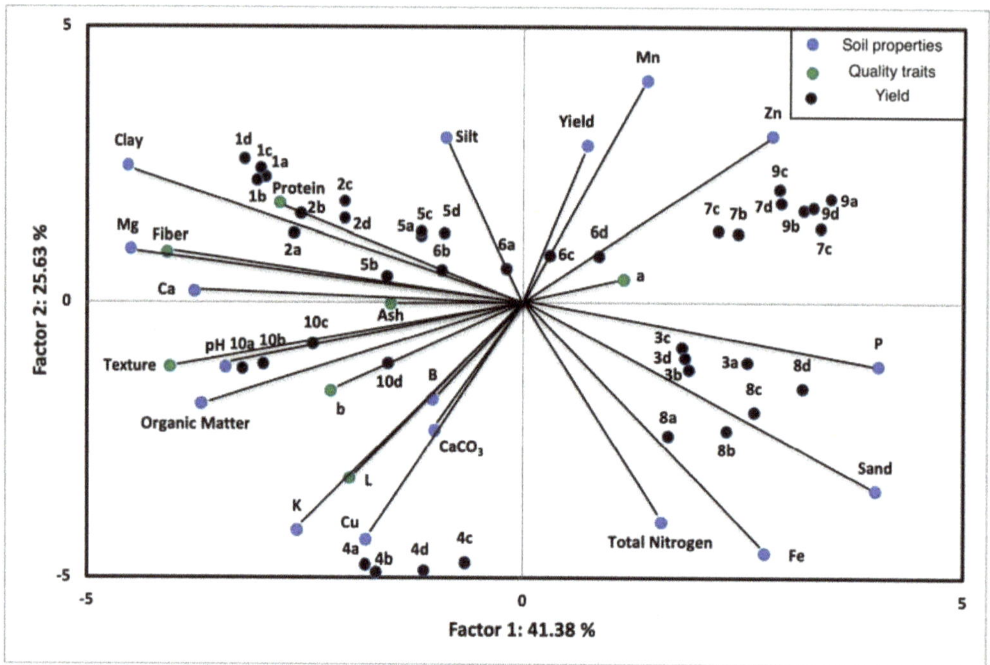

Figure 1. Biplot of principal component analysis of the four different genotypes of maize hybrids at five different environmental conditions. Code of different environments of maize hybrids on different genotypes used on principal components analysis (PCA) is listed as follows: Environments: (1) ENV1, (2) ENV2, (3) ENV3, (4) ENV4, (5) ENV5, (6) ENV6, (7) ENV7 (8) ENV8, (9) ENV9 and (10) ENV10 and Genotypes: (a) GEN1, (b) GEN2, (c) GEN3 and (d) GEN4.

According to the PCA plot, the texture, fiber, protein, clay, Mg, Ca, pH and OM had a negative effect on PC1 and the total nitrogen, silt, $CaCO_3$, Fe and K had a negative effect on PC2, while the sand, P and Zn had a positive effect on PC1 and the Mn, silt and yield had a positive effect on PC2. Furthermore, there are correlations between the Mg and fiber, clay and protein, pH and texture, and between L, B, K, Cu and $CaCO_3$. Based on PCA score plot of the tested samples, four main groups of samples were noted. The groups are (a) 1a, 1b, 1c, 1d, 2a, 2b, 2c, 2d (b) 5a, 5b, 5c, 5d, 6a, 6b, 6c, 6d (c) 10a, 10b, 10c, 10d (d) 3a, 3b, 3c, 3d, 8a, 8b, 8c, 8d and (e) 7a, 7b, 7c, 7d, 9a, 9b, 9c, 9d.

All five testing locations are suitable for maize production in Greece. However, they have different soil conditions and rainfall, influencing the yield and the quality characteristics of corn grains. The samples of group (a) confirmed that ENV1 and ENV2 were the most effective environments for all the tested hybrids, giving corn grains with the highest protein and fiber content. These findings could be attributed to the enhanced soil fertility of ENV1 and ENV2 having the highest concentration of nutrients including Mg, Ca, clay and silt. Many studies demonstrated that the protein content was mostly affected by the environment, indicating its sensitivity to the environment [18,35,36,38] and that the soil nutrient supply affected positively the yield and the quality characteristics of the crop products [39].

The samples of group (b) indicated that ENV5 and ENV6 showed good soil conditions in terms of nutrients resulting in high yields. This finding is in accordance with other studies indicating that any higher nutrient uptake by the plant can result in higher yields [18,40]. The samples of group (c) showed that ENV10 resulted in grains with the maximum hardness concerning their texture essential soil indices such as pH and OM compared with the other environments. Moreover, the samples of group (d) showed that ENV3 and ENV8 had similar characteristics in terms of protein and fiber content growing on a

Fe-, P- and N- and sand-rich environment. The samples of group (e) indicated that ENV7 and ENV9 had similar characteristics in terms of yield and quality properties growing on a Zn-rich environment. Concluding, the results depicted by principal components analysis are in agreement with those discussed above.

3.4. GE Interaction for Yield, Protein and Fiber Content

According to the ANOVA, genotypes (GEN), environments (ENV) and their interaction (G×E) gave statistically significant differences ($p < 0.001$) concerning the yield measurement. Moreover, the highest percentage of variation explained by ENV (68.89%), followed by the G×E (27.15%) effect, while GEN explained (3.95%) of the variation (Table S3).

Figure 2 shows that GEN3 had the highest mean yield, followed by GEN1, GEN4 and GEN2. Among these, GEN4 was the hybrid with the lowest score of the first interaction principal components axis (IPC1). The great score values of IPC1 mean that these genotypes are adapted to certain environments [24]. As for the environments, ENV10 presented the highest mean yield (18,862 kg/ha) with an IPC1 score close to zero, indicating small interactions and ENV4 the lowest yield (13,227 kg/ha). ENV1, ENV6, ENV8 and ENV9 had IPC1 values close to zero and yield higher than the mean value (17,941, 18,839, 17,691 and 18,446 kg/ha respectively).

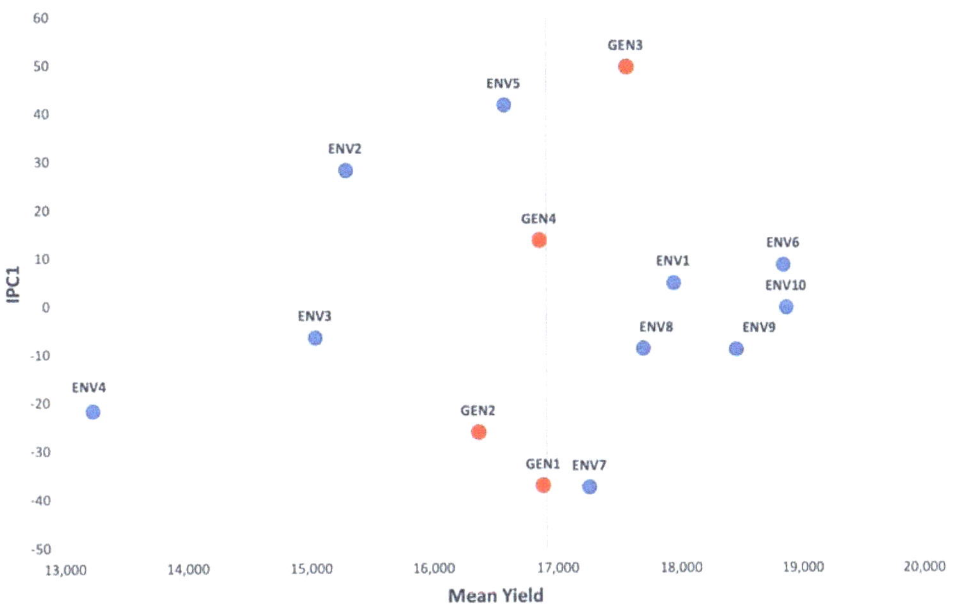

Figure 2. Additive main effects and multiplicative interaction (AMMI) biplot presenting mean grain yield (kg/ha) and the first interaction principal components axis (IPC1) of 4 genotypes (red) evaluated in 10 environments (blue).

Based on AMMI1 model, GEN3 and GEN1 resulted in the highest narrow adaptations and these were the best adapted genotypes of the two mega-environments delineated (Table 2). The first one consisted of ENV5, ENV2, ENV6, ENV1, ENV10 and ENV3 in which GEN3 was the best adapted genotype. The other one consisted of ENV8, ENV9, ENV4 and ENV7 with GEN1 presenting better results. Genotypes and environments had been listed according to their IPC1 order (Table 3), resulting that top and bottom performances have the opposite GE pattern [33]. For instance, GEN3 had positive GE with the environments such as ENV5 and ENV2 and a negative GE with environments like ENV7 and ENV4.

Table 2. AMMI family models for the grain yield, protein content and fiber content dataset, winning genotypes and the numbers of mega-environments.

Yield			AMMI Model Family		
Genotype		0	1	2	F
3	GEN3	10	6	7	6
4	GEN4			2	2
2	GEN2				1
1	GEN1		4	1	1
Mega-environments		1	2	3	4
Protein					
Genotype		0	1	2	F
3	GEN3	10	5	2	3
4	GEN4			4	4
2	GEN2		5	4	3
Mega-environments		1	2	3	3
The other 1 genotype never win and so it's not listed					
Fiber					
Genotype		0	1	2	F
2	GEN2	10	8	5	4
3	GEN3				1
1	GEN1			3	3
4	GEN4		2	2	2
Mega-environments		1	2	3	4

AMMI F denotes the full model.

Table 3. Ranking of the genotypes and the environments based on their IPC1 scores for grain yield, protein content and fiber content.

	Yield		Protein		Fiber	
GENOTYPE	Code	**IPC1 Score**	Code	**IPC1 Score**	Code	**IPC1 Score**
	GEN3	49.60	GEN3	0.79	GEN2	0.65
	GEN4	13.60	GEN4	0.57	GEN3	0.45
	GEN2	−26.06	GEN1	−0.62	GEN1	0.04
	GEN1	−37.13	GEN2	−0.73	GEN4	−1.14
ENVIRONMENT						
	ENV5	41.74	ENV7	0.72	ENV10	0.76
	ENV2	28.32	ENV3	0.43	ENV7	0.51
	ENV6	8.50	ENV9	0.34	ENV5	0.28
	ENV1	4.81	ENV8	0.33	ENV2	0.15
	ENV10	−0.33	ENV1	0.20	ENV6	−0.04
	ENV3	−6.36	ENV10	−0.27	ENV8	−0.09
	ENV8	−8.78	ENV5	−0.29	ENV9	−0.09
	ENV9	−8.95	ENV4	−0.41	ENV4	−0.15
	ENV4	−21.51	ENV2	−0.41	ENV1	−0.50
	ENV7	−37.45	ENV6	−0.64	ENV3	−0.83

The first two principal components in AMMI analysis were significant ($p < 0.001$), explaining 82.13% of GE (44.49% IPC1 and 37.64% IPC2) of interaction variation (Table S3). According to the biplot of the first (IPC1) and the second (IPC2) interaction principal components (Figure 3), GEN2 had a positive interaction with five out of ten environments (ENV3, ENV4, ENV8, ENV9 and ENV10). Generally, all genotypes were located far from the biplot origin and contribute to the G×E interaction for yield. AMMI analysis is widely used for the evaluation of maize hybrids yield in multienvironment field trials [20,24,25,41–43], and grain yield of wheat varieties [22,36,44–46], seed yield of oilseed rape [23], nutritional composition of sweet potato [38], yield of sugarcane [47] and yield of chickpea [48].

As for the protein content, according to the ANOVA, genotypes (G), environments (E) and their interaction (GxE) gave statistically significant values ($p < 0.001$). Moreover, the highest percentage of variation explained by ENV (66.90%) and G×E (29.32%) effects, while G explained the rest of variation (3.79%) (Table S3). Figure 4 shows that all genotypes had protein content percentages close to the mean value, with IPC1 values far from zero. ENV1 was the environment that presented the best results for the protein content (8.63%) of corn grains, while ENV9 presented the lowest one (6.54%). ENV10, ENV8 and ENV9 were the environments that presented the most stable results in terms of protein content.

According to AMMI1 model, GEN3 and GEN2 resulted in the highest narrow adaptations delineating two mega-environments (Table 2). The first one consisted of ENV7, ENV3, ENV9, ENV8 and ENV1, in which GEN3 was the best adapted genotype and the other one consisted of ENV10, ENV5, ENV4, ENV2 and ENV6, where GEN2 was the better suited genotype.

The first two principal components in AMMI analysis were significant ($p < 0.001$), explaining 83.93% of GE (48.61% IPC1 and 35.32% IPC2) of interaction variation (Table S3). According to the biplot of the first (IPC1) and the second (IPC2) interaction principal components (Figure 5), GEN1 had a large positive interaction with ENV4 and ENV2, GEN2 had a large positive interaction with ENV6, while GEN4 and GEN3 had a positive interaction with ENV8 and ENV9 and ENV3 respectively. Likewise the yield, all genotypes are located far from the biplot origin and contribute to the G×E interaction for yield.

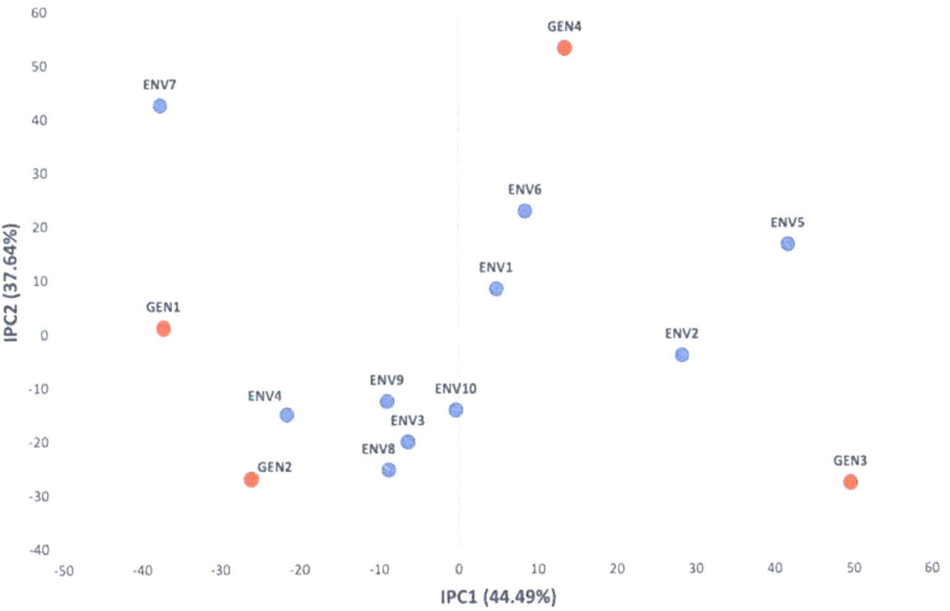

Figure 3. AMMI biplot presenting the second interaction principal components axis (IPC2) against the first interaction principal components axis (IPC1) scores for grain yield (kg/ha) of 4 genotypes (red) evaluated in 10 environments (blue).

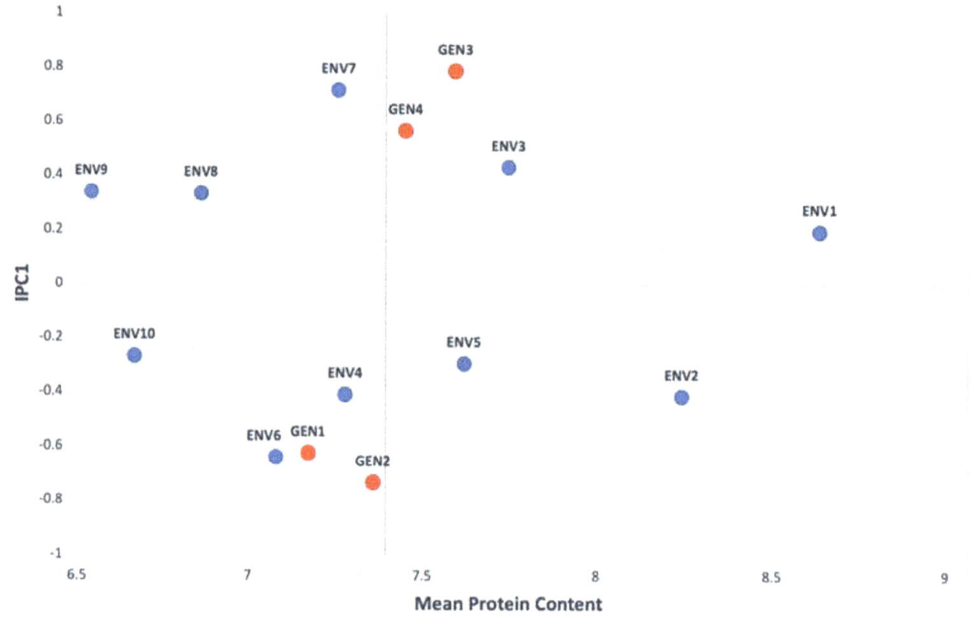

Figure 4. AMMI biplot presenting mean protein content and the first interaction principal components axis (IPC1) of 4 genotypes (red) evaluated in 10 environments (blue).

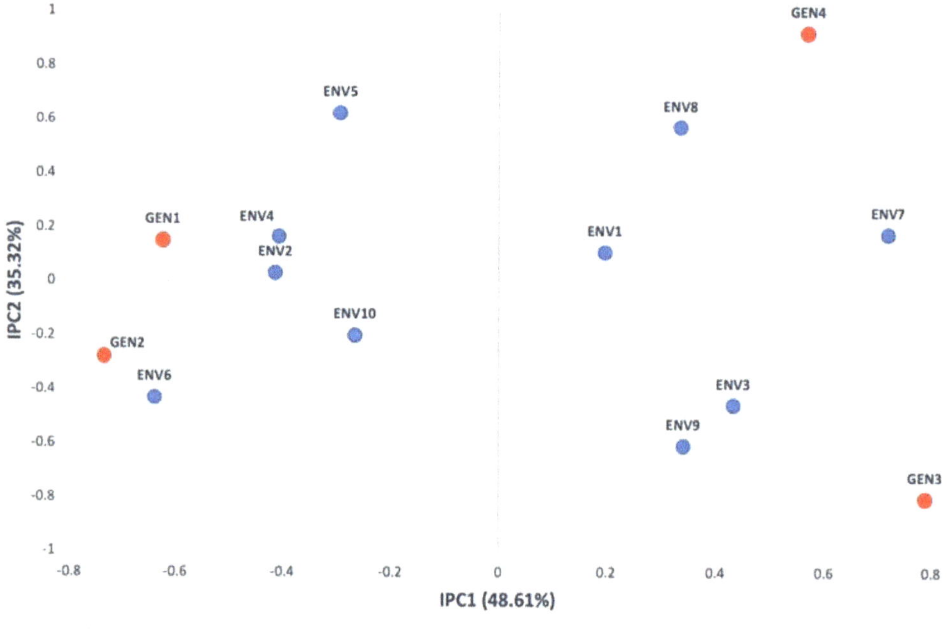

Figure 5. AMMI biplot presenting the second interaction principal components axis (IPC2) against the first interaction principal components axis (IPC1) scores for protein content of 4 genotypes (red) evaluated in 10 environments (blue).

Regarding fiber content, according to the ANOVA, genotypes (G), environments (E) and their interaction (GxE) gave statistically significant values ($p < 0.001$). Moreover, the highest percentage of variation explained by ENV (56.43%) and G×E (39.15%) effects, while G explained the rest of variation (4.42%) (Table S3). Figure 6 presents that GEN2 had the highest value, followed by GEN1, GEN3 and GEN4. GEN1 had an IPC1 score close to zero, indicating small interactions. ENV6 scored an IPC1 value near 0, and at the same time had fiber content slightly higher than the mean value. ENV2 was the environment that presented the best results for the fiber content (3.65%) of corn grains, while ENV7 presented the lowest one (2.01%).

According to AMMI1 model, GEN2 and GEN4 resulted in the highest narrow adaptations defining two mega-environments (Table 2). The first one consisted of ENV10, ENV7, ENV5, ENV2, ENV6, ENV8, ENV9 and ENV4 in which GEN2 was the genotype that presented better results and the other one consisted of ENV1 and ENV3 that had GEN4 as the better suited genotype. The AMMI statistical model has been used to evaluate the effects of genotypes, environments and their interaction for quality characteristics, like iron and zinc concentrations in the grain of maize [49], protein and tryptophan in maize [50], nutritional composition (protein, β-carotene, iron, zinc, starch and sucrose) in sweet potato [38] and vitreousness, SDS sedimentation test, yellow pigment index, protein content and test weight in durum wheat [46].

The first two principal components in AMMI analysis were significant ($p < 0.001$), explaining 90.79% of GxE (55.20% IPC1 and 35.59% IPC2) of interaction variation (Table S3). According to the biplot of the first (IPC1) and the second (IPC2) interaction principal components (Figure 7), GEN2 had a large positive interaction with ENV7, GEN4 had a large positive interaction with ENV3, while GEN3 had a positive interaction with ENV10.

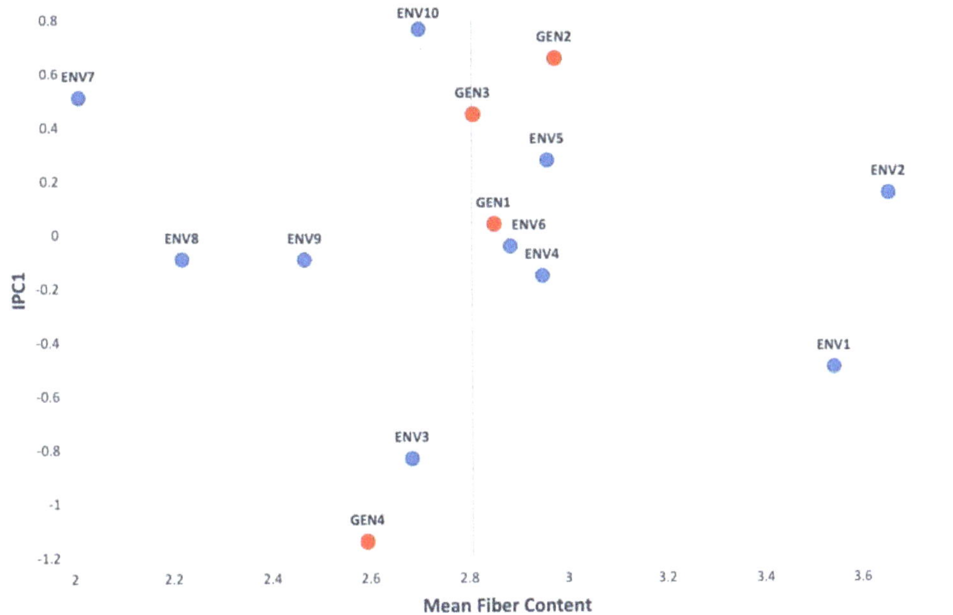

Figure 6. AMMI biplot presenting mean fiber content and the first interaction principal components axis (IPC1) of 4 genotypes (red) evaluated in 10 environments (blue).

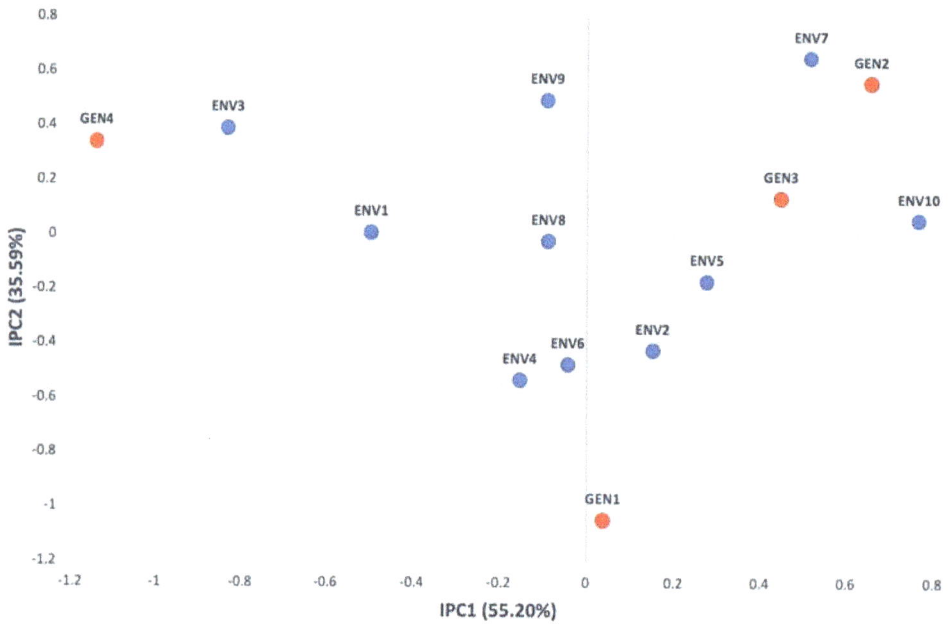

Figure 7. AMMI biplot presenting the second interaction principal components axis (IPC2) against the first interaction principal components axis (IPC1) scores for fiber content of 4 genotypes (red) evaluated in 10 environments (blue).

4. Conclusions

The evaluation of yield results of different genotypes, under different environmental conditions is a complicated issue, as a lot of parameters have to be considered, in order to lead to reliable results. In such experiments, often the AMMI analysis was used, which provides valuable information that contributes to the understanding of the G×E interaction. In this study, the GEN effect explained a low percentage of the variation and could not lead to the selection of a stable genotype for all environments. However, the results of the AMMI analysis contributed to dividing the region into mega-environments and introduce the most suitable genotype for every environment.

Concerning the yield, based on the AMMI1 model, GEN3 and GEN1 resulted in the highest narrow adaptations and these were the best adapted genotypes of the two mega-environments delineated (the first consisted of ENV5, ENV2, ENV6, ENV1, ENV10 and ENV3 and the second consisted of ENV8, ENV9, ENV4 and ENV7, respectively). A group of five environments (ENV1, ENV8, ENV6 ENV10 and ENV9) gave higher yields than the mean value and at the same time had low IPC1 scores, which indicated that they gave high yield with all the genotypes used. Regarding the grain quality, GEN3 and GEN2 for the protein content and GEN2 and GEN4 for the fiber content resulted in the highest narrow adaptations delineating two mega-environments. Specifically for the protein, the results of the analysis indicated that in order to obtain higher protein content, certain genotypes should be used in certain environments. It is important to note that ENV1 and ENV2 (location Giannouli for years 2019 and 2020 respectively, gave the highest values concerning protein and fiber content. These findings could be attributed to the high concentration rates of nutrients like Mg, Ca and the soil texture (C).

The results of this study suggest that the target to increase the quantity and quality of grain yield of maize hybrids is a very challenging issue, due to the high G×E interac-

Agronomy **2021**, *11*, 357

tion, which can be implemented by exploiting positive GE interactions, by dividing the environment into mega-environments.

Supplementary Materials: The following are available online at https://www.mdpi.com/2073-439 5/11/2/357/s1, Table S1. Coordinates, altitude, soil texture and cultivation information for the ten environments. Table S2. Climatic conditions of the 10 examined environments during the cultivation period (March-September). Table S3. AMMI analysis of variance for grain yield, protein content and fiber content of 4 genotypes evaluated in 10 environments.

Author Contributions: Conceptualization, N.K. and A.E.; Methodology, N.K., P.S., S.C., M.G., D.L., M.V.C., G.K. and A.E.; Software, N.K., P.S., S.C. and M.G.; Validation, D.L., G.K. and A.E.; Formal Analysis, N.K., P.S. and S.C.; Investigation, N.K., S.C., M.G., D.L. and M.V.C.; Resources, N.K., P.S., S.C., M.G., D.L. and M.V.C.; Writing—Original Draft Preparation, N.K., P.S., S.C. and M.G.; Writing— Review and Editing, N.K., P.S., S.C., M.G., D.L., M.V.C., G.K. and A.E.; Supervision, A.E.; Project Administration, N.K., D.L. and G.K. All authors have read and agreed to the published version of the manuscript.

Funding: This research received no external funding.

Data Availability Statement: The data presented in this study are available on request from the corresponding author.

Conflicts of Interest: The authors declare no conflict of interests.

References

1. Duvick, D.N. The Contribution of Breeding to Yield Advances in maize (*Zea mays* L.). *Adv. Agron.* **2005**, *86*, 83–145. [CrossRef]
2. Bruulsema, T.W.; Tollenaar, M.; Heckman, J.R. Boosting crop yields in the next century. *Better Crops* **2000**, *84*, 9–13.
3. Frei, O.M. Changes in yield physiology of corn as a result of breeding in northern Europe. *Maydica* **2000**, *45*, 173–183.
4. Rosegrant, M.W.; Cai, X.; Cline, S.A. *Global Water Outlook to 2025: Averting an Impending Crisis*; International Food Policy Research Institute and International Water Management Institute: Washington, DC, USA; Colombo, Sri Lanka, 2002.
5. Tardieu, F. Plant response to environmental conditions: Assessing potential production, water demand, and negative effects of water deficit. *Front. Physiol.* **2013**, *4*, 17. [CrossRef] [PubMed]
6. McMichael, B.; Quisenberry, J. The impact of the soil environment on the growth of root systems. *Environ. Exp. Bot.* **1993**, *33*, 53–61. [CrossRef]
7. Pardo, A.; Amato, M.; Chiarandà, F.Q. Relationships between soil structure, root distribution and water uptake of chickpea (*Cicer arietinum* L.). Plant growth and water distribution. *Eur. J. Agron.* **2000**, *13*, 39–45. [CrossRef]
8. Kang, M.S. Using Genotype-by-Environment Interaction for Crop Cultivar Development. *Adv. Agron.* **1997**, *62*, 199–252. [CrossRef]
9. Kang, M.S. Breeding: Genotype-by-environment interaction. In *Encyclopedia of Plant and Crop Science*; Goodman, R.M., Ed.; Marcel-Dekker: New York, NY, USA, 2004; pp. 218–221.
10. Kang, M.S. Genotype-environment interaction: Progress and prospects. In *Quantitative Genetics, Genomics, and Plant Breeding*; Kang, M.S., Ed.; CABI Publ.: Wallingford, DC, USA; Oxon, UK,, 2002; pp. 221–243.
11. Annicchiarico, P. *Genotype × Environment Interaction—Challenges and Opportunities for Plant Breeding and Cultivar Recommendations*; FAO Plant Production and Protection Papers: Rome, Italy, 2002.
12. Fan, X.-M.; Kang, M.S.; Chen, H.; Zhang, Y.; Tan, J.; Xu, C. Yield Stability of Maize Hybrids Evaluated in Multi-Environment Trials in Yunnan, China. *Agron. J.* **2007**, *99*, 220–228. [CrossRef]
13. Flores, F.; Moreno, M.; Cubero, J. A comparison of univariate and multivariate methods to analyze G×E interaction. *Field Crop. Res.* **1998**, *56*, 271–286. [CrossRef]
14. Baker, R.J. Tests for Crossover Genotype-Environmental Interactions. *Can. J. Plant Sci.* **1988**, *68*, 405–410. [CrossRef]
15. Kang, M.S.; Balzarini, M.G.; Guerra, J.L.L. Genotype-by-environment interaction. In *Genetic Analysis of Complex Traits Using SAS*; Saxton, A.M., Ed.; SAS Publ., SAS Inst.: Cary, NC, USA, 2004; pp. 69–96.
16. Lu'Quez, J.E.; Aguirrezabal, L.A.N.; Aguero, M.E.; Pereyra, V.R. Stability and Adaptability of Cultivars in Non-balanced Yield Trials. Comparison of Methods for Selecting 'High Oleic' Sunflower hybrids for Grain Yield and Quality. *J. Agron. Crop. Sci.* **2002**, *188*, 225–234. [CrossRef]
17. Signor, C.-L.; Dousse, S.; Lorgeou, J.; Denis, J.-B.; Bonhomme, R.; Carolo, P.; Charcosset, A. Interpretation of Genotype × Environment Interactions for Early Maize Hybrids over 12 Years. *Crop. Sci.* **2001**, *41*, 663–669. [CrossRef]
18. Nehe, A.; Akin, B.; Sanal, T.; Evlice, A.K.; Ünsal, R.; Dinçer, N.; Demir, L.; Geren, H.; Sevim, I.; Orhan, Ş.; et al. Genotype x environment interaction and genetic gain for grain yield and grain quality traits in Turkish spring wheat released between 1964 and 2010. *PLoS ONE* **2019**, *14*, e0219432. [CrossRef]
19. Gauch, H.G. A Simple Protocol for AMMI Analysis of Yield Trials. *Crop. Sci.* **2013**, *53*, 1860–1869. [CrossRef]

20. Mitrović, B.; Stanisavljevi, D.; Treski, S.; Stojaković, M.; Ivanović, M.; Bekavac, G.; Rajković, M. Evaluation of experimental maize hybrids tested in multi-location trials using ammi and gge biplot analyses. *Turk. J. Field Crops* **2012**, *17*, 35–40.

21. Kaya, Y.; Palta, C.; Taner, S. Additive Main Effects and Multiplicative Interactions Analysis of Yield Performances in Bread Wheat Genotypes across Environments. *Turk. J. Agric. For.* **2002**, *26*, 275–279.

22. Li, W.; Yan, Z.-H.; Wei, Y.-M.; Lan, X.-J.; Zheng, Y.-L. Evaluation of Genotype x Environment Interactions in Chinese Spring Wheat by the AMMI Model, Correlation and Path Analysis. *J. Agron. Crop. Sci.* **2006**, *192*, 221–227. [CrossRef]

23. Agahi, K.; Ahmadi, J.; Oghan, H.A.; Fotokian, M.H.; Orang, S.F. Analysis of genotype × environment interaction for seed yield in spring oilseed rape using the AMMI model. *Crop. Breed. Appl. Biotechnol.* **2020**, *20*, 26502012. [CrossRef]

24. Mafouasson, H.N.A.; Gracen, V.; Yeboah, M.A.; Ntsomboh-Ntsefong, G.; Tandzi, L.N.; Mutengwa, C.S. Genotype-by-Environment Interaction and Yield Stability of Maize Single Cross Hybrids Developed from Tropical Inbred Lines. *Agronomy* **2018**, *8*, 62. [CrossRef]

25. Hongyu, K.; García-Peña, M.; De Araújo, L.B.; Dias, C.T.D.S. Statistical analysis of yield trials by AMMI analysis of genotype × environment interaction. *Biom. Lett.* **2014**, *51*, 89–102. [CrossRef]

26. International Standard Organisation (ISO). *ISO 11260:1994. Soil Quality—Determination of Effective Cation Exchange Capacity and Base Saturation Level Using Barium Chloride Solution*; ISO: Geneva, Switzerland, 1994.

27. International Standard Organisation (ISO). *ISO 14870:2001. Soil Quality—Extraction of Trace Elements by Buffered DTPA Solution*; ISO: Geneva, Switzerland, 2001.

28. Bingham, F.T. Boron. In *Methods of Soil Analysis, Part 2. Chemical and Microbiological Properties*, 2nd ed.; Agronomy Monograph No. 9; Page, A.L., Miller, R.H., Keeney, D.R., Eds.; SSSA: Madison, WI, USA, 1982; pp. 431–447.

29. International Standard Organisation (ISO). *ISO 11261:1995. Soil Quality—Determination of Total Nitrogen—Modified Kjeldahl Method*; ISO: Geneva, Switzerland, 1995.

30. International Standard Organisation (ISO). *ISO 14235:1998. Soil Quality—Determination of Organic Carbon by sulfochromic Oxidation*; ISO: Geneva, Switzerland, 1998.

31. International Standard Organisation (ISO). *ISO 11263:1994. Soil Quality—Determination of Phosphorus—Spectrometric Determination of Phosphorus Soluble in Sodium Hydrogen Carbonate Solution*; ISO: Geneva, Switzerland, 1994.

32. Bouyoucos, G.J. Hydrometer method improved for making particle and size analysis of soils. *Agron. J.* **1962**, *54*, 464–465. [CrossRef]

33. Mult, Z.; Akay, H.; Köse Özge, D.E. Grain yield, quality traits and grain yield stability of local oat cultivars. *J. Soil Sci. Plant Nutr.* **2018**, *18*, 269–281. [CrossRef]

34. Peterson, D.M.; Wesenberg, D.M.; Burrup, D.E.; Erickson, C.A. Relationships among Agronomic Traits and Grain Composition in Oat Genotypes Grown in Different Environments. *Crop. Sci.* **2005**, *45*, 1249–1255. [CrossRef]

35. Williams, R.M.; O'Brien, L.; Eagles, H.A.; Solah, V.; Jayasena, V. The influences of genotype, environment, and genotype × environment interaction on wheat quality. *Aust. J. Agric. Res.* **2008**, *59*, 95–111. [CrossRef]

36. Kaya, Y.; Ii, I.; Akcura, M. Effects of genotype and environment on grain yield and quality traits in bread wheat (*T. aestivum* L.). *Food Sci. Technol.* **2014**, *34*, 386–393. [CrossRef]

37. Doehlert, D.C.; McMullen, M.S.; Hammond, J.J. Genotypic and Environmental Effects on Grain Yield and Quality of Oat Grown in North Dakota. *Crop. Sci.* **2001**, *41*, 1066–1072. [CrossRef]

38. Gurmu, F.; Shimelis, H.; Laing, M.; Mashilo, J. Genotype-by-environment interaction analysis of nutritional composition in newly-developed sweetpotato clones. *J. Food Compos. Anal.* **2020**, *88*, 103426. [CrossRef]

39. Juhos, K.; Szabó, S.; Ladányi, M. Explore the influence of soil quality on crop yield using statistically-derived pedological indicators. *Ecol. Indic.* **2016**, *63*, 366–373. [CrossRef]

40. Nehe, A.; Misra, S.; Murchie, E.; Chinnathambi, K.; Foulkes, M. Genetic variation in N-use efficiency and associated traits in Indian wheat cultivars. *Field Crop. Res.* **2018**, *225*, 152–162. [CrossRef]

41. Mohammed, A. Genotype by environment interaction and yield stability analysis of open pollinated maize varieties using AMMI model in Afar Regional State, Ethiopia. *J. Plant Breed. Crop. Sci.* **2020**, *12*, 8–15. [CrossRef]

42. Júnior, L.A.Y.B.; Da Silva, C.P.; De Oliveira, L.A.; Nuvunga, J.J.; Pires, L.P.M.; Von Pinho, R.G.; Balestre, M. AMMI Bayesian Models to Study Stability and Adaptability in Maize. *Agron. J.* **2018**, *110*, 1765–1776. [CrossRef]

43. De Oliveira, R.L.; Von Pinho, R.G.; Ferreira, D.F.; Pires, L.P.M.; Melo, W.M.C. Selection Index in the Study of Adaptability and Stability in Maize. *Sci. World J.* **2014**, *2014*, 1–6. [CrossRef] [PubMed]

44. Tarakanovas, P.; Ruzgas, V. Additive main effect and multiplicative interaction analysis of grain yield of wheat varieties in Lithuania. *Agron. Res.* **2006**, *4*, 91–98.

45. Brancourt-Hulmel, M.; Lecomte, C. Effect of Environmental Variates on Genotype × Environment Interaction of Winter Wheat. *Crop. Sci.* **2003**, *43*, 608–617. [CrossRef]

46. Taghouti, M.; Gaboun, F.; Nsarellah, N.; Rhrib, R.; El-Haila, M.; Kamar, M.; Abbad-Andaloussi, F.; Udupa, S.M. Genotype x environment interaction for quality traits in durum wheat cultivars adapted to different environments. *Afr. J. Biotechnol.* **2010**, *9*, 3054–3062. [CrossRef]

47. Ramburan, S.; Zhou, M.; Labuschagne, M. Interpretation of genotype × environment interactions of sugarcane: Identifying significant environmental factors. *Field Crops Res.* **2011**, *124*, 392–399. [CrossRef]

48. Farshadfar, E.; Sabaghpour, S.H.; Zali, H. Comparison of parametric and non-parametric stability statistics for selecting stable chickpea (*Cicer arietinum* L.) genotypes under diverse environments. *AJCS* **2012**, *6*, 514–524.
49. Oikeh, S.O.; Menkir, A.; Maziya-Dixon, B.; Welch, R.M.; Glahn, R.P.; Gauch, G. Environmental stability of iron and zinc concentrations in grain of elite early-maturing tropical maize genotypes grown under field conditions. *J. Agric. Sci.* **2004**, *142*, 543–551. [CrossRef]
50. Pixley, K.V.; Bjarnason, M.S. Stability of Grain Yield, Endosperm Modification, and Protein Quality of Hybrid and Open-Pollinated Quality Protein Maize (QPM) Cultivars. *Crop. Sci.* **2002**, *42*, 1882–1890. [CrossRef]

Article

Nematicidal Activity and Phytochemistry of Greek Lamiaceae Species

Nikoletta G. Ntalli [1,*], Efstathia X. Ozalexandridou [2], Konstantinos M. Kasiotis [3], Maria Samara [1] and Spyros K. Golfinopoulos [2,4,*]

[1] Laboratory of Biological Control of Pesticides, Department of Pesticides Control and Phytopharmacy, Benaki Phytopathological Institute, 8 S. Delta Str., 14561 Athens, Greece; m.samara@bpi.gr

[2] School of Sciences and Technology, Hellenic Open University, 18 Parodos Aristotelous st., 26335 Patra, Greece; e.ozalexandridou@gmail.com

[3] Laboratory of Pesticides' Toxicology, Department of Pesticides Control and Phytopharmacy, Benaki Phytopathological Institute, 8 S. Delta Str., 14561 Athens, Greece; k.kasiotis@bpi.gr

[4] Department of Financial and Management Engineering, School of Engineering, University of the Aegean, 41 Kountourioti st., 82132 Chios, Greece

* Correspondence: nntali@agro.auth.gr (N.G.N.); s.golfinopoulos@fme.aegean.gr (S.K.G.); Tel.: +30-210-818-0343 (N.G.N.); +30-227-103-5459 (S.K.G.)

Received: 23 June 2020; Accepted: 30 July 2020; Published: 1 August 2020

Abstract: Natural pesticides are in the forefront of interest as ecofriendly alternatives to their synthetic ancestors. In the present study, we evaluated the nematicidal activity of seven Greek Lamiaceae species and discerned among principal components for activity according to GC-MS analysis. Care was taken that all botanicals used were easily prepared without employing elaborate procedures and toxic solvents. We established the *in vitro* EC_{50} values of the hydrosols of *Origanum vulgare* L., *Mentha piperita* L. and *Melissa officinalis* L. and the water extracts of *Origanum vulgare*, *Thymus vulgaris* L., *Thymus citriodorus* (Schreb), *Rosmarinus officinalis* (Spenn), and *Ocimum basilicum* L. against *Meloidogyne javanica* (Treub) and *Meloidogyne incognita* (Kofoid & White). Furthermore, we amended nematode-infested soil with powdered leaves and flowers of *O. vulgare* to assess for efficacy. According to *in vitro* studies, the most active botanical preparations against both nematode species was *O. vulgare*, as regards its hydrosol and water extract. *Thymus citriodorus* was proved very potent against *M. javanica*, provoking 100% paralysis at 4 µL/mL after 96 h, but was only nematostatic against *M. incognita* since the second-stage juveniles (J2s) recovered movement 48 h after immersion in test solutions. Interestingly, *O. vulgare* was also proved nematicidal in pot bioassays but at test concentrations over 50 g/kg was phytotoxic for tomato plants. According to GC-MS analysis, the principal components sustaining activity of *O. vulgare* are carvacrol and thymol. The nematicidal activity of *O. vulgare* seems promising in the forms of essential oil leftovers (i.e., hydrosol), self-prepared water extract that can be of consideration as a "basic substance", and powder for soil amendment.

Keywords: natural substances; nematicidal; root-knot nematodes; oregano; soil amendments; basic substances

---

1. Introduction

Nematodes are among the most complex and numerous organisms on the planet. The name Nematoda, or Nematelminthes, is derived from the Greek word "νήμα" (thread or threadworms). They belong to the kingdom Animalia, phylum Nematoda [1]; and after arthropods, they form the second most numerous group of Metazoa. They have the form of a worm, with a cylindrical, elongated body and a circular cross-section. Their diffusion on earth is wide due to their ability to adapt easily due to their inner and outer morphology [2]. Nematode species that cause damage to cultivated plant species are called plant-parasitic nematodes. Root-knot nematodes (*Meloidogyne* sp.) cause considerable damage to more than 5000 plant species and use their stylets to feed on the roots of the plants. [3]. *Meloidogyne incognita*, *M. javanica*, and *M. arenaria* infect *Solanaceae* and *Malvaceae*, have broad host ranges, and are in the list of the most economically damaging root-knot nematodes [4]. On the other hand, tomato is one of the most significant crop hosts of *Meloidogyne* spp. and in the presence of disease complexes with soil pathogens the subsequent damages may even lead to total crop loss [5].

Plant nematodes are not controlled using just one method but are tackled by a combination of methods in the context of an integrated system for management of harmful organisms [6]. Although many synthetic pesticides have been used in the past as chemical nematicides, only a few are still authorized according to European legislation [7]. This fact creates an urgent need for discovering less toxic and environmentally friendly substitutes for commercial use [8]. The use of plant secondary metabolites, as well as the reutilization of culture's debris can potentially constitute a powerful weapon for plant protection. Efforts for a transition to an environmentally friendly crop protection pave a new way towards pest control. The essential oils (EOs) of aromatic plants are used in many fields because of their antimicrobial, antifungal, antioxidant, and antibacterial activities [9]. The biological activity of EOs is related to their chemical composition, which is influenced by the specific climatic, seasonal, and geographic conditions affecting the aromatic species from which EOs derive [10]. It is accepted that the EOs exhibit efficacy against insects [11] and nematodes [12]. We have previously shown that the EOs of the family Lamiaceae, including the species *Mentha pulegium* L., *Origanum vulgare*, *Origanum dictamnus* L. and *Melissa officinalis* display powerful *in vitro* nematicidal activity [13]. In another study using material from different botanical families, a consistent suppression of *M. incognita* population on tomato roots was evident after soil drench treatments with water emulsions of EOs from *Schinus molle* L., *Cinnamomum camphora* L., *Eugenia caryophillata* L., *Cinnamomum zeylanicum* (J. Presl), and *Citrus aurantium* L. [14]. Likewise, the EOs of *Eucalyptus citriodora* Hook, *Eucalyptus globulus* (), *Mentha piperita*, *Pelargonium asperum* (L'Hér), and *Ruta graveolens* L. were proved to be nematicidal against *M. incognita* [15], while *Dysphania ambrosioides* L., *Filipendula ulmaria* L., *Ruta graveolens*, *Satureja montana* L. and *Thymbra capitata* L. EOs revealed EC_{50} values lower than 0.15 μL/mL against root-knot nematodes according to Faria et al. [16]. The bioactive EOs' principal components are the terpenes, playing an essential protective role for the plants [17]; their significant synergistic and/or antagonistic activities compose the overall efficacy against target organisms [18,19].

According to SANCO draft working document 10472, reduced data requirements are described for plant protection products made from all edible parts of plants of human and animal feed. The document refers to plant extracts made with water or ethanol from parts of plants currently authorized as herbal drugs according to European Pharmacopoeia. This manufacturing process (e.g., crushing, drying, and water and/or ethanol extraction) is not considered to modify the toxicity or ecotoxicity profiles [20,21]. Moreover, the regulation 1107/2009 (EC2009) contains some facilitation for "low-risk active substances" and for "basic substances", that is, "nematicidal recipes" prepared by the farmer with low risk of harmfulness for soil, water, air, plants, or animals [22]. In this frame, water extracts prepared from *O. vulgare* and *T. citriodorus* may plausibly be developed as "low-risk plant protection products" or "basic substances".

The aim of this study was to examine the nematicidal effect of botanical extracts that are easily prepared from seven Greek Lamiaceae species without employing elaborate procedures and toxic solvents. Interestingly, we used *T. citriodorus* in a recent study to test for nematicidal activity against *M. incognita* and *M. javanica*, but with a different extraction protocol [23], thus herein we evaluate any subsequent differences in activity. Specifically we study the *in vitro* nematicidal activity of the hydrosols of *Origanum vulgare*, *Mentha piperita*, and *Melissa officinalis* and the water extracts of *Origanum vulgare*, *Thymus vulgaris*, *Thymus citriodorus*, *Rosmarinus officinalis* and *Ocimum basilicum* against *Meloidogyne javanica* and *Meloidogyne incognita*. Furthermore, we evaluate *in planta* the most effective botanical species, i.e., *O. vulgare*, in the form of powdered leaves and flowers amended into soil that has been artificially infested by nematodes and assess for efficacy. The outcomes of the study are self-prepared water-based extracts that can be of consideration as "basic substances" or "low-risk" active ingredients for plant protection products, as well as soil bio-amendment practices that can plausibly be integrated into IPM schemes and/or organic farming.

2. Materials and Methods

2.1. Aromatic Materials

The aromatic materials tested for the nematicidal activity against *M. javanica* and *M. incognita* include: (a) hydrosols obtained as aquatic leftovers after EO obtainment from Clevenger distillation of *Origanum vulgare*, *Mentha piperita*, and *Melissa officinalis* from the Aitheria Company based in Velvento, Kozani, Greece, and (b) water extracts from powdered plant parts of *Origanum vulgare*, *Thymus vulgaris*, *Thymus citriodorus*, *Rosmarinus officinalis*, and *Ocimum basilicum* obtained from the Ethericon Greek Herbs company based in Larisa, Greece.

2.2. Plant Extraction

The extraction of plant powders was performed with the Sonicator Branson 2510 ultrasound device. Initially, 5 g of plant residue powder was weighed and placed in a beaker with 25 mL of water, forming a ratio of 1/25 (*w/v*). The mixture was carefully stirred in order to ensure even distribution and placed in the Sonicator device. Afterwards, distilled water was added to the volume needed to surpass the surface of the solvent in the beaker, and the mixture was sonicated for 15 min. Finally, the water extract was obtained by filtering through cotton (filtering had no effect on the recovery of the constituents assessed by GC-MS). To proceed to GC-MS chemical analysis, water extracts and hydrosols were subjected to subsequent extraction with two portions of petroleum ether (25 mL each). The combined organic phases were dried ($MgSO_4$), evaporated to dryness using N_2 stream, filtered with Chromafil syringe Nylon filter (0.22 µm, Macherey Nagel GmbH & Co. KG, Düren, Germany), and injected into the GC-MS system.

2.3. Nematode Culture

A population of *M. javanica* and another of *M. incognita* were obtained upon a single eggmass per species and were reared on tomato (*Solanum lycopersicum* L.) cv Belladonna, a variety that is particularly sensitive to root-knot nematodes. The tomato plants used for the development of the population of root-knot nematodes were maintained in a growth chamber at 25–28 °C, 60% relative humidity and 16 h photoperiod in plastic plant pots (18 cm diameter) containing a 10:1 (*v/v*) mixture of peat and pearlite. These conditions remained stable at these levels throughout the whole duration of the experiments. Plants used for inoculations were 7 weeks old, at the five-leaf stage. After 40 days, the plants were uprooted, and the roots were washed to remove any soil residue and cut into 2 cm pieces. The roots were placed in a solution of 1% sodium hypochlorite, and the suspension was shaken for 5 min. Then, it was washed in running water through nested sieves of a 250 and 38 µm cross-section, and the eggs of the nematodes were recovered and finally transferred in Baermann modified funnels at 28 °C [24]. The water suspension with the nematode eggs was placed in filter paper inside a sieve (2 mm hole size) and secured in a plastic tray with distilled water. All second-stage juveniles (J2s) hatching in the first 3 days were discarded. After an additional 24 h, J2s were collected and used in the experiments.

2.4. J2 Paralysis

Appropriate amounts of hydrosol and extract solutions were used in order to achieve the concentration range for EC_{50} calculation. The solutions were mixed with the suspension of nematodes, in the wells of a polystyrene plate of 96-well plates at a ratio 1:1 (*v/v*), and the final volume per well was 140 µL. Each test solution was 2 × so as to reach the expected concentration after mixing with the nematode suspension.

Five replicates were performed on five concentration levels covering the range of 25 to 200 µL/mL for all treatments. Regarding the best effective ones, which were the water extracts of *O. vulgare* and *T. citriodorus*, the EC_{50} values were finally established using the test concentration range of 3.9, 7.8, 15.6, 31.2, and 62.5 µL/mL. Distilled water was used as control, and the nematode number per experimental well was 15–20 J2s. The plates were covered with a lid, so as to prevent evaporation which would differentiate final test concentrations, and intermediate wells were used to immerse J2 in water so as to control cross-contamination between treatments due to the volatility. The plates were placed in a chamber with stable conditions at 28 °C. J2 paralysis was assessed by observation in a reverse microscope (Euromex, Holland) at 40× zoom at the timepoints of 24, 48, and 96 h after establishment of the experiment. The J2s were classified into two categories: motile and paralyzed. Paralysis assessed after 96 h of immersion was characterized as death if J2s never regained motility after significantly augmenting the water volume in immersion solutions for an additional 24 h.

2.5. Soil Amendment with O. vulgare Powder for M. incognita Control and Phytotoxicity to Tomato Plants

The sandy loam soil (18% clay, 22% silt, 60% sand), with pH 6.5, 3.3% organic carbon, and 1.9 mg g^{-1} total N, was collected from a noncultivated field of the Benaki Phytopathological Institute. Initially, it was sieved through a 3 mm sieve and partially air dried overnight; then, a mixture with sand at a ratio of 2:1 was prepared to form the hereafter referred to soil. Six plastic bags represented the experimental treatments, 1 kg of soil each receiving a nematode inoculation of 2500 J2s per kg. After appropriate mixing and overnight incubation at room temperature, according to Ntalli et al. [23], the plastic bags were spiked with appropriate amounts of *O. vulgare* powder to reach the test concentrations of 1, 5, 10, 50, and 100 g kg^{-1} soil. A water control was also included in the experiment. Seven-week-old tomato plants, cv. Belladonna, were transplanted into the treated soil, separated in five different pots containing 200 g of soil each, and the bioassay was kept in an incubator at 27 °C and 60% relative humidity at a 16 h photoperiod for 40 days. Every pot received 20 mL of water every 3 days for 40 days; afterwards, plants were uprooted and gently washed. Shoots were separated from roots and the latter were stained with acid fuchsin, according to Byrd et al. [25], and the following parameters were assessed: (a) *M. incognita* females per g of root at 10× magnification, (b) fresh stem weight, and (c) fresh root weight. The experiment was performed twice, and the treatments were arranged in a completely randomized design with five replicates.

2.6. Gas Chromatography–Mass Spectrometry Analyses of Water Extracts (WEs) and Hydrosols (Hs)

The GC-MS analysis was conducted on a Chromtech Evolution 3 MS/MS triple quadrupole mass spectrometer built on an Agilent 5975 B inert XL EI/CI MSD system that was operated in full scan data acquisition mode, within the mass range from m/z 50 to 500. Samples were injected with a Gerstel MPS-2 autosampler using a 10 µL syringe. Component separations were performed on the chromatographic column HP-5 ms Ultra-Inert (UI), with a length of 30 m, inner diameter (ID) of 0.25 mm, and film thickness of 0.25 µm (J&W Scientific, Folsom, CA, USA). Helium (99.9999% purity) was used as the carrier gas at a flow rate of 1.2 mL min^{-1}. The column oven temperature program initiated from 45 °C and stayed there for 1 min before increasing at a rate of 5 °C min^{-1} to 250 °C, where it stayed for 5 min. The transfer line, manifold, and source of ionization temperatures were 300, 40 and 230 °C, respectively. The electron multiplier voltage was set at 2000 V. The total GC analysis time was 47 min. Identified peaks in GC-MS (triplicate analysis) were verified by matching the acquired mass spectra with those in the commercial library of NIST 08.

2.7. Statistical Analysis

Concerning the effect of the hydrosols and the aquatic extracts on J2 mobility, the experiments were repeated five times for each concentration level on an experimental project of completely randomized groups. The experiment was conducted twice.

For each paralysis test, the analysis of all data was correlated to time. The average of the two temporal repetitions for each experiment is presented, as the correlated analysis of variability did not show a significant interaction between the interventions and the execution time of the experiments.

For the statistical analysis and since the paralysis in the carrier did not differ from that measured in water, the paralysis data for both experiments were expressed as percentage rates of the paralysis values corresponding to the water control according to Schneider Orelli's equation [26]: Paralysis increase % = ((paralysis % during treatment − paralysis % in the water control)/(100 − paralysis % in the water control)) × 100.

Following the immersion in hydrosol and extract solutions, a variability analysis (ANOVA) was performed. Then the log-logistic equation of Seefeldt et al. [27] for the calculation of the values EC_{50} was used, according to the following equation: $Y = C + (D − C)/\{1 + \exp[b\ (\log(x) − \log(EC_{50}))]\}$ where C is the lower limit, D is the upper limit, b is the slope of line in value EC_{50}, and EC_{50} is the concentration of the hydrosol or aquatic extract required for the 50% increase of paralyzed J2s compared with those of the water control.

In the particular regression equation, the concentration of the hydrosol or extract (μg/mL) was the independent factor (x), and J2 immobility (percentage increase as compared to the water control) the dependent factor (y).

Concerning the pot bioassays, the means were averaged over experiments, since ANOVAs showed no significant treatment between runs. Statistical analysis was performed using SPSS 20 (IBM, Armonk, NY, USA). Both ANOVA and Duncan's test were set at $p \leq 0.05$.

3. Results and Discussion

3.1. In Vitro Nematicidal Activity

Among all the seven species tested for their nematicidal activity, only *O. vulgare* and *T. citriodorus* have nematicidal activity at the concentration range tested (3.9 to 200 μL/mL). Table 1 presents the effect on the paralysis of J2s after their immersion for 24, 48, and 96 h in test solutions. Paralysis reported after 96 h was irreversible in all cases. In most cases, the rate of J2 paralysis was proportionate to the increase of concentration and time of immersion in test solutions. Instead, plant extracts of *M. piperita*, *M. officinalis*, *T. vulgaris*, *R. officinalis*, and *O. basilicum* did not show activity in the test concentration range (25 to 200 μL/mL). *Thymus citriodorus* water extract (WE) was the most effective against *M. javanica*, and the observed paralysis for J2s immersed at 3.9 μL/mL was 100% at 96 h after establishment of the experiment. In contrast, *T. citriodorus* was not found equally active on *M. incognita* since paralysis obtained at 24 h was reversible 48 h after establishment of the experiment.

In the same context, previously we studied *T. citriodorus* WE obtained with a different extraction protocol applying lower eluent volume 1/10 (*w/v*), and the $EC_{50/48h}$ values were calculated to be 84.19 and 61.97 μL/mL against *M. incognita* and *M. javanica*, respectively [23]. In this study, *T. citriodorus* was extracted with water, using a ratio of 1/25 (*w/v*), and the respective $EC_{50/48h}$ value against *M. javanica* was lower than the value previously reported; meanwhile it was found nematostatic against *M. incognita*. Consequently, it appears that the solvent volume affects extraction efficiency, influencing the equilibrium constant of the analytes' partitioning between the two phases [28]. In this case, the higher amount of solvent seems to yield higher extraction recoveries of the particular nematicidal components (i.e., a more concentrated extract, not affected under these conditions by the dilution), leading to higher activity against *M. javanica*.

In fact, the WE of *O. vulgare* was the best effective against *M. incognita*, followed by *T. citriodorus*. Interestingly, *O. vulgare* hydrosol (H) only achieved paralysis against *M. javanica*, but was found inactive against *M. incognita*. *Origanum vulgare* WE achieved better paralysis of *M. incognita* than *T. citriodorus* did. It should be noted that there was no cross-contamination between the treatments due to volatility, as the mobility of the J2s in the peripheral wells, where J2s were immersed in water, did not change.

Table 1. EC_{50} (μL/mL) values of water extracts (WE) and hydrosols (H) against *M. incognita* and *M. javanica*, calculated after immersion of second-stage juveniles (J2s) in test solutions for 24, 48, and 96 h. The test concentrations used for the H of *O. vulgare* were 25 to 200 μL/mL, while for the water extracts of *O. vulgare* and *T. citriodorus* 3.9, 7.8, 15.6, 31.2, and 62.5 μL/mL.

H of *O. vulgare* on *M. javanica*								
24 h			48 h			96 h		
EC_{50} (μL/mL)		95% Conf Int	EC_{50} (μL/mL)		95% Conf Int	EC_{50} (μL/mL)		95% Conf Int
173.580	7.127	158.837–188.324	151.196	8.550	133.510–168.883	118.26	5.996	105.856–130.663
WE of *O. vulgare* on *M. javanica*								
24 h			48 h			96 h		
EC_{50} (μL/mL)		95% Conf Int	EC_{50} (μL/mL)		95% Conf Int	EC_{50} (μL/mL)		95% Conf Int
na			27.041	3.084	20.661–33.421	37.844	5.091	27.312–48.375
WE of *O. vulgare* on *M. incognita*								
24 h			48 h			96 h		
EC_{50} (μL/mL)		95% Conf Int	EC_{50} (μL/mL)		95% Conf Int	EC_{50} (μL/mL)		95% Conf Int
30.686	1.426	27.735–33.636	23.531	11.241	0.276–46.781	27.626	9.134	8.731–46.521
WE of *T. citriodorus* on *M. javanica*								
24 h			48 h			96 h		
EC_{50} (μL/mL)		95% Conf Int	EC_{50} (μL/mL)		95% Conf Int	EC_{50} (μL/mL)		95% Conf Int
38.057	3.736	30.329–45.781	27.914	2.557	22.582–33.246	na	-	-
WE of *T. citriodorus* on *M. incognita*								
24 h			48 h			96 h		
EC_{50} (μL/mL)		95% Conf Int	EC_{50} (μL/mL)		95% Conf Int	EC_{50} (μL/mL)		95% Conf Int
35.974	1.066	33.769–38.180	na	-	-	na	-	-

EC_{50} values were not calculated (na) they were outside the test concentration range.

3.2. In Planta Nematicidal Activity and Secondary Effects on Tomato Plants

When different *O. vulgare* powder quantities were used in the pot bioassay to assess for efficacy, a clear dose–response relationship was established and efficacy was evident at 5 g/kg of soil (Figure 1) both in terms of root galls and female counts. Considering stem and root weights, no statistical differences were evident until the test concentration of 50 g/kg of soil, which was proved to be phytotoxic for the tomato plants.

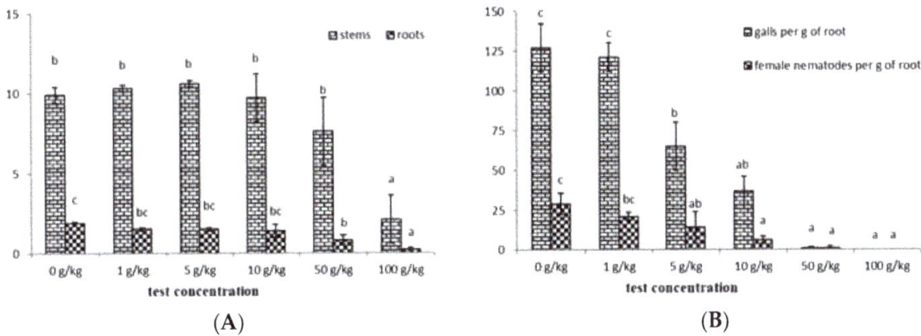

(A)   (B)

Figure 1. Tomato stem and root weights (g), as assessed after treatment with *O. vulgare* powder for *M. incognita* control in pot bioassays 40 days after experiment establishment (**A**). Root galls and female counts per g of tomato root 40 days after experiment establishment (**B**). The data are means of ten replicates with standard deviations. Means followed by the same letter are not significantly different according to Duncan's test ($p \leq 0.05$). Within each graph, letters correspond to statistical differences amongst same pattern bars.

3.3. Chemical Composition Analysis of Water Extracts and Hydrosols of Lamiaceae Species—Nematicidal Activity Implication

The GC-MS chemical analysis of the petroleum ether extracts of the WE and hydrosols unveiled several substances. More specifically, numerous constituents related to EOs of the respective species were identified. According to the GC-MS, the principal components in respective extracts were as follows: thymol in *T. vulgaris* WE (89.15%), carvacrol in *O. vulgare* WE (86.77%), eugenol and linalool in *O. basilicum* WE (50.75 and 34.45%, respectively), carvacrol in *O. vulgare* H (93.00%), levomenthol in *M. piperita* H (54.35%), and thymol in *M. officinalis* H (39.43%) (Table 2). Therefore, the predominant detections in the most active species (*Thymus* spp. and *O. vulgare*) were carvacrol and thymol. Carvacrol in *O. vulgare* WE and H displayed a profound difference in abundance from thymol.

Previously, we have demonstrated the significant individual nematicidal activity of carvacrol, along with its synergic potency with other terpenes, against *Meloidogyne* sp. [13,18]. Similarly, carvacrol has been proved to be a nematicidal component against *M. incognita* by others [29]. Interestingly, although *O. vulgare* H was richer in terms of number of constituents compared to its WE, it exhibited higher EC_{50} values—a fact that reveals the complexity of bioactivity interactions amongst plant secondary metabolites within an extract.

In specific, based on the abundances of the key constituents of the GC-MS chromatograms (same plant material quantity used), the H contains higher concentrations of carvacrol and thymol than the WE; however, the WE is more potent than the H. Likewise, *M. piperita* H did not exhibit significant activity, although it yielded numerous ingredients considered active on *Meloidogyne* sp., like pulegone and geraniol [13,18].

Lamiaceae species' EOs, extracts, hydrosols, and contained constituents are reported to exhibit nematicidal activity [13,30–34]. It is worth mentioning that, in this work, emphasis was given to the volatile and semivolatile constituents of the specific species, since many of these constituents display nematicidal activity. Nevertheless, it is expected that some of the semipolar and polar liquid-chromatography-amenable compounds (such as phenolic acids and flavonoids) can contribute to the nematicidal properties exhibited in this work. Therefore, the enhanced activity of *O. vulgare* WE and *T. citriodorus* WE might be attributed to the potential high content of phenolic acids (such as rosmarinic acid and oleanolic acid) that exhibit nematicidal properties [32,35,36] and other phenolic compounds, including their glycosides and hexosides.

Table 2. Chemical and relative composition of water extract (WE) and hydrosol (H) of Lamiaceae species *.

Retention Time (min)	RI **	Analyte	WE *** (Relative Amount (%))			H ** (Relative Amount (%))		
			Thymus vulgaris	Origanum vulgare	Ocimum basilicum	Origanum vulgare	Mentha piperita	Melissa officinalis
4.52	838 (841)	Tyranton					(0.80 ± 0.13)	(2.06 ± 0.18)
9.48	-	Trifluoro-acetyl-α-terpineol			(4.60 ± 0.27)			
9.49	1032 (1031)	Eucalyptol				(0.08 ± 0.02)	(10.19 ± 0.93)	
10.81	1086 (1091)	trans-Linalool-oxide (furanoid)				(0.03 ± 0.01)	(1.22 ± 0.14)	
11.24	1090 (1090)	Ethyl-2-(5-methyl-5-vinyltetrahydrofuran-2-yl)-propan-2-yl-carbonate					(0.34 ± 0.07)	
11.67	1099 (1100)	Linalool			(34.45 ± 2.27)	(0.23 ± 0.04)	(5.64 ± 0.35)	(3.89 ± 0.42)
12.27	1116 (1118)	trans-p-Mentha-2,8-dienol						(1.81 ± 0.20)
12.28	1122 (1123)	cis-2p-Menthen-1-ol					(0.35 ± 0.06)	
12.76	1123 (1122)	1R,4R-p-Mentha-2,8-dien-1-ol						(2.41 ± 0.27)
13.01	1140 (-)	(E)-p-2-Menthen-1-ol)						(1.07 ± 0.10)
13.02	1146 (1146)	Isopulegol					(0.84 ± 0.09)	
13.23	1164 (1164)	cis-p-Menthan-3-one					(1.57 ± 0.23)	
13.54	1166 (1166)	p-Menthan-3-one					(3.24 ± 0.30)	
13.68	1168 (1173)	(+)-Borneol					(4.61 ± 0.38)	
13.71	1167 (-)	p-Mentha-1,5-dien-8-ol						(4.95 ± 0.54)
13.77	1167 (-)	endo-Borneol				(0.46 ± 0.10)		
14.01	1177 (1177)	Terpinen-4-ol				(0.73 ± 0.11)		(5.25 ± 0.31)
14.01	1175 (1172)	Levomenthol					(54.35 ± 4.17)	
14.02	1182 (1182)	(-)-Terpinen-4-ol			(10.20 ± 0.80)			
14.31	1180 (1180)	m-Cymen-8-ol				(0.27 ± 0.03)		
14.45	1190 (1187)	L-α-Terpineol					(5.70 ± 0.44)	
14.46	1183 (1182)	p-Cymen-8-ol				(0.42 ± 0.10)		
14.63	1189 (1189)	α-Terpineol				(0.15 ± 0.04)		(12.45 ± 0.75)
14.72	1217 (1217)	trans-Carveol						(3.73 ± 0.19)
15.08	1222 (1222)	2-Thiaadamantan-4-ol					(1.88 ± 0.19)	
15.25	1228 (1228)	D-Verbenone						(3.07 ± 0.22)
15.67	1240 (-)	Isogeraniol						(2.97 ± 0.33)
15.78	1240 (1242)	β-Citral						(5.22 ± 0.40)
15.80	1237 (1236)	Pulegone					(2.02 ± 0.28)	
16.14	-	unidentified				(0.12 ± 0.03)		
16.22	1250 (1249)	Thymoquinone		(2.24 ± 0.19)		(0.16 ± 0.04)		
16.24	1253 (1252)	Piperitone					(4.74 ± 0.14)	
16.40	1255 (1258)	Geraniol						(6.68 ± 0.57)
16.63	-	p-Menthan-1,2,3-triol					(1.35 ± 0.22)	
16.69	1276 (-)	Citral						(5.02 ± 0.17)
17.57	1291 (1292)	Thymol	(89.15 ± 5.10)	(10.98 ± 0.75)		(4.36 ± 0.68)		(39.43 ± 2.09)
18.04	1299 (1298)	Carvacrol	(11.85 ± 2.13)	(86.77 ± 3.15)		(93.00 ± 3.17)		
19.72	1357 (1358)	Eugenol			(50.75 ± 1.56)			

* *Thymus citriodorus* WE displayed traces of thymol, and *Rosmarinus officinalis* WE had no essential oil components; ** RI, retention index on HP5-MS UI column (relative to *n*-alkanes), identification based on mass spectra comparison with the reference databases and comparison with literature RIs obtained in equivalent column (depicted in parentheses); *** positive findings are indicated.

4. Conclusions

The Mediterranean basin constitutes a chemical arsenal owing to its wealth of self-sown, aromatic plants which could be used for the development of "low-risk" plant protection products and "basic substances". As the EO manufacturing industry grows, producers are imposing plans, strategies, and technology for waste management and reutilization. The reutilization of this waste for the production of a new generation of nematicides constitutes an utterly successful practice not only for plant protection but also for the environment. Our results show that the easily prepared water extracts of *T. citriodorus* and *O. vulgare*, along with the distillation waste of *O. vulgare*, can be alternatives for the control of *Meloidogyne*. Additionally, plant powder of these two species incorporated into nematode-infested soil blocks infestation augmentation and can thus be an additional practice for farmers to incorporate into an integrated nematode management frame.

Author Contributions: Conceptualization, N.G.N.; methodology, N.G.N., E.X.O., K.M.K. and M.S.; software, N.G.N. and K.M.K.; validation, N.G.N., E.X.O., K.M.K. and S.K.G.; formal analysis, N.G.N. and K.M.K.; investigation, N.G.N., E.X.O., K.M.K. and M.S.; resources, M.S.; data curation, N.G.N. and K.M.K.; writing—original draft preparation, N.G.N. and S.K.G.; writing—review and editing, N.G.N., K.M.K. and S.K.G.; supervision, N.G.N. All authors have read and agreed to the published version of the manuscript.

Funding: This research received no external funding.

Acknowledgments: The authors are grateful to T. Koufakis and AGRIS SA for providing seeds and seedlings. We are also grateful to Aitheria Company, based in Velvento, Kozani, Greece, and Ethericon Greek Herbs Company, based in Larisa, Greece, for the kind offer of aromatic material.

Conflicts of Interest: The authors declare no conflict of interest.

References

1. Chitwood, B.G. The English word «Nema» revised. *Syst. Zool.* **1957**, *6*, 184–186. [CrossRef]
2. Hirschmann, H. Gross Morphology on Nematodes. In *Nematology*; Sasser, J.N., Jenkins, W.R., Eds.; California Univ. Press: Berkeley, CA, USA, 1960; pp. 125–129.
3. Trudgill, D.L.; Blok, V.C. Apomictic polyphagous root-knot nematodes; exceptionally successful and damaging biotrophic root pathogens. *Annu. Rev. Phytopathol.* **2001**, *39*, 53–77. [CrossRef] [PubMed]
4. Jones, J.T.; Haegeman, A.; Danchin, E.G.J.; Gaur, H.S.; Helder, J.; Jones, M.G.K.; Kikuchi, T.; Manzanilla-López, R.; Palomares-Rius, J.E.; Wesemael, W.M.L.; et al. Top 10 plant-parasitic nematodes in molecular plant pathology. *Mol. Plant Pathol.* **2013**, *14*, 946–961. [CrossRef] [PubMed]
5. Back, M.A.; Haydock, P.P.J.; Jenkinson, P. Disease complexes involving plant-parasitic nematodes and soilborne pathogens. *Plant Pathol.* **2002**, *51*, 683–697. [CrossRef]
6. Lambert, K.; Bekal, S. Introduction to plant-parasitic nematodes. *Plant Health Instr.* **2002**, *10*, 1094–1218. [CrossRef]
7. European Commission. Regulation (EC) No 1107/2009 of the European 515 Parliament and of the Council of 21 October 2009 concerning the placing of plant 516 protection products on the market and repealing Council Directives 79/117/EEC 517 and 91/414/EEC. *Off. J. Eur. Union* **2009**, *309*, 1–50.
8. Isman, M.B. Plant essential oils for pest and disease management. *Crop Prot.* **2000**, *19*, 603–608. [CrossRef]
9. Karamanoli, K.; Menkissoglu-Spiroudi, U.; Bosabalidis, A.M.; Vokou, D.; Constantinidou, H.I.A. Bacterial colonization of the Phyllosphere of nineteen plant species and antimicrobial activity of their leaf secondary metabolites against leaf associate bacteria. *Chemoecology* **2005**, *15*, 59–67. [CrossRef]
10. Figueiredo, A.C.; Barroso, G.J.; Pedro, L.G.; Scheffer, J.J.C. Factors affecting secondary metabolite production in plants: Volatile components and essential oils. *Flavour Fragr. J.* **2008**, *23*, 213–226. [CrossRef]
11. Papachristos, D.P.; Karamanoli, K.I.; Stamopoulos, D.C.; Menkissoglu-Spiroudi, U. The relationship between the chemical composition of three essential oils and their insecticidal activity against *Acanthoscelides obtectus* (Say). *Pest Manag. Sci.* **2004**, *60*, 514–520. [CrossRef]
12. Al-Banna, L.; Darwish, R.M.; Aburjai, T. Effect of plant extracts and essential oils on root-knot nematodes. *Phytopathol. Mediterr.* **2003**, *42*, 123–128.

13. Ntalli, N.G.; Ferrari, F.; Giannakou, I.; Menkissoglu-Spiroudi, U. Phytochemistry and nematicidal activity of the essential oils from 8 greek lamiaceae aromatic plants and 13 terpene components. *J. Agric. Food Chem.* **2010**, *58*, 7856–7863. [CrossRef] [PubMed]

14. Laquale, S.; Sasanelli, N.; D'Addabbo, T. Attività Biocida Di Olii Essenziali di Specie di Eucalyptus nei Confronti del Nematode Galligeno *Meloidogyne Incognita*. In Proceedings of the I Congresso Nazionale della Società Italiana per la Ricerca sugli Oli Essenziali (S.I.R.O.E.), Roma, Italy, 15–17 November 2013.

15. Laquale, S.; Candido, V.; Avato, P.; Argentieri, M.P.; D'Addabbo, T. Essential oils as soil biofumigants for the control of the root-knot nematode *Meloidogyne incognita* on tomato. *Ann. Appl. Biol.* **2015**, *167*, 217–224. [CrossRef]

16. Faria, J.M.S.; Sena, I.; Ribeiro, B.; Rodrigues, A.M.; Maleita, C.M.N.; Abrantes, I.; Bennett, R.; Mota, M.; Figueiredo, A.C.S. First report on *Meloidogyne chitwoodi* hatching inhibition activity of essential oils and essential oils fractions. *J. Pest Sci.* **2016**, *89*, 207–217. [CrossRef]

17. Lahlou, M. Methods to study the phytochemistry and bioactivity of essential oils. *Phytother. Res.* **2004**, *18*, 435–448. [CrossRef] [PubMed]

18. Ntalli, N.G.; Ferrari, F.; Giannakou, I.; Menkissoglu-Spiroudi, U. Synergistic and antagonistic interactions of terpenes against *Meloidogyne Incognita* and the nematicidal activity of essential oils from seven plants indigenous to Greece. *Pest Manag. Sci.* **2011**, *67*, 341–351. [CrossRef]

19. Mohan, S.Z.; Haider, H.C.; Andola, V.K. Essential oils as green pesticides: For sustainable agriculture. *Res. J. Pharm. Biol. Chem. Sci.* **2011**, *2*, 100–106.

20. Ehlers, R.-U. *Regulation of Biological Control Agents*; Springer: Cham, Switzerland, 2011; pp. 293–294.

21. *Concerning the Data Requirements for Active Substances of Plant Protection Products Made from Plants or Plant Extracts*; Sanco/10472/2003-rev.5; European Commission, Health & Consumer Protection Directorate-General: Brussels, Belgium, 2014.

22. *Working Document on the Procedure for Application of Basic Substances to Be Approved in Compliance with Article 23 of Regulation (EC) No 1107/2009*; SANCO/10363/2012-rev.9; European Commission, Health & Consumers Directorate-General: Brussels, Belgium, 2014.

23. Ntalli, N.; Bratidou Parlapani, A.B.; Tzani, K.; Samara, M.; Boutsis, G.; Dimou, M.; Menkissoglu-Spiroudi, U.; Monokrousos, N. *Thymus Citriodorus* (Schreb) Botanical Products as Ecofriendly Nematicides with Bio-Fertilizing Properties. *Plants* **2020**, *9*, 202. [CrossRef]

24. Hussey, R.S.; Barker, K.R. A comparison of methods of collecting Inocula of *Meloidogyne* spp. including a new technique. *Plant Dis. Rep.* **1973**, *57*, 1025–1028.

25. Byrd, D.W.; Krikpatrick, T.; Barker, K.R. An improved technique for cleaning and staining plant tissue for detection of nematodes. *J. Nematol.* **1983**, *15*, 142–143.

26. Puntener, W. Manual for Field Trials. In *Plant Protection*; Ciba-Geigy: Basel, Switzerland, 1981; p. 205.

27. Seefeldt, S.S.; Jensen, J.E.; Fuerst, E.P. Log-logistic analysis of herbicide rate response relationships. *Weed Technol.* **1995**, *9*, 218–227. [CrossRef]

28. Said, K.A.M.; Yakub, I.; Alipah, N.A.M. Effects of Solvent/Solid Ratio and Temperature on the Kinetics of Vitamin C Extraction from *Musa Acuminata*. *J. Appl. Sci. Process Eng.* **2015**, *2*, 107–115.

29. Nasiou, E.; Giannakou, I.O. The potential use of carvacrol for the control of *Meloidogyne javanica*. *Eur. J. Plant Pathol.* **2017**, *149*, 415–424. [CrossRef]

30. Andres, M.F.; Gonzalez-Coloma, A.; Munoz, R.; De la Pena, F.; Julio, L.F.; Burillo, J. Nematicidal potential of hydrolates from the semi industrial vapor-pressure extraction of Spanish aromatic plants. *Environ. Sci. Pollut. Res.* **2018**, *25*, 29834–29840. [CrossRef]

31. Barbosa, P.; Lima, A.S.; Vieira, P.; Dias, L.S.; Tinoco, M.T.; Barroso, J.G.; Pedro, L.G.; Figueiredo, A.G.; Mota, M. Nematicidal activity of essential oils and volatiles derived from Portuguese aromatic flora against the pinewood nematode, *Bursaphelenchus xylophilus*. *J. Nematol.* **2010**, *42*, 8–16.

32. Caboni, P.; Saba, M.; Tocco, G.; Casu, L.; Murgia, A.; Maxia, A.; Menkissoglu-Spiroudi, U.; Ntalli, N. Nematicidal Activity of Mint Aqueous Extracts against the Root-Knot Nematode *Meloidogyne incognita*. *J. Agric. Food Chem.* **2013**, *61*, 9784–9788. [CrossRef]

33. Andres, M.F.; Gonzalez-Coloma, A.; Sanz, J.; Burillo, J.; Sainz, P. Nematicidal activity of essential oils: A review. *Phytochem. Rev.* **2012**, *11*, 371–390. [CrossRef]

34. Pandey, A.K.; Singh, P.; Tripathi, N.N. Chemistry and bioactivities of essential oils of some *Ocimum* species: An overview. *Asian Pac. J. Trop. Biomed.* **2014**, *4*, 682–694. [CrossRef]

35. Wang, J.; Pan, X.; Han, Y.; Guo, D.; Guo, Q.; Li, R. Rosmarinic Acid from Eelgrass Shows Nematicidal and Antibacterial Activities against Pine Wood Nematode and Its Carrying Bacteria. *Mar. Drugs* **2012**, *10*, 2729–2740. [CrossRef]

36. Qamar, F.; Begum, S.; Raza, S.M.; Wahab, A.; Siddiqui, B.S. Nematicidal natural products from the aerial parts of *Lantana camara* Linn. *Nat. Prod. Res.* **2005**, *19*, 609–613. [CrossRef]

Review

Zeolites Enhance Soil Health, Crop Productivity and Environmental Safety

Mousumi Mondal [1], Benukar Biswas [1], Sourav Garai [1], Sukamal Sarkar [1,2], Hirak Banerjee [1], Koushik Brahmachari [1], Prasanta Kumar Bandyopadhyay [3], Sagar Maitra [4], Marian Brestic [5,6,*], Milan Skalicky [6], Peter Ondrisik [7] and Akbar Hossain [8,*]

[1] Department of Agronomy, Bidhan Chandra Krishi Viswavidyalaya (BCKV), Nadia 741252, West Bengal, India; mou.mousumi98@gmail.com (M.M.); kripahi@yahoo.com (B.B.); garai.sourav93@gmail.com (S.G.); sukamalsarkarc@yahoo.com (S.S.); hirak.bckv@gmail.com (H.B.); brahmacharis@gmail.com (K.B.)

[2] Office of the Assistant Director of Agriculture, Bhagwangola-II Block, Directorate of Agriculture, Government of West Bengal, Murshidabad 742135, West Bengal, India

[3] Department of Agricultural Chemistry and Soil Science, Bidhan Chandra Krishi Viswavidyalaya (BCKV), Nadia 741252, West Bengal, India; pkb_bckv@rediffmail.com

[4] Department of Agronomy, Centurion University of Technology and Management, Paralakhemundi 761211, Odisha, India; sagar.maitra@cutm.ac.in

[5] Department of Plant Physiology, Slovak University of Agriculture, Nitra, Tr. A. Hlinku 2, 949 01 Nitra, Slovakia

[6] Department of Botany and Plant Physiology, Faculty of Agrobiology, Food and Natural Resources, Czech University of Life Sciences Prague, Kamycka 129, 165 00 Prague, Czech Republic; skalicky@af.czu.cz

[7] Department of Environment and Zoology, Slovak University of Agriculture, Nitra, Tr. A. Hlinku 2, 949 01 Nitra, Slovakia; peter.ondrisik@uniag.sk

[8] Bangladesh Wheat and Maize Research Institute, Dinajpur 5200, Bangladesh

* Correspondence: marian.brestic@uniag.sk (M.B.); akbarhossainwrc@gmail.com (A.H.)

Citation: Mondal, M.; Biswas, B.; Garai, S.; Sarkar, S.; Banerjee, H.; Brahmachari, K.; Bandyopadhyay, P.K.; Maitra, S.; Brestic, M.; Skalicky, M.; et al. Zeolites Enhance Soil Health, Crop Productivity and Environmental Safety. *Agronomy* **2021**, *11*, 448. https://doi.org/10.3390/agronomy11030448

Academic Editors: Nikolaos Monokrousos and Efimia M. Papatheodorou

Received: 4 February 2021
Accepted: 24 February 2021
Published: 28 February 2021

Publisher's Note: MDPI stays neutral with regard to jurisdictional claims in published maps and institutional affiliations.

Abstract: In modern days, rapid urbanisation, climatic abnormalities, water scarcity and quality degradation vis-à-vis the increasing demand for food to feed the growing population necessitate a more efficient agriculture production system. In this context, farming with zeolites, hydrated naturally occurring aluminosilicates found in sedimentary rocks, which are ubiquitous and environment friendly, has attracted attention in the recent past owing to multidisciplinary benefits accrued from them in agricultural activities. The use of these minerals as soil ameliorants facilitates the improvement of soil's physical and chemical properties as well as alleviates heavy metal toxicity. Additionally, natural and surface-modified zeolites have selectivity for major essential nutrients, including ammonium (NH_4^+), phosphate (PO_4^{2-}), nitrate (NO_3^-), potassium (K^+) and sulphate (SO_4^{2-}), in their unique porous structure that reduces nutrient leaching. The slow-release nature of zeolites is also beneficial to avail nutrients optimally throughout crop growth. These unique characteristics of zeolites improve the fertilizer and water use efficiency and, subsequently, diminish environmental pollution by reducing nitrate leaching and the emissions of nitrous oxides and ammonia. The aforesaid characteristics significantly improve the growth, productivity and quality of versatile crops, along with maximising resource use efficiency. This literature review highlights the findings of previous studies as well as the prospects of zeolite application for achieving sustenance in agriculture without negotiating the output.

Keywords: soil amelioration; resource use efficiency; water conservation; nutrient retention; heavy metal toxicity

1. Introduction

The increasing pressure of the population leads to a higher food demand, and at least 50% more food production is required to meet the demand of people by 2050, without any

115

scope of horizontal land diversification [1,2]. Therefore, intensive agricultural practices in food and nutritional security force the use of irrational chemical inputs, water and heavy machinery. More than two-thirds of the renewable water resources are exclusively used by agricultural activities, resulting in uneven water sharing with the other sectors [2–4]. Furthermore, the consequences of intensive practices are the degradation of soil and water qualities, such as depletion of soil organic carbon and inherent soil nutrient status, heavy metal contamination and residual fertilizer and/or pesticide mixing with groundwater vis-à-vis surface water resources, that dwindle crop productivity and ultimately the per capita food grain availability [5]. Long-term intensive farming activities make the agricultural land unproductive, resulting in low soil retention capacity. The most important element, nitrogen, is widely used in agricultural systems, although its use efficiency in nitrogenous fertilizers rarely exceeds 50% as it is mostly lost through denitrification, leaching and volatilisation [6]. Moreover, irrational application of nitrogenous fertilizers facilitates easy NO_3^+ discharge from soil to groundwater, causing negative anthropogenic impacts on the groundwater quality and public health hazards such as methemoglobinemia, cancer of digestive organs, eutrophication in water bodies and production of greenhouse gases such as nitrous oxide (N_2O) through the denitrification process [7–10]. Phosphate (PO_4^{3+}) is another major nutrient in fertilizer, also responsible for eutrophication in water bodies [11]. Therefore, soil nutrient retention is a major concern in modern agriculture to account for maximum nutrient use efficiency, improve the soil nutrient status and prevent groundwater contamination [12–14]. Nutrient use efficiency and better plant growth are highly related to soil's physical and chemical properties. In this context, the application of soil amendments, more particularly natural or organic amendments, has great importance for the long-term reclamation of soil's physicochemical properties [15–17]. Zeolites are naturally occurring, alkaline-hydrated aluminosilicates with more than 50 different forms [18,19] and a wide range of applications such as soil-binding agents and nutrient supplements for animal and aquatic lives. Additionally, they can be used as heat storage materials and solar refrigerators, both absorber and adsorber; ion-exchanging elements; molecular sieving agents; and catalysing agents in various chemical reactions [20,21]. In agriculture, the importance of zeolites has been realised to a greater extent with their varying applicability (Figure 1) [20]. Natural zeolites are being considered as good soil ameliorating substances, having good water and nutrient holding capacity (WHC); it improves infiltration rate, saturated hydraulic conductivity, cation exchange capacity, and prevents water losses from deep percolation [22–26]. Moreover, zeolites could be used as fertilizer and chelating agent [27]. Zeolites minimize the rate of nutrient release from both organic and inorganic fertilizers and enable better nutrient availability throughout the crop growth stages [27]. The improvement of the wide range of agronomic and horticultural crops in respect to growth, yield and quality traits with the application of zeolites has been well reported by various researchers [28–33]. Additionally, zeolite can effectively absorb heavy metals such as cadmium (Cd), lead (Pb), nickel (Ni), anions like chromate (CrO_4^{2-}) and arsenate (AsO_4^{-3}), and organic pollutants such as volatile organic compounds (VOCs) including benzene, toluene, ethylbenzene, and xylene (BTEX) from soil or water body [34–36]. Acknowledging all the aforesaid advantages, the applications of zeolites in the agricultural research field have been widely gained importance since the last two decades (Figure 2), evidenced by the chronological ascending trend of the publication rate accessed from "Scopus" online database with the keywords of "zeolit", "soil remediation", "water retention", "nutrient retention", "crop production" and "heavy metal toxicity". Several earlier findings reported the applicability of zeolites on soil properties along with water and nutrient retention capacity, crop yield and heavy metal toxicity. Therefore, it is high time to give importance to zeolites application in agricultural activities and this review article gives a comprehensive assessment on the sources of zeolites, their structure and properties, and wide application in agriculture with the special consideration of soil properties, resource conservation, pest management, pollution control and crop productivity.

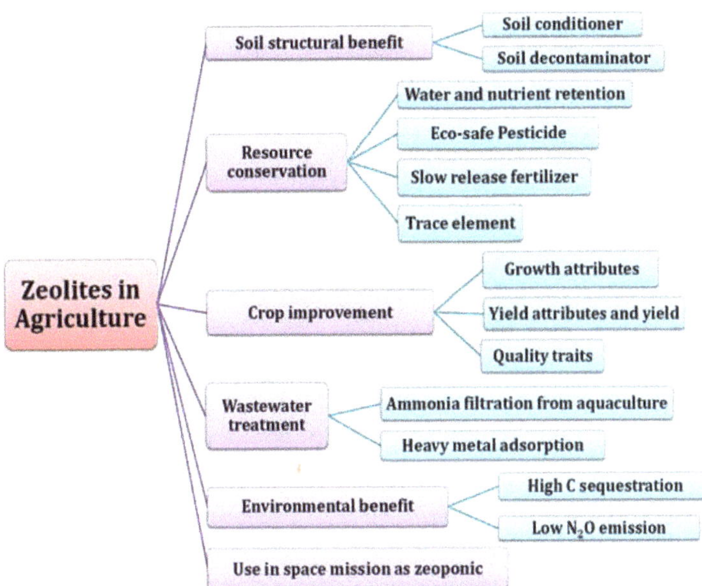

Figure 1. Multidimensional Uses of Zeolites in Agriculture.

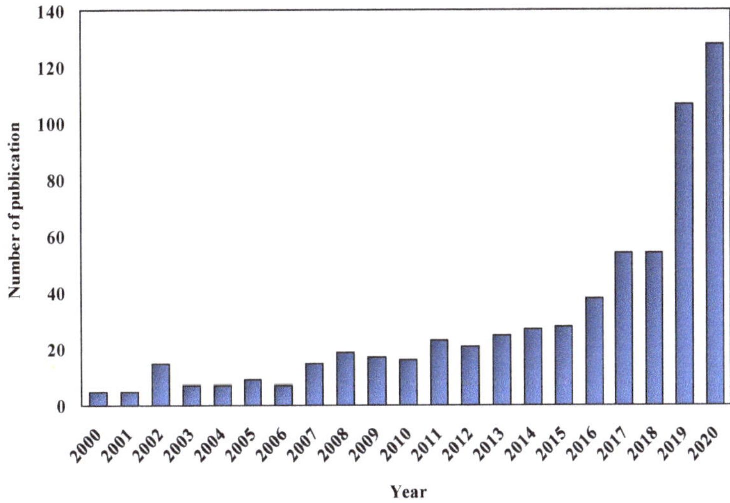

Figure 2. The Trend of Annual Publications on Zeolite Applications in Agriculture for the Last Two Decades. Source: Scopus Preview [37].

2. Origin, Structure and Properties of Zeolites

The word zeolites refer to 'boiling stones' because of their ability to froth when heated to about 200 °C. The first time, the mineral zeolites are identified by a Swedish mineralogist Alex Fredrik Cronstedt in 1756 [38]. However, zeolites production was started commercially in the 1960s [38]. China contributes ~75% market share of total zeolites production, followed by Korea (8%), the United States (3%), and Turkey (2%) [39]. In India, the maximum zeolitic enriched soil is found in the state of Maharashtra followed by Karnataka, Gujrat, Andhra Pradesh and West Bengal (Figure 3).

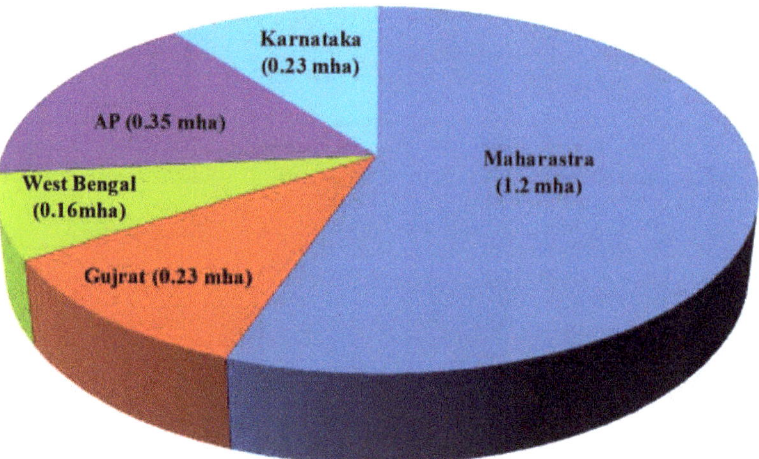

Figure 3. Distribution of Zeolitic Soil in India. Modified from Bhattacharyya et al. [40].

Structurally, zeolite is comprised of aluminosilicate (AlO_4 and SiO_4) tetrahedrons, joined into three-dimensional frameworks and seems like a honeycomb structure (Figure 4) [41]. The cages in the porous structure of zeolite are approximately 12 Å in diameter, interlinked through the channels of 8 Å diameter, includes 12 tetrahedrons rings [42]. Depending on the minerals, the pores are interlinked to form long wide channels which facilitate easy molecular movement into and out of the zeolite structure. The negative charge of aluminum ions in the zeolite structure is balanced by positively charged cations.

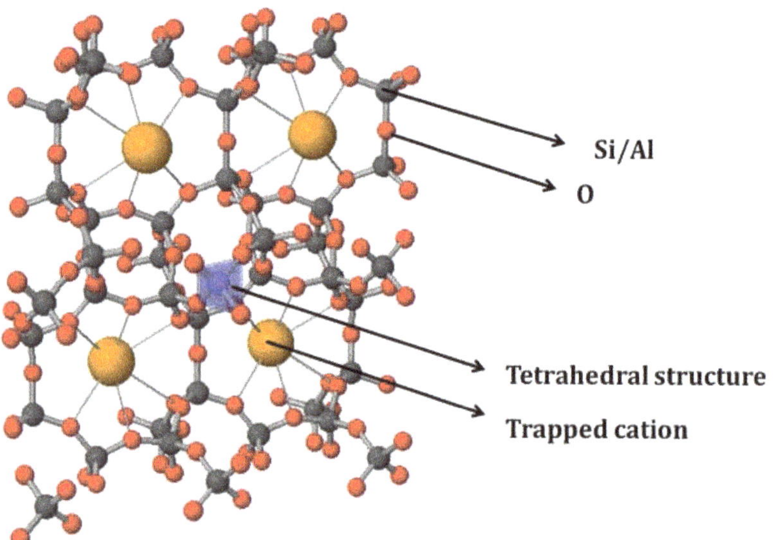

Figure 4. The Tetrahedral Framework of Clinoptilolite Zeolite. Modified from IZA [43].

The general empirical formula that refers to a zeolite structure is $M_{2n}O \cdot Al_2O_3 \cdot xSiO_2 \cdot yH_2O$. M refers to any alkali or alkaline earth cation; the valence of the cation is indicated by n, x ranges between 2 and 10, and y ranges between 2 and 7, with structural cations comprising Si^{2+}, Al^{3+} and Fe^{3+}, and exchangeable cations K^+, Na^+ and Ca^{2+} [44]. The spacious porous structure with large channels in zeolite structure makes it unique in nature as compared to other silicate minerals [45]. Natural zeolites are loaded with aforesaid cations with various considerable properties such as higher cation exchange capacity (CEC) than normal soil, ranges between 100 and 200 centimol (+) kg^{-1} [46], free water storage within their structural channels, and also have a great ability of ion adsorption in large surface area. Zeolites can adsorb or exchange various cations viz. strontium (Sr) and cesium (Cs); heavy metals like zinc (Zn), cadmium (Cd), lead (Pb), manganese (Mn), nickel (Ni), chrome (Cr), iron (Fe), and copper (Cu) [34]; anions such as chromate (CrO_4^{2+}) and arsenate (AsO_4^{3+}) [35]; and numerous organic pollutants mentioned earlier [36]. Other useful physical and chemical properties of zeolites include high void volume (~50%), low density (2.1–2.2 g/cm^3), excellent molecular sieving properties and high cation selectivity exclusively for ammonium, potassium, and cesium ions [40]. Physical characteristics of some naturally occurring zeolites are summarized in Table 1. In respect to pore diameter Zeolites have been classified by Flanigen [47], viz. (i) Small-pore (0.3–0.45 nm diameter with 8 rings), (ii) Medium-pore (0.45–0.6 nm diameter with 10 rings), (iii) Large-pore (0.6–0.8 nm diameter with 12 rings), and (iv) Extra-large pore zeolites (0.8–1.0 nm diameter with 14 rings).

Table 1. Physical Characteristics of Some Naturally Occurring Zeolites.

Zeolites	Porosity (%)	Channel Dimensions (Å)	Heat Stability	Ion Exchange Capacity ($meq\ g^{-1}$)	Specific Gravity ($g\ cm^{-3}$)	Bulk Density ($g\ cm^{-3}$)	References
Analcine $Na_{10}(Al_{16}Si_{32}O_{96})\cdot16H_2O$	18	2.6	High	4.55	2.24–2.29	1.85	Sangeetha and Baskar [42]
Chabezite$(Na_2Ca)_6$ $(Al_{12}Si_{24}O_{72})\cdot40H_2O$	47	3.7 × 4.2	High	3.85	2.50–2.10	1.45	IZA [43]
Clinoptilolite $(Na_3K_3)(A_{16}Si_{30}O_{72})\cdot24H_2O$	34	3.9 × 5.4	High	2.17	2.15–2.25	1.15	IZA [43]
Erionite $(AlCaH_{60}KNaO_{36}Si_2^{+3})$	35	-	High	3.12	2.02–2.08	1.51	Hemingway and Robie [44]
Heulandite $(Ca_4)(Al_8Si_{28}O_{72})\cdot24H_2O$	39	4.0 × 5.5	Low	2.90	2.18–2.20	1.69	Sangeetha and Baskar [42]
Mordenite$(Na_8)(Al_8Si_{40}O_{96})\cdot24H_2O$	28	2.9 × 5.7	High	4.30	2.12–2.15	1.70	Chmielewska and Lensỳ [45]
Philipsite$(NaK)_5(Al_5Si_{11}O_{32})\cdot20H_2O$	31	4.2 × 4.4	Moderate	3.32	2.15–2.20	1.58	Chmielewska and Lensỳ [45]
Faujasite $(Na_{58})(Al_{58}Si_{134}O_{384})\cdot240H_2O$	47	7.4	High	3.38	-	-	Hemingway and Robie [44]
Laumonitte$(Ca_4)(Al_8Si_{16}O_{48})\cdot16H_2O$	34	4.6 × 6.3	Low	4.25	-	-	Sangeetha and Baskar [42]
Linde $A(Na_{12})(Al_{12}Si_{12}O_{48})\cdot27H_2O$	47	4.4	High	5.47	-	-	Sangeetha and Baskar [42]
Linde $X(Na_{86})(Al_{86}Si_{106}O_{384})\cdot264H_2O$	50	7.4	High	4.72	-	-	Sangeetha and Baskar [42]

3. Impacts of Zeolite Application in Agriculture

3.1. Improvement of Soil Physical Properties

Soil physical properties include bulk density, particle density, aeration, soil porosity, water holding capacity in which bulk density is the basic soil property that influences the total porosity and topsoil stability [48]. The application of zeolites in light texture soil reduces the bulk density that modifies the water holding capacity and soil air porosity [49]. However, total porosity is not influenced significantly [49]. In a previous study, Xiubin and Zhanbin [3] opined the natural zeolite mainly mordenite with less than 0.25 mm size to the fine-grained calcareous loess which had low WHC. Result revealed that after 25 h of water addition to treated and normal soils, the zeolites applied soil resulted in higher water content (Figure 5). They also reported that water holding capacity in zeolites treated soil increased 0.4–1.8% in drought condition while 5–15% in normal situation as compared to non-treated soil.

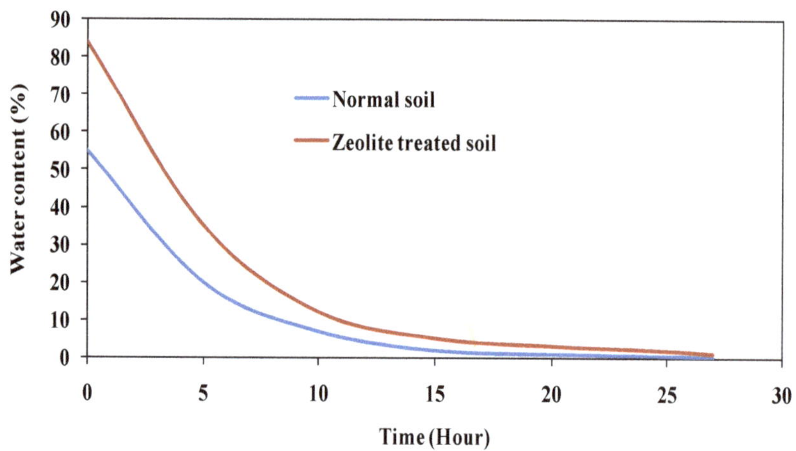

Figure 5. Soil Water Content as Influenced by Zeolitic Soil. Modified from Xiubin and Zhanbin [3].

In another study, the effect of modified Ca^{+2} type zeolite on sand dune soil was determined where irrigated was given with saline water. Sand dune soil samples were treated with the three different rates of zeolite i.e., 5 kg m^{-2}, 1 kg m^{-2} and no zeolite (control) and irrigated with seawater diluted to electrical conductivity (EC) levels of 3 and 16 deciSiemens per metre (dSm^{-1}). Results showed that soil with 5 kg zeolites m^{-2} enhanced soil water as well as salt content, accounting for 20 and 1.4% higher than no application of zeolite [16]. The concentration of cations namely Ca^{2+}, K^+, Na^+, Mg^{2+} is increased with the increasing soil salinity. The findings were attributed to the fact that zeolite increases the cation exchange capacity, and subsequent cations holding on the surface soil, and release them at the expense of salts in the saline water [22]. Thus, the low salt accumulation in subsurface soil facilitates low salt stress on plants and creates a better environment for plant growth. Lowering of particle size with the application of zeolites in sandy soil might be another reason for higher water holding capacity. Higher pore volumes in zeolites facilitate greater water holding in their structures [49]. Such structures are not damaged by water particles during surface evaporation and/or reabsorption. Zeolites may be considered as the permanent water reservoir. Retention of soil moisture in longer duration, particularly during dry periods helps to mitigate drought-induced abiotic stresses and enable plants to withstand in dry spell; zeolites also facilitate to rapid rewetting and the lateral water spreading throughout the root zone during the time of irrigation that reduces the timing of water application [41]. Soil amelioration with zeolites increases the water availability to plants by 50% [42]. Application of zeolite @ 10 g kg^{-1} soil could maintain maximum water percentage (8.4%) at field capacity and delay in permanent wilting point in sandy loam soils [50]. Al–Busaidi et al. [16] reported that the existence of fine particles and micropores in zeolites slowed down the deep percolation of soil water. The infiltration rate is inversely proportional to zeolites application (Figure 6) indicating the higher soil water residence and subsequent restriction in nutrient and salt leaching. Xiubin and Zhanbin [3] observed that the mixing of zeolites with fine grain calcareous loess soil increased the infiltration rate by 7–30% and 50% in a gentle and steep slope respectively. Furthermore, run-off and subsequent soil erosion were reduced with the zeolites application and the sedimentation also found to be decreased by 85% and 50% in a gentle and steep slope respectively. Interestingly, a combination of zeolites and selenium application check the water deficit oxidative damages in plants [51]. Colombani et al. [52]

quantified the changes in flow and transport parameters induced by the addition of zeolites in a silty-clay soil and reported that NH_4^+ enriched zeolites enhanced the capacity of water retention in silty-clay soil, thus diminishing the water and solute losses. Maximum irrigation water productivity (0.81 kg m^{-3}) under limited irrigation supply was registered with the supplemental application of zeolites (21% ww^{-1}) along with urea, while the minimum water productivity (0.48 kg m^{-3}) was observed under full irrigation supply and exclusive urea application [53]. Bernardi et al. [54] also observed that concentrated zeolites as a sand-soil amendment increase at least 10% of soil-water retention and 15% of available water to plants. Zeolite increases the periods between the starting of rainfall and runoff occurrence. Rainfall intensity with 10 mm per hr. results in the beginning of runoff within 15 min in normal soil while in zeolites (20%) treated soil runoff starts after 30 min of rainfall occurrence [55].

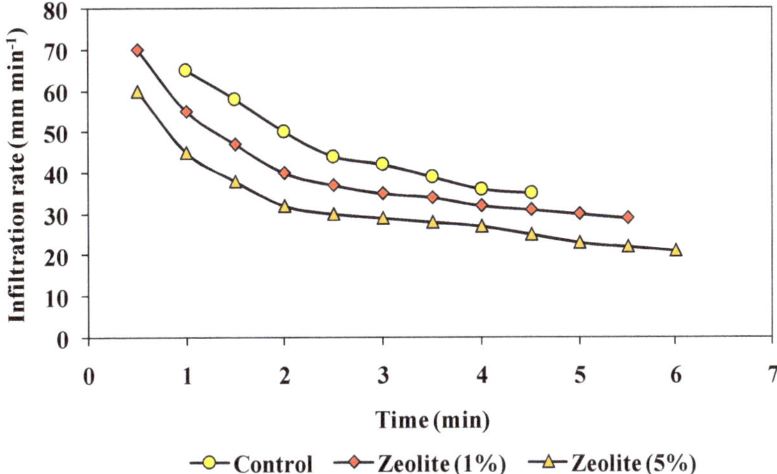

Figure 6. Soil Infiltration Rate as Influenced by Different Rates of Zeolite Application. Modified from Al-Busaidi et al. [16].

Zeolites help to improve the water-stable aggregates in soil. As per example, nano zeolite with 30% concentration increased the mean weight diameter of water-stable aggregates by 0.735 mm [56]. With the use of this property, Moritani et al. [57] reported that the incorporation of 10% artificial zeolites in sodic soils resulted in improved wet aggregate stability ranged between 22.4% and 59.4% depends on the soil textural classes. Cario et al. [58] categorized the soil with average assessment ranking 'good' and 'excellent' in terms of water-stable aggregates and degree of soil aggregation in Vertisols and they showed the application of zeolite along with chemical fertilizers or organic manure (Zeolite @ 7.5 t h^{-1} + sugarcane filter cake @ 22.5 t h^{-1}) improved the soil properties from good to excellent. Sepaskhah and Yousefi [59] conducted an experiment to justify the effect of various rate of calcium-potassium zeolite on the pore velocity of water in the soil they observed higher pore water velocity (35 and 74%) with the application of 4 and 8 g zeolite kg^{-1} soil respectively. Changes of soil physical properties with the Zeolite application in thin (heavy) textured medium-thin textured, and medium coarse (light) textured soil was observed by Gholizadeh-Sarabi and Sepaskhah [60] reported that in fine and medium texture soil, zeolites application at the rate of 4 and 8 g kg^{-1} of soil at the low salinity level (0.5 and 1.5 dS m^{-1}) and 16 g zeolites kg^{-1} soil at the high salinity level (3.0 and 5.0 dS m^{-1}) increased saturated hydraulic conductivity significantly while in coarse texture soil similar rate of zeolites application reduced the saturated hydraulic conductivity considerably. They also assumed that zeolites application in the heavy (clay loam) and medium-textured soil (loam) changed the shape and size of the soil pores and resulted in an improvement of soil structure and the water movement in these soils. Zeolites application alleviates the adverse effect of salinity on hydraulic conductivity and thus it would prevent waterlogging in heavy and medium soil textures. In case of sandy soils, zeolites addition would be appropriate to decrease the hydraulic conductivity and the transferability of water that results in low deep percolation and loss of soil water. However, Razmi and Sepaskhah [61] reported that the application of zeolite (8 g kg^{-1}) in silty clay soils significantly improved the hydraulic conductivity. They also established that the soil treated with zeolite resulted in 50% less crack depth in dry puddled soil with pre-application of zeolites in comparison to no zeolites application. A similar observation was also recorded after the first and second irrigation in puddled condition (Figure 7).

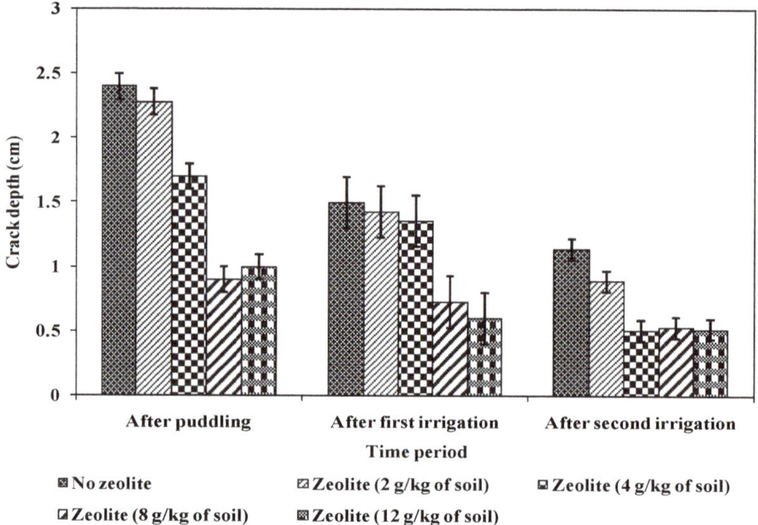

Figure 7. Effect of Zeolite on Crack Depth in Puddled Transplanted Rice. Modified from Razmi and Sepaskhah [61].

Furthermore, the sorptivity of clay-loam soil was reduced with a higher rate of zeolites application as reported by Gholizadeh-Sarabi and Sepaskhah [60]. However, a contrasting result was observed in the case of sandy-loam and loamy soil. Proper use of water is the immediate need in agriculture to ensure food security with available water resources; hence, technologies that enhance water use efficiency are being widespread. The aforesaid discussions indicate that zeolites addition positively influence the inter-particle porosity as well as total porosity, bulk density, hydraulic conductivity, infiltration rate, and cation exchange capacity of soil that ultimately accelerates the soil water content. Additionally, the open pore network channels into zeolites structure mainly play the significant roles' in water retention. The summarization of zeolitic impacts to the wide range of soils in Table 2 indicates that the use of zeolites as a soil ameliorant would be a welcome strategy in agriculture.

Table 2. Physical Properties of Soils as Influenced by Zeolites Application.

Types of Zeolite	Application Rate (ww^{-1})	Soil Textural Classes	Changes in Soil Physical Properties			References
			Water Content	Infiltration Rate	Hydraulic Conductivity	
Clinoptilolite	1–15%	Clay, loamy sand, sand	• 20% increase at 10% zeolite application rate in sandy soil	• Infiltration rate reduction with a higher rate of application	• Decreased in sandy and loamy soils; Increased in clay soil	Mahabadi et al. [15]
Mordenite	–	Calcareous loess	• 0.4–1.8% and 5–15% increase in drought conditions and normal conditions	• 7–30% increase with gentle slopes >50% increase with steep slope	–	Xiubin and Zhanbin [3]
Non-specified natural zeolite	0.4%, 0.8%, 1.6% and no zeolitie	Sandy loam	–	• Highest sorptivity at 0.4% (0.5 dS m^{-1} salinity) • Lowest sorptivity at 1.6%	• Decreased at 0.8 and 1.6%	Gholizadeh-Sarabi and Sepaskhah [60]
		Loam	–	• Highest sorptivity at 1.6% (all salinity levels)	• Maximized at 1.6% and 0.4% at 3–5 dS m^{-1} and 0.5–1.5 salinity respectively	
		Clay loam	–	• Lowest sorptivity at 1.6% and 0.8% at 0.5–3 and 5 dS m^{-1} salinity	• Maximized at 1.6% and 0.8% at 3–5 and 0.5–1.5 dS m^{-1} salinity level	
Synthetic zeolite (Ca^{2+}-type)	1% and 5%	Sand dune soil	• Increased at 5% level	• Reduction in infiltration rate (more for higher application rate)		Al-Busaidi et al. [16]
Stilbite	3.33, 6.67, and 10%	Sandy soil	• 10%, 38% and 67% increase with 3.33%, 6.67% and 10% level respectively	–		Bernardi et al. [54]
Clinolite and Ecolite	15.85%	–	–	–	• Increase (More in Ecolite over Clinolite)	Githinji et al. [62]

3.2. Nutrient Retention

Zeolites positively influence the physical, chemical, and biological properties of soil directly or indirectly which in turns improves the nutrient dynamics as well as nutrient retention capacity. Zeolitic minerals have high CEC which attributes to high NH_4^+ sorption selectivity as a consequence of the electrostatic attraction between positively charged NH_4^+ and negatively charged sites in zeolite structure [63,64]. The effective diffusion coefficient was around 4–5 × 10–12 m^2 s^{-1} for ammonium and sodium ions respectively in clinoptilolite [65,66]. The adsorption capacity of zeolites for these ions is determined by isotherms and kinetics and this adsorption property is used for various purposes such as wastewater treatment, heavy metal removal. Clinoptilolite generally exhibits a high selectivity for NH_4^+ ion, having theoretical CEC of 2.16 cmol (+) kg^{-1} [67]. I on adsorption efficiency of zeolites are mainly depends of the factors like mass, particle size, initial concentration of cations of model solution, contact time, temperature and pH [68,69]. Additionally, modification of zeolites surface with strong acids accelerates the cation sorption capacity [70]. The modification of natural zeolites includes pretreatment by grinding and sieving, mixing with sodium salt and finally, calcinations makes a change in the pore size and surface area of zeolites, and thereby the ammonium ion uptake is increased [71].Soil application of zeolites in combination with chemical fertilizers reduces nitrogen leaching [72–74] and volatilization [75–77] slows down the mineralization process and subsequent reduction in greenhouse gases (GHGs) emission [78], and retards the nutrients release into soil solution [79,80]. In the incubation studies, researchers had clearly seen the difference in the ammonia loss with chemical fertilizers and chemical fertilizers with zeolite and reported low ammonia losses when fertilizer applied with zeolite [81,82]. Omar et al. [83] proved the significant improvement in soil exchangeable ammonium retention by 40–50% in zeolite treated soil. The leaching reduction of NH_4^+ and NO_3^- from different nitrogenous fertilizer with the application of zeolite is depicted in Table 3. The zeourea and nano-zeourea contain 18.5% and 28% of N respectively and capable to release N up to 34 and 48 days, respectively, while from conventional urea the N releases within 4 days after application [84]. The reason behind this may be the urease activity is significantly reduced by zeolite application that lowers the nutrient release from fertilizer [85]. The slow-released nature of fertilizer helps to release their nutrient contents gradually and to coincide with the nutrient requirement of a plant [86].

Table 3. Leaching Reduction Percent of NH_4^+ and NO_3^- from Different Nitrogenous Fertilizer with the Application of Zeolite.

Soil Type	Zeolite Application Rate	Source of N	N Dose (kg h^{-1})	Leaching Reduction		References
				NH_4^+	NO_3^-	
Sand-based putting green	10%	Ammonium Sulphate	293	99%	86%	Huang and Petrovic [87]
Sandy soil	0.8%	Ammonium Sulphate	32	>90%	–	Zwingmann et al. [88]
Loamy sand	5%	Ammonium Sulphate	200 [†]	83%	–	Mackown and Tucker [89]
Sandy loam	9 [*]	Urea	270	–	36%	Golamhoseini et al. [90]
Silty loam	4%	Wastewater	14.2 [‡]	–	54.9%	Taheri-Sodejani et al. [91]

[*] With the unit of t ha^{-1}; [†] With the unit of mg kg^{-1}; [‡] With the unit of mg L^{-1}.

Urea saturated zeolite chips have also been developed elsewhere. Piñón-Villarreal et al. [92] experimented to assess the leaching loss from urea ammonium nitrate solution (UAN32) where 443 mg total N was present per liter of solution. They observed that 82% reduction in leaching loss happened from the pure clinoptilolite zeolite loaded column in comparison to the column of loamy sand. In a sorption experiment, Piñón-Villarreal et al. [92] reported more than 90% NH_4^+ absorption by zeolite incorporated soil in initial several minutes. Very small particle size with a greater surface area of zeolitic minerals accelerates the stabilization of exchange equilibrium in only a few hours. Zeolite minerals also protect the conversion of NH_4^+ to NO_3^+ through the nitrification process. The latter is more prone to leach out into the soil and facilitates to groundwater contaminations [59]. The small pores in zeolite crystal lattice structure (4–5 Å) in which cations like ammonium can easily adsorb, do not give access to the nitrifying microorganisms into the pores [93]; thus, nitrification does not take place easily in zeolites treated soil. One of the most usefulness of zeolite is utilized in compost making a way to convert agricultural farm waste into valuable organic amendments. However, a significant amount of N losses take place during the time of composting [94]. In an experiment, Ramesh and Islam [95] confirmed that the application of 14–21% zeolite in fresh manure resulted in low ammonium loss. Zeolite also could absorb volatile substances such as acetic acid, butanoic acid, skatole and isovaleric acid and also could effectively control the odor released during composting [96,97].

The extent of reduction in total nitrogen and even phosphorus losses with the application of zeolite into organic manure was successfully reported by Murnane et al. [98]. The reason behind the low N losses from manure is the high specific selectivity of zeolites to ammonium (NH_4^+) that helps in holding this ion during volatilization. Moreover, the existences of small internal channels protect NH_4^+ from rapid nitrification by microbes [99]. Interestingly, zeolites not only help to protect the N loss but also reduces P leaching; however, it helps in reducing NO_3^- leaching greater than P leaching [53,100,101]. Being alkaline in nature and the presence of negative charges, zeolite ameliorated soil improves soil P availability through lowering of soil acidity, soil exchangeable Al, and Fe [101–103]. These help in less P fixation by metal oxyhydroxides. Moreover, zeolites supplementation triggers more P uptake by enhancing the exchange- induced dissolution mechanisms as follows [102]:

$$RP(rock\ phosphate) + NH_4^+ + zeolite \rightarrow Ca - zeolite + NH_4^+ + H_2PO_4^- \tag{1}$$

In this reaction, released Ca is adsorbed on the zeolite surface due to high CEC and as a result, more rock phosphate will be dissolved with lowering Ca^{2+} activity in the solution. This system releases the NH_4^+ and PO_4^{3-} ions. The addition of clinoptilolite zeolites with a 75% recommended rate of fertilizers showed comparable total and available P with the existing recommended dose without any zeolite application [31]. In this experiment, the addition of clinoptilolite zeolites also helped to reduce Al as well as soil acidity that resulted in low P fixation to soil colloid. A similar trend of observation was recorded by Zheng et al. [104], accounted for 14.1% higher available P with the application of zeolite relative to non-zeolite treatment. Antoniadis et al. [105] also reported an increase in P recovery efficiency of 4.02% due to zeolite application in acidic soil as compared to no zeolite application. The slow-release nature of zeolite in P release was observed by Bansiwal et al. [106] resulted in the continuous phosphate release even after 1080 h of continuous percolation from zeolite loaded modified phosphorous surface, while within only 264 h phosphate from potassium dihydrogen phosphate (KH_2PO_4) was exhausted.

Rather than N and P zeolites have strong selectivity on K^+ than Na^+, Ca^{2+}, and Mg^{2+} that makes it difficult to remove K^+ from exchange sites, facilitating greater absorption of K^+ by plant root hairs through the ion exchange within root and zeolite [107]. The losses of K^+ by surface runoff and groundwater leaching can be reduced by supplementing the zeolites as slow-release fertilizer [108]. For example, the application of zeolites in municipal compost to investigate the K^+ release pattern resulted in six times less leaching loss from the zeolitic compost as compared to normal compost [109]. Additionally, Williams and Nelson [110] observed that in a soil-less medium K^+ saturated clinoptilolite recorded 23% less leaching of K^+ over-controlled control substrate. Moraetis et al. [109] reported that there was 18-fold increase in bioavailable K when zeolites were added through kinetic experiment to the soil-compost mixture, suggesting high potassium affinity in the soil-compost-zeolite mixture. Zeolite is considered as nano-enhanced green application as it adsorbs molecules at relatively low pressure [111,112]. Zeolite coated fertilizers have higher potential in water absorption and retention, and this coating materials retard the nutrient release rate from soil applied fertilizers, especially in sandy and sandy loam soil [113].

Similar nutrient retention ability of zeolites in secondary nutrients such as S was registered by Li and Zhang [114] who revealed that after leaching with 50 pore volumes, 85% of the pre-loaded SO_4^{2+} remained on the zeolite modified S fertilizer. Moreover, the initial SO_4^{2+} concentration in the leachate of S-loaded surfaced modified zeolite was found to be lowered, in comparison with the non-zeolitic sulfur sources. In addition to clinoptilolite, nano-zeolite based S fertilizer is also comprised of epistilbite zeolite. The findings from an experiment conducted by Thirunavukkarasu and Subramanian [115] exhibited that SO_4^{2+} was available even after 912 h of continuous percolation from S loaded modified nano-zeolite, while SO_4^{2+} from $(NH_4)_2SO_4$ was depleted within 384 h. The presence of a huge number of channels, pores, and cages in the structure of the zeolite which helps in holding the SO_4^{2+} tightly might be the reason behind the slow release of this secondary nutrient from surface modified nano-zeolite [115].

The increase in micronutrient use efficiency with zeolites supplementation was also registered in previous literatures [33,116–118]. Sheta et al. [116] reported the ability of five natural zeolites and bentonite minerals to adsorb and release of zinc and iron as natural zeolites have a greater affinity to these micronutrients. Iskander et al. [117] found 74.7% and 84.63% are readily extractable by DTPA (diethylene-triamine pentaacetic acid) extractant (0.005 M DTPA + 0.01 M CaCI$_2$ + 0.1 M triethanolamine, adjusted to pH 7.30) after three successive extractions of Zn and Mn, respectively and rest were retained by zeolite. Yuvaraj and Subramaniannano [119] reported that nano-zeolite adsorbed more Zn and the adsorption rate obtained with the nano-zeolite appeared to be efficient adsorbents for Zn. They also observed that $ZnSO_4$ released the Zn up to 200 hours whereas micronutrients from nano-zeolite were releasing even after 800 h (Figure 8). The better availability of micronutrients in soil with zeolite application ultimately facilitates to greater micronutrients contents in plants. Ozbahce et al. [33] resulted in significantly higher Zn, Mn and Cu content in bean leaves with the maximization of zeolite application up to 90 kg ha^{-1} (Figure 9). From the above-mentioned discussions, it can be concluded that the zeolite application accelerates the availability of primary, secondary and micronutrients in soils and subsequent plant uptake (Figure 10), and its application is most significant in arid and semi-arid regions that suffer from high water and nutrient scarcity all-time.

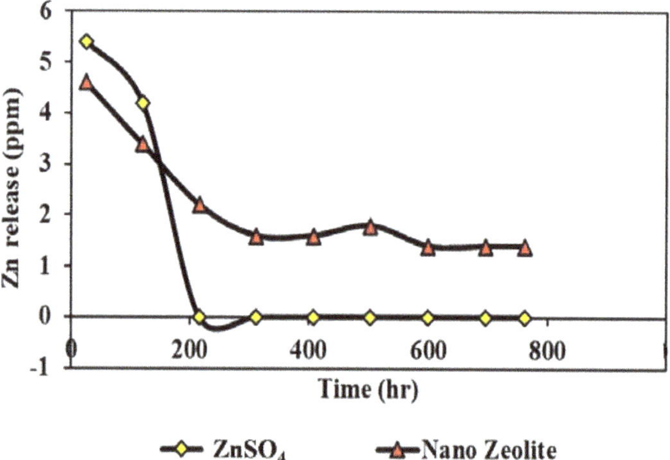

Figure 8. Zn Release Pattern with Time Duration as Influenced by Nano Zeolite Application. Modified from Yuvaraj and Subramaniannano [119].

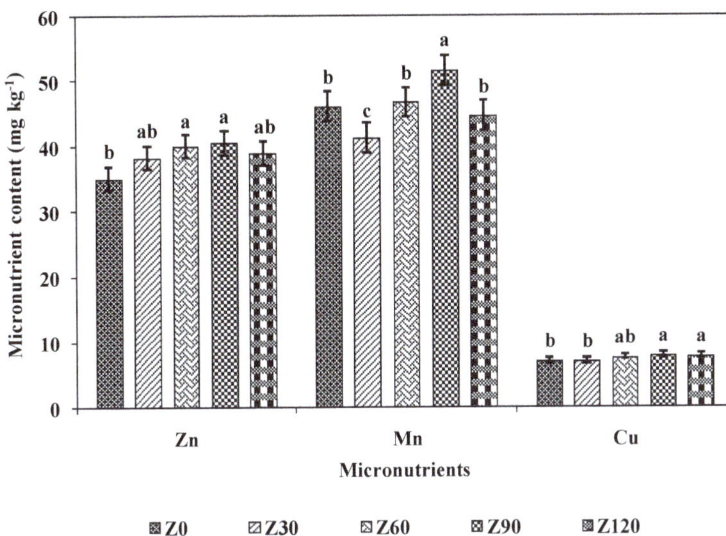

Figure 9. Micronutrient content in bean leaves with different levels of zeolite application (Z_0: 0; Z_{30}: 30; Z_{60}: 60; Z_{90}: 90; Z_{120}: 120 t ha^{-1}); within treatments, different letters indicate significant differences at $p \leq 0.05$(otherwise statistically at par); error bars represent the least significant difference value. Modified after Ozbahce et al. [33].

Figure 10. Effectiveness of Zeolite on Water and Nutrient Retention in Soil. Modified from Nakhli et al. (2014).

3.3. Environmental Impact

The addition of zeolites increases the C sequestration and subsequent soil C stock as compared to untreated soil [120,121]. According to Aminiyan et al. [56], application of zeolite (30%) along with crop residues (5%) to wheat could maintain the highest amount of organic carbon in light and heavy fractions. Soil organic matter even in the light fraction is highly correlated with N mineralization and subsequent soil management practices. The light fraction of soil organic matter (SOM) is not only sensitive to changes in management practices but also correlates well with the rate of N mineralization. Periodical measurements of N_2O and N_2 emissions in fields from the applied cow urine or potassium nitrate (KNO_3)

each at 200 kg N ha^{-1} with and without the addition of zeolite (clinoptilonite) showed that zeolite significantly lowered the total N_2O emissions by 11% from urine treated soils.

Specific channel size enables zeolite to act as molecular gas sieves. Wang et al. [122] recommended the use of zeolite as an amendment to reduce GHGs emission from duck manure as they found almost 27% of GHG emissions reduction from zeolite treated soil than no zeolite application. Additionally, low NO_3^- and PO_4^{2-} leaching from zeolite amended soil helps to prevent groundwater pollution as well as surface water contamination and subsequent eutrophication [59]. They claimed that the better retention of anions in zeolite structure might be the reason for less leaching loss. Zeolite prevents rapid mineralization by preventing the entry of nitrifying bacteria into its structure and thus reduces the emission of N_2O [99].

3.4. Slow Release of Herbicides

Being porous in nature along with a well-ordered structure, zeolites are considered as potential substances for storage and release of organic guest molecules. The most hydrophobic solid form of zeolite 'ZSM 5' adsorbs triazine group of herbicides in the compartmentalized intra-crystalline void space and release them slowly [123]. Furthermore, ZSM-5 was found to be restricted to the mobility of post-emergence herbicide such as paraquat [124,125]. Humic acid zeolites act as a sorbent of the herbicides belongs to the phenylurea group [126]. Clinoptilolitic turf has the potential to remove atrazine from soil and water [127,128]. Application of 2, 4–D herbicide along with zeolites results in a gradual temporal release pattern and keeps the active ingredient of herbicide in upper 0–5 cm of soil layer [129,130]. This slow-release nature of herbicide when used with zeolites improves the herbicide efficiency to control the weed floras and the prolonged effect of herbicide keeps the weed-free crop field throughout the entire crop weed competition period. Zeolite-rich nanocapsule is used as an herbicide carrier, adsorbent and retaining agent [130]. A longer retention period of zeolite added herbicide on weed leaves helps in maximizing the efficacy of the herbicidal mode of action. Interestingly, the synergistic effect between zeolite-loaded catalysts with isoproturon accelerates the visible light absorption and moreover better adsorption of recalcitrant molecules by the porous structure of zeolites [131].

3.5. Remediation of Contaminated Soil

Heavy metals induced soil pollution is one of the major concerns in modern agriculture. The anthropogenic activities of human, rapid industrialization and injudicious use of fertilizers without proper precaution make the soil toxic with heavy metal contamination. The solubility of heavy metal in soil is depending on complex chemical degradation and numerous factors. Among them, low soil pH is one of the major determining factors. In an acidic environment, oxides of iron, aluminum and manganese are slowly solubilized, and the primary and secondary minerals release the heavy metal into soil [132]. Soil sorption capacity is another determining factor for the retention of heavy metal ions. The ongoing concern in relation to the purity of the soil and the need to restore its original properties forced us to seek new and alternative ways of soil cleansing. Zeolite additions increase the soil pH significantly which facilitates to the heavy metal adsorption on its surface; thus, the solubility and bioavailability of heavy metals are ultimately reduced [133]. Chen et al. [134] observed that the cadmium and lead accumulation in wheat is significantly reduced with soil application of zeolite in soil. Moreover, it has been well reported that the clinoptilolite zeolite effectively controlled the heavy metal solubility including cadmium and lead up to 72% and 81% respectively [135,136]. However, this area of research needs extensive studies to find out heavy metal-specific appropriate dose and methods of zeolite application [85].

3.6. Wastewater Treatment

Industrial development with fast urbanization produces large quantities of wastewater that contains heavy metals, oils and organics that badly affect the aqueous environment [137]. Various efficient techniques such as solvent extraction, ion exchange and adsorption are often used to remove those contaminants. Among them, the use of zeolites as adsorbents is most popular due to low-cost involvement, eco-friendly and poses good selectivity for toxic cations [138]. It also prevents the generation of new waste materials [139]. Furthermore, zeolites more specifically clinoptilolite could adsorb dyes, humic acid, phenols and phenol derivatives from the water body [140–142]. The clinoptilolite is mostly effective against metallic cations such as Al^{3+}, Cd^{2+}, Cu^{2+}, Ni^{2+}, Pb^{2+}, and Zn^{2+} from copper mine wastewater [143]. The selectivity by clinoptilolite for heavy metals following the order: $Pb^{2+} > Cd^{2+} > Cu^{2+} > Co^{2+} > Cr^{3+} > Zn^{2+} > Ni^{2+}$ [144]. The most advantage of clinoptilolite use in wastewater treatment is it can adsorb the heavy metals at a wide range of temperature (25–60 °C), pH (1–4) and different agitation speed (0, 100, 200, 400 rpm) [145]. The greater surface area along with high cation exchange capacity makes zeolite as a good adsorbent of cations [142]. The ability of heavy metals uptake by clinoptilolite zeolite was investigated by Baker et al. 2009 and opined the high selectivity of zeolite for the discharge of Pb^{2+} (98%), followed by Cr^{3+}, Cu^{2+} and Cd^{2+} with 96% selectivity within 90 min. Morkou et al. (2015) [146] reported that wastewater nutrients can be recycled and used for microalgal and cyanobacterial biomass production by using zeolite as a medium.

3.7. Crop Management Practices

Zeolites have been used in a wide range of field crops production such as rice (*Oryza sativa* L.), corn (*Zea mays* L.), wheat (*Triticum aestivum* L.), potato (*Solanum tuberosum* L.), soybeans (*Glycine max* L.), and other upland crops in all types of soil to improve their productivity, water, and nutrient use efficiency (NUE), also maintaining the soil ecology and environment [83,147,148]. In an experiment, Chen et al. [29] estimated the effect of different rates of zeolite in combination with different N levels on transplanted rice and concluded that the highest yield was achieved consistently when rice plant was treated with a maximum dose of N (157.50 kg ha^{-1}) along with zeolite supplementation (15 t ha^{-1}), accounting 14.90% higher than the exclusive application of N. They also revealed that yield attributing characters namely effective tillers per plant, number of grains per panicle, grain filling percentage, and 1000-grain weight were positively influenced by the higher dose of N; however, zeolite consistently increased the number of effective tillers (Figure 11). A possible explanation of these results is the slow-release characteristics of zeolite amendment that makes the essential plant nutrients available throughout the crop growth within 0–30 cm soil depth. Furthermore, the supplementary application of zeolite significantly influenced the quality traits like protein content and tasting score of rice but did not influence the head rice recovery and chalkiness of rice grain [29].

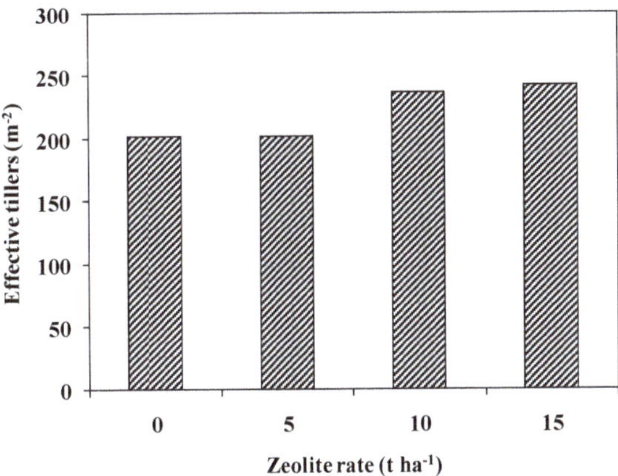

Figure 11. Effect of Zeolite Application on Tillering Pattern of Rice. Modified from Chen et al. [29].

In another experiment, Zheng et al. [32] evaluated the effect of zeolite application on rice under limited water condition and they confirmed that the zeolite treatment (15 t ha^{-1}) improved the LAI, transpiration rate and stomatal conductance (Figure 12). They observed that chalky rice rate and chalkiness were decreased by 29.6% and 41.2% respectively in zeolite treated plants as compared to the non-zeolite control. There was no significant difference in zeolite application on the starch viscosity properties. As rice quality is thought to be determined both genetically and environmentally, any improvements with zeolite application may result from better nitrogen and water availability to plants. The better crop performance and N partitioning in different parts of the rice plant with higher levels of zeolite application were depicted by Wu et al. [149]. Kavoosi et al. [150] resulted in both rice grain and straw yield increment with the application of zeolite at a certain level and thereafter decreased (Figure 13).

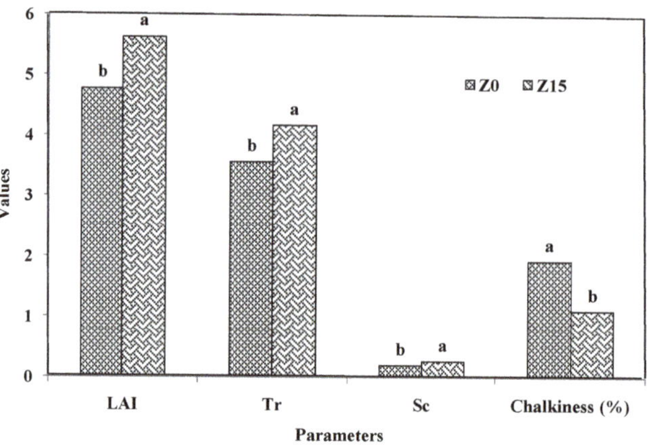

Figure 12. Effect of zeolite application on LAI, Tr, Sc and Chalkiness of rice. Tr: Transpiration rate (mmol m^{-2} s^{-1}); Sc: Stomatal conductance (mol m^{-2} s^{-1}); Z$_0$: No Zeolite; Z$_{15}$: Zeolite at 15 t ha^{-1}. Within treatments, different letters indicate significant differences at $p \leq 0.05$(otherwise statistically at par). Modified from Zheng et al. [32].

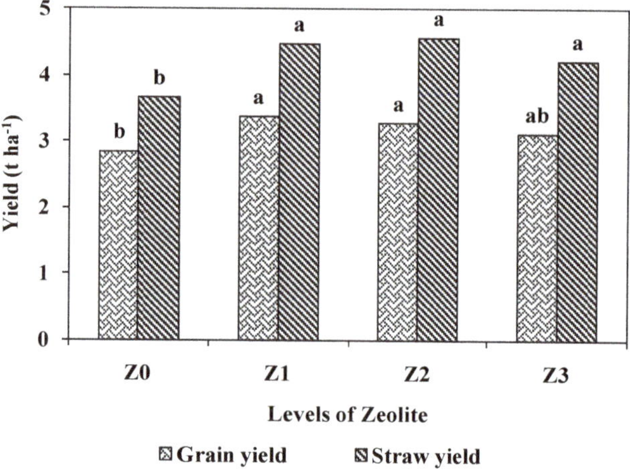

Figure 13. Effect of zeolite application on grain and straw yield of rice (Z$_0$: 0; Z$_8$: 8; Z$_{16}$: 16; Z$_{24}$: 24 t ha^{-1}) Within treatments, different letters indicate significant differences at $p \leq 0.05$(otherwise statistically at par). Modified from Kavoosi [150].

According to Wu et al. [151], the zeolites amendment significantly improved the root characteristics in terms of root length, dry weight, root diameter and volume, total root surface area, root bleeding intensity in rice plant over no zeolite application. Developed root traits may enhance nutrient transportation from the root to the above-ground parts and result in higher biomass and grain yield [152]. In previous studies, researchers confirmed that additional zeolites supply maximized the leaf area index (LAI) as well as leaf SPAD values and photosynthetic efficiency in rice plant, which might be attributed to its better ammonium retention capacity and slow-release nature that increase the better N availability to plants [53,72]. In a lowland rice production system, Sepaskhah and Barzegar [153] established the positive correlation between zeolites application and N retention in the upper soil profile. This higher availability favours better N uptake and subsequently higher nitrogen use efficiency (Figure 14). Zeolites induced rice cultivation resulted in greater apparent N recovery (65%) while 40% recovery was observed in exclusive N fertilization [150,154].

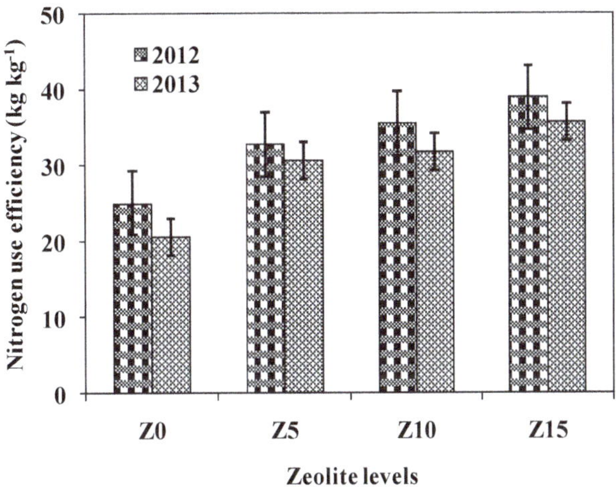

Figure 14. Nitrogen use Efficiency of Rice with Different Rates of Zeolite Application (Z_0: 0; Z_5: 5; Z_{10}: 10; Z_{15}: 15 t ha^{-1}). Modified from Chen et al. [29].

Zheng et al. [104] evaluated the consequence of zeolite and phosphorus applications in rice under different irrigation regimes and resulted in 15.2% higher water use efficiency (WUE) as well as greater leaf and stem P concentration by 20.3% and 32.7% respectively than no-zeolite control. The better water use efficiency may be attributed to higher soil water retention in the porous structure of zeolites and thus better water availability to plant [3,53]. Additionally, restriction in deep percolation and leaching beyond the crop root zone in zeolite loaded soil are major reasons for better water use efficiency [23,155].

The application of zeolite in maize cultivation was reported by Malekian et al. (2011) [156] who opined that maize plants resulted in better response to zeolite when used as a fertilizer carrier at the rate of 60 g kg^{-1} of soil. The application of clinoptilolite zeolite (CZ) with a 75% recommended dose of fertilizer resulted in significantly similar cobs yield in maize as compared to the full recommended dose of fertilizer [31]. A similar trend of observation was recorded regarding dry matter production and nutrient uptake, especially N and K uptake. It is possible due to the higher cation exchange capacity and affinity of CZ to NH_4^+ and K^+ ions. More specifically, reduced nitrification, prevention of leaching and volatilization by inhibiting ureolytic activity of microorganisms in the presence of CZ facilitate better nutrients availability [157]. Moreover, the cation selectivity of the CZ in the order to $K^+ > NH_4^+ > Na^+ > Ca^{2+} > Mg^{2+}$ supports to the aforesaid observation [31,158]. Increased nitrogen-use efficiency with the application of zeolites and ensured good retention of soil-exchangeable cations, available P and NO_3^- within the soil have been found by Rabai et al. [159] in maize cultivation. Low fertilizer requirement with zeolites application not only gives a similar yield but also reduces the environmental pollution in respect to nitrous oxide emission, with maintaining the economic viability. Andronikashvilf et al. [147] also suggested that the zeolite application facilitates a reduction in the recommended dose of fertilizer by 25% and maintains a positive effect for 2–3 years in upland crops production systems.

In high saline condition zeolite amendment in soil responded well in Barley crop and it was reported that zeolite at 5% level produced taller plants; accumulated maximum plant biomass and more grain yield over 1% and no zeolite application [16]. Similarly, in alkaline condition soil application of zeolites for French bean (*Phaseolus vulgaris* L.) cultivation maximized the nutrient accumulation in plant tissues. Additionally, better crop performance as well as greater water use efficiency, water productivity and crop yield were recorded from the zeolites treated plots [154]. Usually, the higher Na^+ content in alkaline and saline soils disturbs the soil nutritional balance and osmotic regulations in plant tissues. Zeolite provides additional Ca^{2+} cations in the soil to reduce the Na^+/Ca^{2+} ratio. The provision of Ca^{2+} from zeolite in the growing media would alleviate the toxic Na^+ ions accumulation and helps in the improvement of soil structure by aggregating the soil particles [16].

Not only in cereals and pulses zeolites have significant importance in oilseed crops. An additional supply of 10-ton zeolites ha^{-1} with recommended fertilizer significantly increased the seed and oil yield in safflower, accounted for 2.7 and 9.38 t ha^{-1} respectively [160]. Zahedi et al. [161] evaluated the effects of zeolite and selenium applications on some agronomic traits of three Canola cultivars under drought stress. They opined that stem diameter significantly decreased due to water stress, while the application of zeolite along with selenium improved stem diameter may be attributed to better water and nutrients availability from zeolites induced soil. They reported that 10 t zeolite ha^{-1} significantly improved the growth, yield attributes, and yield (Figure 15). They also observed reduced N leaching along with higher water holding capacity and CEC in alkaline soil when supplemented with 10 t zeolite as compared to normal soil. The oil yield and oil qualities such as palmitic acid, Oleic acid, Linoleic acid, Linolenic acid and Erucic acid of canola significantly improved with zeolite application (15 t ha^{-1}) rather than no zeolite use [28].

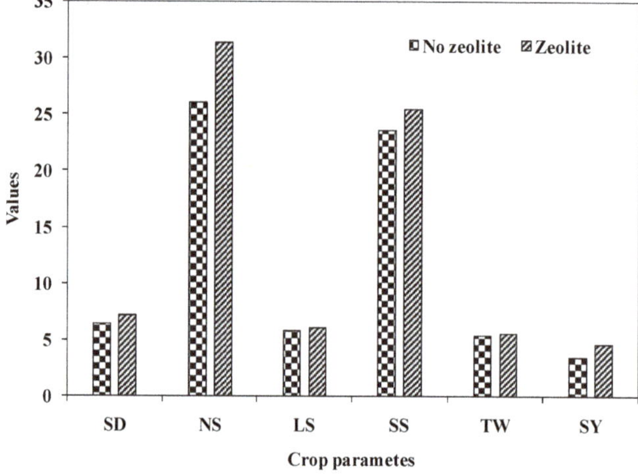

Figure 15. Effects of zeolite on some agronomic traits in canola. SD—Stem diameter (mm); NS—Number of siliquae; LS—Length of siliqua (cm); SS—Seeds per siliqua; TW—Test weight (g); SY—Seed yield (t ha^{-1}). Adapted and modified from Zahedi et al. [162].

From an experiment, the findings recorded by Ozbahce et al. [162] revealed that application of 60 t zeolite ha^{-1} along with proper irrigation and nutrient management, potato yielded (39.1 t ha^{-1}) maximum tubers (Figure 16). They also recorded superior crop performance even under limited water supply when treated with zeolite while non-zeolite traditional practices sharply decreased the tuber yield. The interaction of zeolites and irrigation regimes was found to be significant for tuber weight, tuber diameter and crude protein percentage.

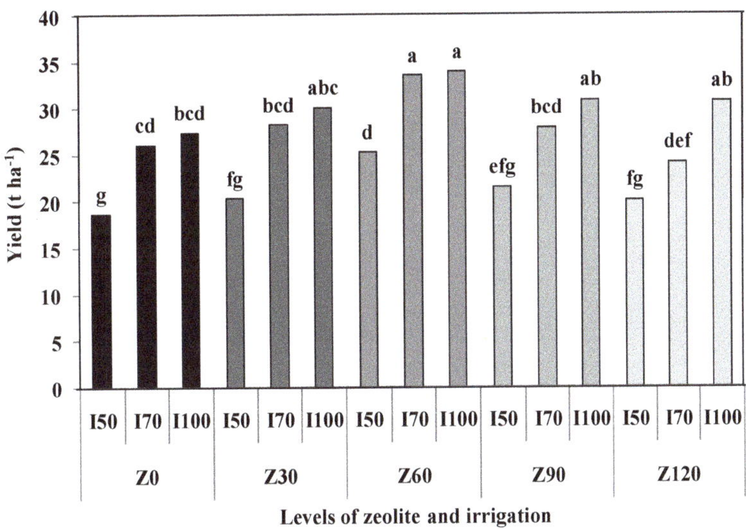

Figure 16. Effect of Zeolite Rates Eighth Different Levels of Irrigation on Tuber Yield of Potato (Z_0: 0; Z_{30}: 30; Z_{60}: 60; Z_{90}: 90; Z_{120}: 120 t ha^{-1}) Within treatments, different letters indicate significant differences at $p \leq 0.05$(otherwise statistically at par). (Adapted and modified from Ozbahce et al. [163]).

The effectiveness of zeolite on Peppermint (*Mentha piperita* L.) cultivation was reported by Ghanbari and Ariafar [30]. They opined that zeolite treatment significantly improved the fresh and dry leaf weight of mint and the highest value of fresh dry leaf weight was observed in 2.5 g zeolite application per kg of soil even under—water scare situation. They also observed that drought intensity was decreased with increasing the zeolite application. In 30% field capacity, zeolite application maximized the leaf dry weight from 18.54 to 32.76 g and fresh leaf weight from 41.7 to 67.14 g. Interestingly, zeolite helps to keep the essential components of mint oil such as menthol, menthone, methyl acetal, menthofuran and palegone [163,164]. Actually, these essential components are adversely affected by drought and salinity stress whereas, zeolites consist of alkali and alkaline materials and crystalline alumino—silicate which act as a water reservoir in their internal surface area during drought situation [165,166].

Numerous scientific reports were also concluded that significant positive influence on cocoa fruiting [167], eye numbers in potato tubers [160], pod and siliqua number in pulses and oilseeds [33,168], and overall development of soybean, sweet potato, wheat, bean and safflower with the application of soil—applied zeolites [160–171]. The use of Clinoptilolite—rich tuff as soil conditioner was found to be effective to improve the productivity of wheat, eggplant, carrots, and apples by 13–15%, 19–55%, 13–38% and 63% respectively [172]. Not only in field crops or vegetables, zeolite induced soil significantly improved the production as well as qualitative traits of mycelium mushroom [173]. The treatment with 30% zeolite + 70% urea resulted in a positive effect on the microbiological community in spring barley, soybean and maize [174]. Andronikashvili et al. [175] opined that the introduction of clinoptilolite containing tuffs into soils enhanced the soil microbial population viz. bacteria, fungi and actinomycetes.

Another interesting dimension of zeolite application was introduced by the National Aeronautics and Space Administration (NASA), which developed a special type of clinoptilolite loaded plant growth media including synthetic apatite, dolomite, and several essential trace nutrients mainly for vegetable production (10% higher than non—zeolite application) in space missions, known as 'zeoponic' [176]. Life support system for regenerating and recycling the air, water and food are essentially required for the long duration Mars mission and only the growing of plants could be fulfilled this aim. The ultimate objective of zeoponic research is to develop a solid substrate that can supply all essential macro and micronutrients slowly for a long duration in a space habitat. In an experiment, Gruener et al. [176] resulted in higher biomass accumulation, root and leaf development and nutrient uptake by radish when cultivated in zeoponic as compared to normal soil. Rodriguez—Fuentes et al. [177] reported that root architecture, plant growth and yield of different vegetables, spices and strawberries, were significantly improved by zeoponic substrates without further fertilization. Researchers confirmed that the native clinoptilolite in zeoponic acts as a good source of N and K as the clinoptilolite cations are exchanged for NH_4^+ and K^+ ions [102]. Additionally, apatite and dolomite dissolution supplies Ca^{2+} into soil solution. This Ca^{2+} rich solution removes the NH_4^+ and K^+ ions from zeolite exchange complex and makes them more available to plants [176]. Sometimes, nitrifying bacteria are supplemented to zeoponic substrates prior to plant growth to augment the nitrification process [178]. Since most zeolites are advantageous in the growth and development of crops, however, erionite (one type of zeolite) was found to be detrimental to the proper growth of plants [179]. Therefore, the selection of an appropriate form of zeolites should be taken into consideration.

3.8. Used as a Pesticide

Zeolites that contain silica gel and alumina silicate crystals have been successfully tested against some stored grain pests such as lesser grain borers (Rhyzopertha dominica), rice weevils (Sitophilus oryzae), and saw—toothed grain beetles (Oryzaephilus surinamensis) [180]. Natural zeolites application at the rate of 50 g kg^{-1} of maize grain were also found to be effective against maize weevil (*Sitophilus zeamais*) in accordance with Haryadi et al. [181]. Clinoptilolite was successfully investigated on organic oilseed rape fields against the pollen beetle (*Meligethes* sp.). Daniel et al. [182] observed that under dry and sunny weather condition, pollen beetles were significantly reduced by 50 to 80% with zeolite application while in rainy weather zeolite did not perform against pollen beetles. Zeolites loaded organophosphorus compound was used with success against the *Aedes aegyptii* [183]. Clinoptilolite is gaining importance as possible sorbents because it acts as a slow—release carrier and retard water contamination [184]. Clinoptilolite riched metalaxyl application on turfgrass against *Phythium* sp. resulted that the active ingredient of fungicide was prevented from groundwater contamination by clinoptilolite zeolite [185]. Actually, the adsorption of pesticide molecules is happened due to polar chemical bonds with the external surface of the microporous zeolitic minerals [108]. Additionally, the dusting of natural zeolites has been successfully tested to control the aphid population in fruit orchard [186]. Moreover, in herbicide application, pest control and in nano—sensing for pest detection, the nano—porous zeolites have been implicated as nano—capsules [187–189]. Stadler et al. [190] examined the insecticidal effect of nanostructured zeolites on two stored—grain insect species, *S. oryzae* and *R. dominica*, and found 80–100% mortality rate within 14 days after application to wheat grain. In this regard, natural zeolites may provide a cheap and reliable alternative to commercial insecticides in pest management. The insecticidal efficacy of natural zeolite on different stored grain pests is summarized in Table 4. Additional research is needed to investigate the mode of action, non—target toxicity, and the potential use in integrated pest control strategies.

Table 4. Efficacy of Natural Zeolites on Stored—Product Pests.

Tested Crop	Type of the NZ	Affecting Insects	Reference
Rice		*Oryzaephilus mercator*	Eroglu et al. [191]
Wheat	Minazel plus	*Rhyzopertha dominica* *Sitophilus oryzae* *Tribolium castaneum*	Kljajic et al. [192]
Maize		*Sitophilus zeamais* *Sitophilus oryzae*	Haryadi et al. [181]
Chickpea		*Lasioderma serricorne*	Perez et al. [193]
Oilseed (Rapeseed)	Klinofeed (dust)	*Meligethes* sp.	Daniel et al. [182]

3.9. Mycotoxin Control

The use of aluminosilicates such as zeolites has emerged as a mycotoxin—binding agent in the feed and food industry to effectively adsorb mycotoxin [194]. Clinoptilolite has the capacity to adsorb aflatoxins by chelating of the β—dicarbonylmoiety in aflatoxin with uncoordinated metal ions [195]. There are some well—established criteria to evaluate the function of any binding additive, such as low inclusion rate, stability over a wide range of pH, huge capacity, and affinity to absorb various concentrations of mycotoxins [194]. The supplementation of mycotoxin binders in contaminated foods has been suggested as the most advantageous dietary approach to lower the mycotoxins efficacy [196]. Hydrated sodium calcium alumino—silicates—zeolite powder (HSCAS) has been identified as "aflatoxin—selective clay", but it does not adsorb other mycotoxins such as cyclopiazonic acid which may coexist with aflatoxin [197] while responses seem to be dose—dependent [198]. Parlat et al. [199] observed that clinoptilolite could successfully minimize the effects of aflatoxin in quail. Natural zeolites with high clinoptilolite content (over 80%) effectively adsorbed aflatoxin B1, aflatoxin B2, and aflatoxin G2 [200]. On the contrary, surface modified zeolites with NH_4^+ showed very well adsorption of ochratoxin A, T—2 toxin, zearalenone and aflatoxin B1 [201]. According to Adamovic et al. [202], the application of zeolites at 2 g kg^{-1} of silage accelerates the fermentation and reduction of T—2 toxin, mould and zearalenone. Zeolites application as mycotoxin binder is impressive against aflatoxicosis, however, their effectiveness against trichothecenes, zearalenone and ochratoxin is restricted. At the same time, these compounds show high inclusion rates for vitamins and minerals, which are considered as one of the major disadvantages [203].

4. Limitation of Zeolites

Rather than the huge applicability of zeolites in agriculture, it should be considered that the zeolites are not without disadvantages. The fine—grained synthetic zeolites are highly dispersive in nature which creates worrisome problems during their use. After mining the usable form of natural zeolites is obtained via isolation procedures like crushing and pellet generation while the application of the synthetic form of zeolites are limited into hard, wear—resistant granular forms. The practical use of granular zeolites is not yet discovered [204]. The distribution of the zeolites sources is very limited such asthezeolitic soil is confined to only 1% of the total geographic area of India and more than 50% of natural zeolites are produced in China among all over the world [40] that may increase the price and the gap between demand and supply. Therefore, the uninterrupted availability of zeolites for farming purposes in worldwide is another major constraint.

5. Future Scope

The significant application of zeolites in agricultural activities has been well established by various researchers. However, systematic and comprehensive efforts are further needed for future research, including (a) precision mapping of the available zeolite deposits in each country, (b) determination of the physical stability of zeolites in various agro-climatic conditions, (c) economically viable organo-zeolitic manure or fertilizer development, (d) evaluation of the risk of leaching of a toxic surfactant that is loosely attached to the zeolite surface, (e) assessment of the long-term impact of zeolite application on rhizospheric microflora and fauna, (f) understanding of the mechanisms of zeolite-mediated heavy metal stabilisation in contaminated soil and (h) development of zeolite-rich herbicides to minimise the residual risk hazard.

6. Conclusions

In the situation of rapid urbanisation and over-increasing population where resources are limited, there is no choice for us but to depend on agricultural productivity. In this context, various researchers suggest that farming with zeolites may be an option to improve soil's physical environments in terms of decreasing bulk density, increasing total porosity and increasing water-holding capacity. Furthermore, the existence of open networks in the zeolite structure leads to the formation of new routes for water movement, subsequently improving the infiltration rate and saturated hydraulic conductivity. Zeolites also show a strong affinity to various essential nutrient ions by modifying their surface chemistries using cationic surfactants, multifunctional adsorbents that have the capacity to trap anions and non-polar organics. Thus, the application of zeolite-loaded fertilizer improves the nutrient retention in soil and releases nutrients slowly throughout the crop life; otherwise, rapid mineralisation would take place, leading to nutrient loss. Zeolites are very much effective in remediation of heavy metal toxicity and wastewater treatment, and they could help to improve soil's biological properties. Zeolite application in space missions as zeoponic substrates opens a new dimension of zeolites. The aforesaid positive impacts ultimately enhance crop growth, productivity and even quality attributes of various agronomic and horticultural crops. The higher input use efficiency significantly reduces greenhouse gas emissions and energy involvement and facilitates better carbon sequestration. However, the impact of zeolite application varies with the agro-climatic location, the nature of zeolites, their availability and application strategies, and soil textural classes. Further studies are needed to identify zeolite resources and the long-term impact on the soil environment and to develop new, cost-effective zeolite-based nutrient resources for sound agricultural practices.

Author Contributions: Conceptualization, M.M.; B.B.; and S.G.; writing—original draft preparation, M.M.; S.G.; S.S.; H.B.; and A.H.; writing—review and editing, B.B.; S.M.; K.B.; P.K.B.; H.B., A.H., M.S., P.O.; and M.B.; funding acquisition, P.O., M.S., A.H. and M.B. All authors have read and agreed to the published version of the manuscript.

Funding: This research was funded by the 'Slovak University of Agriculture', Nitra, Tr. A. Hlinku 2,949 01 Nitra, Slovak Republic under the project 'APVV—18—0465 and EPPN2020—OPVaI—VA—ITMS313011T813'.

Conflicts of Interest: The authors declare no conflict of interest.

References

1. Alexandratos, N.; Bruinsma, J. World Agriculture Towards 2030/2050. Food and Agriculture Organization of the United Nations. 2012. Available online: www.fao.org/economic/esa (accessed on 10 February 2021).
2. Bhatt, R.; Singh, P.; Hossain, A.; Timsina, J. Rice–wheat system in the northwest Indo-Gangetic plains of South Asia: Issues and technological interventions for increasing productivity and sustainability. *Paddy Water Environ.* **2021**, 1–21. [CrossRef]
3. Xiubin, H. and Zhanbin, H. Zeolite application for enhancing water infiltration and retention in loess soil. *Resour. Conserv. Recycl.* **2001**, *34*, 45–52. [CrossRef]
4. Wu, Q.; Chi, D.; Xia, G.; Chen, T.; Sun, Y.; Song, Y. Effects of Zeolite on Drought Resistance and Water–Nitrogen Use Efficiency in Paddy Rice. *J. Irrig. Drain. Eng.* **2019**, *145*, 04019024. [CrossRef]
5. Garai, S.; Mondal, M.; Mukherjee, S. Smart Practices and Adaptive Technologies for Climate Resilient Agriculture. In *Advanced Agriculture*, 1st ed.; Maitra, S., Pramanick, B., Eds.; New Delhi Publishers: Kolkata, India, 2020; pp. 327–358.
6. Ming, D.W.; Mumpton, F.A. Zeolites in soils. *Miner. Soil Environ.* **1989**, *1*, 873–911. [CrossRef]
7. Peña-Haro, S.; Llopis-Albert, C.; Pulido-Velazquez, M.; Pulido-Velazquez, D. Fertilizer standards for controlling groundwater nitrate pollution from agriculture: El Salobral-Los Llanos case study, Spain. *J. Hydrol.* **2010**, *392*, 174–187. [CrossRef]

8. Hosono, T.; Tokunaga, T.; Kagabu, M.; Nakata, H.; Orishikida, T.; Lin, I.; Shimada, J. The use of δ 15 N and δ 18 O tracers with an understanding of groundwater flow dynamics for evaluating the origins and attenuation mechanisms of nitrate pollution. *Water Res.* **2013**, *47*, 2661–2675. [CrossRef] [PubMed]

9. Aschebrook-Kilfoy, B.; Shu, X.; Gao, Y.; Ji, B.; Yang, G.; Li, H.L.; Rothman, N.; Chow, W.; Zheng, W.; Ward, M.H. Thyroid cancer risk and dietary nitrate and nitrite intake in the Shanghai women's health study. *Int. J. Cancer* **2013**, *132*, 897–904. [CrossRef]

10. Mondal, M.; Garai, S.; Banerjee, H.; Sarkar, S.; Kundu, R. Mulching and nitrogen management in peanut cultivation: An evaluation of productivity, energy trade-off, carbon footprint and profitability. *Energy Ecol. Environ.* **2020**. [CrossRef]

11. Delkash, M.; Al-Faraj, F.A.; Scholz, M. Comparing the export coefficient approach with the soil and water assessment tool to predict phosphorous pollution: The Kan watershed case study. *Water Air Soil Pollut.* **2014**, *225*, 1–17. [CrossRef]

12. Bakhshayesh, B.E.; Delkash, M.; Scholz, M. Response of vegetables to cadmium-enriched soil. *Water* **2014**, *6*, 1246–1256. [CrossRef]

13. Nakhli, S.A.; Ahmadizadeh, K.; Fereshtehnejad, M.; Rostami, M.H.; Safari, M.; Borghei, S.M. Biological removal of phenol from saline wastewater using a moving bed biofilm reactor containing acclimated mixed consortia. *SpringerPlus* **2014**, *3*, 112. [CrossRef]

14. Leggo, P.J. The efficacy of the organo-zeolitic bio-fertilizer. *Agrotechnology* **2015**, *4*, 1000130. [CrossRef]

15. Mahabadi, A.A.; Hajabbasi, M.; Khademi, H.; Kazemian, H. Soil cadmium stabilization using an Iranian natural zeolite. *Geoderma* **2007**, *137*, 388–393. [CrossRef]

16. Al-Busaidi, A.; Yamamoto, T.; Inoue, M.; Eneji, A.E.; Mori, Y.; Irshad, M. Effects of zeolite on soil nutrients and growth of barley following irrigation with saline water. *J. Plant. Nutr.* **2008**, *31*, 1159–1173. [CrossRef]

17. Sarkar, B.; Naidu, R. Nutrient and water use efficiency in soil: The influence of geological mineral amendments. In *Nutrient Use Efficiency: From Basics to Advances*; Rakshit, A., Singh, H.B., Sen, A., Eds.; Springer: New Delhi, India, 2015; pp. 29–44.

18. Jha, B.; Singh, D.N. *Fly Ash Zeolites: Innovations, Applications, and Directions*; Springer: Singapore, 2016; pp. 33–51.

19. Virta, R.L. Zeolites. In *Us Geological Survey Minerals Yearbook*; US Geological Survey: Reston, VA, USA, 2002; pp. 84.1–84.4.

20. Elliot, A.D.; Zhang, D. Controlled release zeolite fertilisers: A value-added product produced from fly ash. *Bioprocess. Technol* **2005**, *4*, 39–47.

21. Ober, J.A. Mineral commodity summaries. *US Geol. Surv.* **2017**, 202. [CrossRef]

22. Inglezakis, V.; Elaiopoulos, K.; Aggelatou, V.; Zorpas, A.A. Treatment of underground water in open flow and closed—loop fixed bed systems by utilizing the natural minerals clinoptilolite and vermiculite. *Desalination Water Treat.* **2012**, *39*, 215–227. [CrossRef]

23. Talebnezhad, R.; Sepaskhah, A.R. Effects of bentonite on water infiltration in a loamy sand soil. *Arch. Agron. Soil Sci.* **2013**, *59*, 1409–1418. [CrossRef]

24. Chmielewska, E. Zeolitic adsorption in course of pollutants mitigation and environmental control. *J. Radioanal. Nucl. Chem.* **2014**, *299*, 255–260. [CrossRef]

25. Ebrazi, B.; Banihabib, M.E. Simulation of Ca^{2+} and Mg^{2+} removal process in fixed-bed column of natural zeolite. *Desalination Water Treat.* **2015**, *55*, 1116–1124. [CrossRef]

26. Enamorado-Horrutiner, Y.; Villanueva-Tagle, M.; Behar, M.; Rodríguez-Fuentes, G.; Dias, J.F.; Pomares-Alfonso, M. Cuban zeolite for lead sorption: Application for water decontamination and metal quantification in water using nondestructive techniques. *Int. J. Environ. Sci. Technol.* **2016**, *13*, 1245–1256. [CrossRef]

27. Perez-Caballero, R.; Gil, J.; Benitez, C.; Gonzalez, J.L. The effect of adding zeolite to soils in order to improve the N-K nutrition of olive trees, preliminary results. *Am. J. Agric. Biol. Sci.* **2008**, *2*, 321–324.

28. Shahsavari, N.; Jais, H.M.; Shirani Rad, A.H. Responses of canola oil quality characteristics and fatty acid composition to zeolite and zinc fertilization under drought stress. *Int. J. Agric. Sci.* **2014**, *4*, 49–59.

29. Chen, T.; Guimin, X.; Qi, W.; Zheng, J.; Jin, Y.; Sun, D.; Wang, S.; Chi, D. The Influence of Zeolite Amendment on Yield Performance, Quality Characteristics, and Nitrogen Use Efficiency of Paddy Rice. *Crop. Sci.* **2017**, *57*, 1–11. [CrossRef]

30. Ghanbari, M.; Ariafar, S. The effect of water deficit and zeolite application on Growth Traits and Oil Yield of Medicinal Peppermint (*Mentha piperita* L.). *Int. J. Med. Arom. Plants* **2013**, *3*, 33–39.

31. Aainaa, H.N.; Ahmed, O.H.; Majid, N.M.A. Effects of clinoptilolite zeolite on phosphorus dynamics and yield of *Zea Mays* L. cultivated on an acid soil. *PLoS ONE* **2018**, *13*, e0204401. [CrossRef]

32. Zheng, J.; Chen, T.; Wu, Q.; Yu, J.; Chen, W.; Chen, Y. Effect of zeolite application on phenology, grain yield and grain quality in rice under water stress. *Agric. Water Manag.* **2018**, *206*, 241–251. [CrossRef]

33. Hazrati, S.; Tahmasebi-Sarvestani, Z.; Mokhtassi-Bidgoli, A.; Modarres-Sanavy, S.A.M.; Mohammadi, H.; Nicola, S. Effects of zeolite and water stress on growth, yield and chemical compositions of *Aloe vera* L. *Agric. Water Manag.* **2017**, *181*, 66–72. [CrossRef]

34. Kazemian, H.; Mallah, M.H. Elimination of Cd2 and Mn2 from wastewaters using natural clinoptilolite and synthetic zeolite P. *Iran. J. Chem. Chem. Eng.* **2006**, *25*, 91–94.

35. Kazemian, H.; Mallah, M. Removal of chromate ion from contaminated synthetic water using MCM-41/ZSM-5 composite. *J. Environ. Health Sci. Eng.* **2008**, *5*, 73–77.

36. Youssefi, S.; Waring, M.S. Indoor transient SOA formation from ozone α-pinene reactions: Impacts of air exchange and initial product concentrations, and comparison to limonene ozonolysis. *Atmos. Environ.* **2015**, *112*, 106–115. [CrossRef]

37. Scopus Preview. Available online: https://www.scopus.com/home.uri (accessed on 7 November 2020).

38. Polat, E.; Karaca, M.; Demir, H.; Naci-Onus, A. Use of natural zeolite (clinoptilolite) in agriculture. *J. Fruit Ornam. Plant. Res.* **2004**, *12*, 183–189.

39. USGS. Mineral Commodity Summaries: Zeolites (Natural). 2016. Available online: http://minerals.usgs.gov/minerals/pubs/commodity/zeolites/index.html (accessed on 31 December 2020).

40. Bhattacharyya, T.; Chandran, P.; Ray, S.K.; Pal, D.K.; Mandal, C.; Mandal, D.K. Distribution of zeolitic soils in India. *Curr. Sci.* **2015**, *109*, 1305–1313. [CrossRef]

41. Nakhli, S.A.A.; Delkash, M.; Bakhshayesh, B.E.; Kazemian, H. Application of Zeolites for Sustainable Agriculture: A Review on Water and Nutrient Retention. *Water Air. Soil Pollut.* **2017**, *228*, 464. [CrossRef]

42. Sangeetha, C.; Baskar, P. Zeolite and its potential uses in agriculture: A critical review. *Agric. Rev.* **2016**, *37*, 101–108. [CrossRef]

43. International Zeolite Association. Available online: http://www.iza-online.org (accessed on 7 November 2020).

44. Hemingway, B.S.; Robie, R.A. Thermodynamic properties of zeolites: Low-temperature heat capacities and thermodynamic functions for phillipsite and clinoptilolite. Estimates of the thermo-chemical properties of zeolitic water at low temperature. *Am. Miner.* **1984**, *69*, 692–700.

45. Chmielewska, E.; Lesný, J. Selective ion exchange onto Slovakian natural zeolites in aqueous solutions. *J. Radioanal. Nucl. Chem.* **2012**, *293*, 535–543. [CrossRef]

46. Stylianou, M.; Inglezakis, V.; Loizidou, M. Comparison of Mn, Zn, and Cr removal in fluidized-and fixed-bed reactors by using clinoptilolite. *Desalination Water Treat.* **2015**, *53*, 3355–3362. [CrossRef]

47. Flanigen, E.M. *Introduction to Zeolite Science and Practice*; Elsevier: Amsterdam, The Netherlands, 2001; pp. 11–35.

48. Bittelli, M.; Campbell, G.S.; Tomei, F. *Soil Physics with Python: Transport in the Soil-Plant. Atmosphere System*; OUP: Oxford, UK, 2015. [CrossRef]

49. Ramesh, K.; Reddy, D.D.; Biswas, A.K.; Subbarao, A. Potential uses of zeolite in agriculture. *Adv. Agron.* **2011**, *113*, 215–236.

50. Torkashvand, A.M.; Shadparvar, V. Effect of some organic waste and zeolite on water holding capacity and PWP delay of soil. *Curr. Biot.* **2013**, *6*, 459–465.

51. Valadabadi, S.A.; Shiranirad, A.H.; Farahani, H.A. Ecophysiological influences of zeolite and selenium on water deficit stress tolerance in different rapeseed cultivars. *J. Ecol. Nat. Environ.* **2010**, *2*, 154–159.

52. Colombani, N.; Mastrocicco, M.; Di Giuseppe, D.; Faccini, B.; Coltorti, M. Variation of the hydraulic properties and solute transport mechanisms in a silty-clay soil amended with natural zeolites. *Catena* **2014**, *123*, 195–204. [CrossRef]

53. Gholamhoseini, M.; Ghalavand, A.; Khodaei-Joghan, A.; Dolatabadian, A.; Zakikhani, H.; Farmanbar, E. Zeolite amended cattle manure effects on sunflower yield, seed quality, water use efficiency and nutrient leaching. *Soil Tillage Res.* **2013**, *126*, 193–202. [CrossRef]

54. Bernardi, A.C.C.; Oliveria, P.P.A.; Barros, F. Brazilian sedimentary zeolite use in agriculture. *Microporous Mesoporous Mater.* **2013**, *17*, 16–21. [CrossRef]

55. Ghazavi, R. The application effects of natural zeolite on soil runoff, soil drainage and some chemical soil properties in arid land area. *Int. J. Innov. Appl. Stud.* **2015**, *13*, 172–177.

56. Aminiyan, M.M.; Sinegani, A.A.S.; Sheklabadi, M. Aggregation stability and organic carbon fraction in a soil amended with some plant residues, nanozeolite, and natural zeolite. *Int. J. Recycl. Org. Waste Agric.* **2015**, *4*, 11–22. [CrossRef]

57. Moritani, S.; Yamamoto, T.; Andry, H.; Inoue, M.; Yuya, A.; Kaneuchi, T. Effectiveness of artificial zeolite amendment in improving the physicochemical properties of saline-sodic soils characterised by different clay mineralogies. *Aust. J. Soil Res.* **2010**, *48*, 470–479. [CrossRef]

58. Cairo, P.C.; De Armas, J.M.; Artiles, P.T.; Martin, B.D.; Carrazana, R.J.; Lopez, O.R. Effects of zeolite and organic fertilizers on soil quality and yield of sugarcane. *Aust. J. Crop. Sci.* **2017**, *11*, 733–738. [CrossRef]

59. Sepaskhah, A.R.; Yousefi, F. Effects of zeolite application on nitrate and ammonium retention of a loamy soil under saturated conditions. *Aust. J. Soil Res.* **2007**, *45*, 368–373. [CrossRef]

60. Gholizadeh-Sarabi, S.; Sepaskhah, A.R. Effect of zeolite and saline water application on saturated hydraulic conductivity and infiltration in different soil textures. *Arch. Agron. Soil Sci.* **2013**, *59*, 753–764. [CrossRef]

61. Razmi, Z.; Sepaskhah, A.R. Effect of zeolite on saturated hydraulic conductivity and crack behavior of silty clay paddled soil. *Arch. Agron. Soil Sci.* **2012**, *58*, 805–816. [CrossRef]

62. Githinji, L.J.; Dane, J.H.; Walker, R.H. Physical and hydraulic properties of inorganic amendments and modeling their effects on water movement in sand-based root zones. *Irrig. Sci.* **2011**, *29*, 65–77. [CrossRef]

63. Aiyuk, S.; Xu, H.; Van Haandel, A. Removal of ammonium nitrogen from pretreated domestic sewage using a natural ion exchanger. *Environ. Technol.* **2004**, *25*, 1321–1330. [CrossRef]

64. Englert, A.; Rubio, J. Characterization and environmental application of a Chilean natural zeolite. *Int. J. Min. Process.* **2005**, *75*, 21–29. [CrossRef]

65. Sprynskyy, M.; Lebedynets, M.; Terzyk, A.P.; Kowalczyk, P.; Namieśnik, J.; Buszewski, B. Ammonium sorption from aqueous solutions by the natural zeolite Transcarpathian clinoptilolite studied under dynamic conditions. *J. Colloid Interface Sci.* **2005**, *284*, 408–415. [CrossRef] [PubMed]

66. Zabochnicka-Swiatek, M. Factors Affecting Adsorption Capacity and Ion Exchange Selectivity of Clinoptilolite towards heavy metal cations. *Inz. I Ochr.* **2007**, *10*, 27.

67. Jha, V.K.; Hayashi, S. Modification on natural clinoptilolite zeolite for its NH_4^+ retention capacity. *J. Hazard. Mater.* **2009**, *169*, 29–35. [CrossRef] [PubMed]

68. Mazloomi, F.; Jalali, M. Ammonium removal from aqueous solutions by natural Iranian zeolite in the presence of organic acids, cations and anions. *J. Environ. Chem. Eng.* **2016**, *4*, 240–249. [CrossRef]

69. Gorre, K.; Yenumula, S.; Himabindu, V.A. Study on the Ammonium Adsorption by using Natural Heulandite and Salt Activated Heulandite. *Int. J. Innov. Appl. Stud.* **2016**, *14*, 483–488.

70. Shaikh, I.I.; Chendaku, Y.J. Removal of ammonium nitrate from aquaculture by sorption using zeolite. *Int. J. Recent Sci. Res.* **2016**, *7*, 11869–11874.

71. Liang, Z.; Ni, J. Improving the ammonium ion uptake onto natural zeolite by using an integrated modification process. *J. Hazard. Mater.* **2008**, *166*, 52–60. [CrossRef]

72. Aghaalikhani, M.; Gholamhoseini, M.; Dolatabadian, A.; Khodaei-Joghan, A.; Sadat, A.K. Zeolite influences on nitrate leaching, nitrogen-use efficiency, yield and yield components of canola in sandy soil. *Arch. Agron. Soil Sci.* **2012**, *58*, 1149–1169. [CrossRef]

73. Vilcek, J.; Torma, S.; Adamisin, P.; Hronec, O. Nitrogen sorption and its release in the soil after zeolite application. *Bulg. J. Agric. Sci.* **2013**, *9*, 228–234.

74. Omar, L.; Ahmed, O.H.; Majid, N.M.A. Improving Ammonium and Nitrate release from Urea using Clinoptilolite zeolite and compost produced from agricultural wastes. *Sci. World J.* **2015**, *574201*, 1–12. [CrossRef]

75. Haruna, A.O.; Husin, A.; Husni, M.H.A. Ammonia volatilization and ammonium accumulation from urea mixed with zeolite and triple superphosphate. *Acta Agric. Scand B Soil Plant. Sci.* **2008**, *58*, 182–186. [CrossRef]

76. Yore, N.; Kalpana, I.; Saya, T. Reducing ammonia volatile concentration from compound fertilizers modified with zeolite. *Int. J. Manures Fert.* **2013**, *2*, 352–354.

77. Rech, I.; Polidoro, J.C.; Pavinato, P.S. Additives incorporated into urea to reduce nitrogen losses after application to the soil. *Pesq. Agropec. Bras.* **2017**, *52*, 194–204. [CrossRef]

78. Zaman, M.; Nguyen, M.L.; Šimek, M.; Nawaz, S.; Khan, M.J.; Babar, M.N.; Zaman, S. Emissions of nitrous oxide (N_2O) and di-nitrogen (N_2) from the agricultural landscapes, sources, sinks, and factors affecting N_2O and N_2 ratios. In *Greenhouse Gases-Emission, Measurement and Management*; Liu, G., Ed.; InTech Europe: Rijeka, Croatia, 2012; published online 14, March, 2012, published in print edition March, 2012; pp. 1–32.

79. Li, J.; Wee, C.; Sohn, B. Effect of Ammonium- and Potassium-Loaded Zeolite on Kale, Growth and Soil Property. *Amer. J. Plant. Sci.* **2013**, *4*, 1976–1982. [CrossRef]

80. Behzadfar, M.; Sadeghi, S.H.; Khanjani, M.J.; Hazbavi, Z. Effects of rates and time of zeolite application on controlling runoff generation and soil loss from a soil subjected to a freeze-thaw cycle. *Int. Soil Water Conserv. Res.* **2017**, *5*, 95–101. [CrossRef]

81. Ahmed, O.; Aminuddin, H.; Husni, M. Reducing ammonia loss from urea and improving soil exchangeable ammonium retention through mixing triple super-phosphate, humic acid and zeolite. *Soil Use Manag.* **2006**, *22*, 315–319. [CrossRef]

82. Palanivell, P.; Ahmed, O.H.; Ab Majid, N.M. Minimizing ammonia volatilization from urea, improving lowland rice (cv. MR219) seed germination, plant growth variables, nutrient uptake, and nutrient recovery using clinoptilolite zeolite. *Arch. Agron. Soil Sci.* **2016**, *62*, 708–724. [CrossRef]

83. Omar, O.L.; Ahmed, O.H.; Muhamad, A.N. Minimizing ammonia volatilization in waterlogged soils through mixing of urea with zeolite and sago waste water. *Int. J. Phys. Sci.* **2010**, *5*, 2193–2197.

84. Manikandan, A.; Subramanian, K.S. Fabrication and characterisation of nanoporouszeolite based N fertilizer. *Afr. J. Agric. Res.* **2014**, *9*, 276–284.

85. Ramesh, K.; Biswas, A.K.; Somasundaram, J.; Subbarao, A. Nanoporous zeolites in farming: Current status and issues ahead. *Curr. Sci.* **2010**, *99*, 760–765.

86. Ge, J.; Wu, R.; Shi, X.H.; Yu, H.; Wang, M.; Wen, L. Biodegradable polyurethane materials from bark and starch. II. Coating materials for controlled release fertilizer. *J. Appl. Polym. Sci.* **2002**, *86*, 2948–2952. [CrossRef]

87. Huang, Z.T.; Petrovic, A.M. Physical properties of sand as affected by clinoptilolite zeolite particle size and quantity. *J. Turfgrass Manag.* **1994**, *1*, 1–15. [CrossRef]

88. Zwingmann, N.; Singh, B.; Mackinnon, I.D.; Gilkes, R.J. Zeolite from alkali modified kaolin increases NH 4 retention by sandy soil: Column experiments. *Appl. Clay Sci.* **2009**, *46*, 7–12. [CrossRef]

89. MacKown, C.; Tucker, T. Ammonium nitrogen movement in a coarse-textured soil amended with zeolite. *Soil Sci. Soc. Am. J.* **1985**, *49*, 235–238. [CrossRef]

90. Gholamhoseini, M.; AghaAlikhani, M.; Dolatabadian, A.; Khodaei-Joghan, A.; Zakikhani, H. Decreasing nitrogen leaching and increasing canola forage yield in a sandy soil by application of natural zeolite. *Agron. J.* **2012**, *104*, 1467–1475. [CrossRef]

91. Taheri-Sodejani, H.; Ghobadinia, M.; Tabatabaei, S.; Kazemian, H. Using natural zeolite for contamination reduction of agricultural soil irrigated with treated urban wastewater. *Desalin. Water Treat.* **2015**, *54*, 2723–2730. [CrossRef]

92. Piñón-Villarreal, A.R.; Bawazir, A.S.; Shukla, M.K.; Hanson, A.T. Retention and transport of nitrate and ammonium in loamy sand amended with clinoptilolite zeolite. *J. Irrig. Drain. Eng.* **2013**, *139*, 755–765. [CrossRef]

93. Baerlocher, C.; McCusker, L.B.; Olson, D.H. *Atlas of Zeolite Framework Types*; Elsevier: Amsterdam, The Netherlands, 2007.

94. Waldrip, H.M.; Todd, R.W.; Cole, N.A. Can surface applied zeolite reduced ammonia losses from feedyard manure? In Proceedings of the ASA, CSA and SSSA International Annual Meetings, Long Beach, CA, USA, 2–5 November 2014.

95. Ramesh, K.; Islam, R. Zeolites regulate nitrogen release from manure amended soil. In Proceedings of the ASA, CSA and SSSA International Annual Meetings, Cincinnati, OH, USA, 21–24 October 2012.

96. Cai, L.; Koziel, J.A.; Liang, Y.; Nguyen, A.T.; Xin, H. Evaluation of zeolite for control of odorants emission from simulated poultry manure storage. *J. Environ. Qual.* **2007**, *36*, 184–193. [CrossRef]

97. Sharadqah, S.I.; Al-Dwairi, R.A. Control of odorants emissions from poultry manure using Jordanian natural zeolites. *Jordan J. Civ. Eng.* **2010**, *4*, 378–388.

98. Murnane, J.G.; Brennan, R.B.; Healy, M.G.; Fenton, O. Use of Zeolite with Alum and Polyaluminum Chloride amendments to mitigate runoff losses of phosphorus, nitrogen, and suspended solids from agricultural wastes applied to grassed soils. *J. Environ. Qual.* **2015**, *44*, 1674–1683. [CrossRef]

99. Latifah, O.; Ahmed, O.H.; Majid, N.M.A. Short Term Enhancement of Nutrients Availability in Zea mays L. Cultivation on an Acid Soil Using Compost and Clinoptilolite Zeolite. *Compos. Sci. Util.* **2016**, *25*, 22–35. [CrossRef]

100. Moharami, S.; Jalali, M. Phosphorus leaching from a sandy soil in the presence of modified and un—modified adsorbents. *Environ. Monit. Assess.* **2014**, *186*, 6565–6576. [CrossRef]

101. Shokouhi, A.; Parsinejad, M.; Noory, H. Impact of zeolite and soil moisture on P uptake. In Proceedings of the 2nd Berlin-European SustainablePhosphorus Conference (ESPC 2), Berlin, Germany, 5–6 March 2015.

102. Allen, E.R.; Hossner, L.R.; Ming, D.W.; Henninger, D.L. Solubility and cation exchange in phosphate rock and saturated clinoptilolite mixtures. *Soil Sci. Soc. Am. J.* **1993**, *57*, 1368–1374. [CrossRef]

103. Mihajloviae, M.; Perisiae, N.; Pezo, L.; Stojanoviar, M.; Milojkoviae, J.; Petroviae, M.; Petroviae, J. Optimization of process parameters to obtain NH$_4$-Clinoptilolite as a supplement to ecological fertilizer. *Clay Min.* **2014**, *49*, 735–745. [CrossRef]

104. Zheng, J.; Chen, T.; Chi, D.; Xia, G.; Wu, Q.; Liu, G.; Chen, W.; Meng, W.; Chen, Y.; Siddique, K.H.M. Influence of Zeolite and Phosphorus Applications on Water Use, P Uptake and Yield in Rice under Different Irrigation Managements. *Agronomy* **2019**, *9*, 537. [CrossRef]

105. Antoniadis, V.; Damalidis, K.; Koutrobas, S.D. Nitrogen and phosphorus availability to ryegrass in acidic and limed zeolite-amended soil. *Agrochim. Pisa* **2012**, *56*, 309–318.

106. Bansiwal, A.K.; Rayalu, S.S.; Labhasetwar, N.K.; Juwarkar, A.A.; Devotta, S. Surfactant-modified zeolite as a slow release fertilizer for phosphorus. *J. Agric. Food Chem.* **2006**, *54*, 4773–4779. [CrossRef]

107. Rivero, L.; Rodríguez-Fuentes, G. Cuban experience with the use of natural zeolite substrates in soilless culture. *Proc. Int. Congr. Soil. Cult. Wagening. Neth.* **1988**, 405–416.

108. Ming, D.W.; Allen, E.R. Natural Zeolites Occurrence Properties, Applications. In *Reviews in Mineralogy and Geochemistry*; Bish, D.L., Ming, D.W., Eds.; Mineralogical Society of America: Washington, DC, USA, 2001; Volume 45, pp. 619–654.

109. Moraetis, D.; Papagiannidou, S.; Pratikakis, A.; Pentari, D.; Komnitsas, K. Effect of zeolite application on potassium release in sandy soils amended with municipal compost. *Desalin. Water Treat.* **2016**, *57*, 13273–13284. [CrossRef]

110. Williams, K.A.; Nelson, P.V. Using precharged zeolite as a source of potassium and phosphate in a soilless container medium during potted chrysanthemum production. *J. Am. Soc. Hortic. Sci.* **1997**, *122*, 703–708. [CrossRef]

111. Kamrudin, K.S.; Hamdan, H.; Mat, H. Methane adsorption characteristic dependency on zeolitic structures and properties. In Proceedings of the 17 th symposium of Malaysian chemical Engineers, Copthorne Orchid Hotel, Penang, Malaysia, 29–30 December 2003.

112. Lavicoli, I.; Leso, V.; Ricciardi, W.; Hodson, L.L.; Hoover, M.D. Opportunities and challenges of nanotechnology in the green economy. *Environ. Health* **2014**, *13*, 78. [CrossRef]

113. Sulakhudin, A.S.; Sunarminto, B.H. Zeolite and hucalcia as coating material for improving quality of NPK fertilizer in costal sandy soil. *J. Trop. Soil* **2011**, *16*, 99–106. [CrossRef]

114. Li, Z.; Zhang, Y. Use of surfactant—modified zeolite to carry and slowly release sulfate. *Desalin. Water Treat.* **2010**, *21*, 73–78. [CrossRef]

115. Thirunavukkarasu, M.; Subramanian, K. Surface modified nano-zeolite used as carrier for slow release of sulphur. *J. Nat. Appl. Sci.* **2014**, *6*, 19–26. [CrossRef]

116. Yuvaraj, M.; Subramanian, K.S. Development of slow release Zn fertilizer using nano-zeolite as carrier. *J. Plant. Nutr.* **2018**, *41*, 311–320. [CrossRef]

117. Sheta, A.; Falatah, A.; Al-Sewailem, M.; Khaled, E.; Sallam, A. Sorption characteristics of zinc and iron by natural zeolite and bentonite. *Microporous Mesoporous Mater.* **2003**, *61*, 127–136. [CrossRef]

118. Iskander, A.L.; Khald, E.M.; Sheta, A.S. Zinc and manganese sorption behavior by natural zeolite and bentonite. *Ann. Agric. Sci.* **2011**, *56*, 43–48. [CrossRef]

119. Najafi-Ghiri, M.; Rahimi, T. Zinc Uptake by Spinach (*Spinacia oleracea* L.) as Affected by Zn Application Rate, Zeolite, and vermicompost. *Compos. Sci. Util.* **2016**, *24*, 203–207. [CrossRef]

120. Filcheva, E.; Chakalov, K. Soil fertility management with zeolite amendments. I. Effect of zeolite on carbon sequestration: A review. In *Agricultural Practices and Policies for Carbon Sequestration in Soils*; Kimble, J.M., Lal, R., Follet, R.F., Eds.; Lewis publishers: New York, NY, USA, 2002; pp. 223–236.

121. Usman, A.R.A.; Kuzyakov, Y.; Stahr, K. Effect of clay minerals on extractability of heavy metals and sewage sludge mineralization in soil. *Chem. Ecol.* **2004**, *20*, 1–13. [CrossRef]

122. Wang, J.; Wang, D.; Zhang, G.; Wang, C. Effect of wheat straw application on ammonia volatilization from urea applied to a paddy field. *Nutr. Cycl. Agroecosyst.* **2012**, *94*, 73–84. [CrossRef]

123. Corma, A.; Garcia, H. Zeolite based photo catalysts. *Chem. Comm.* **2004**, *13*, 1443–1459. [CrossRef]

124. Walcarius, A.; Mouchotte, R. Efficient in vitro paraquat removal via irreversible immobilization into zeolite particles. *Arch. Environ. Con. Tox.* **2004**, *46*, 135–140. [CrossRef] [PubMed]

125. Zhang, H.; Kim, Y.; Dutt, P.K. Controlled release of paraquat from surface-modified zeolite Y. *Microporous Mesoporous Mater.* **2006**, *88*, 312–318. [CrossRef]

126. Capasso, S.; Coppola, E.; Lovino, P.; Salvestrini, S.; Colella, C. Uptake of phenylurea herbicides by humic acid-zeolite tuff aggregate. *Stud. Surf. Sci. Catal.* **2007**, *170*, 2122–2127.

127. Salvestrini, S.; Sagliano, P.; Lovino, P.; Capasso, S.; Colella, C. Atrazine adsorption by acid activated zeolite-rich tuffs. *Appl. Clay Sci.* **2010**, *49*, 330–335. [CrossRef]

128. Jamil, T.S.; Gad-Allah, T.A.; Ibrahim, H.S.; Saleh, T.S. Adsorption and isothermal models of atrazine by zeolite prepared from Egyptian kaolin. *Solid State Sci.* **2011**, *13*, 198–203. [CrossRef]

129. Bakhtiary, S.; Shirvani, M.; Shariatmadari, H. Adsorption desorption behavior of 2,4-D on NCP-modified bentonite and zeolite: Implications for slow release herbicide formulations. *Chemosphere* **2013**, *90*, 699–705. [CrossRef] [PubMed]

130. Shirvani, M.; Farajollahi, E.; Bakhtiari, S.; Ogunseitan, O.A. Mobility and efficacy of 2,4-D herbicide from slow—release delivery systems based on organo-zeolite and organo bentonite complexes. *J. Environ. Sci. Health B* **2014**, *49*, 255–262. [CrossRef] [PubMed]

131. Reddy, P.A.K.; Srinivas, B.; Durgakumari, V.; Subrahmanyam, M. Solar photocatalytic degradation of the herbicide isoprotuoron on a Bi TiO2/zeolite photocatalyst. *Toxicol. Environ. Chem.* **2012**, *94*, 512–524. [CrossRef]

132. Karczewska, A.; Kabała, C. The soils polluted with heavy metals and arsenic in Lower Silesia-the need and methods of reclamation. *Zesz. We Wrocławiu Rol.* **2010**, *96*, 59–79.

133. Shia, W.; Shaoa, H.; Li, H.; Shao, M.; Sheng, D. Progress in the remediation of hazardous heavy metal-polluted soils by natural zeolite. *J. Hazard. Mater.* **2009**, *170*, 1–6. [CrossRef]

134. Chen, Z.S.; Lee, G.J.; Liu, J.C. The effects of chemical remediation treatments on the extractability and speciation of cadmium and lead in contaminated soils. *Chemosphere* **2000**, *41*, 235–242. [CrossRef]

135. Ghuman, G.S.; Sajwan, K.S.; Alva, A.K. Remediation of an acid soil using hydroxyapatite and zeolites. In Proceedings of the Extended Abstracts from the 5th International Conference on the Biogeochemistry of Trace Elements, Vienna, Austria, 11–15 July 1999; pp. 1018–1019.

136. Chlopecka, A.; Adriano, D.C. Inactivation of metals in polluted soils using natural zeolite and apatite. In Proceedings of the 4th International Conference on the Biogeochemistry of Trace Elements, Berkeley, CA, USA, 23–26 June 1997.

137. Wang, S.; Peng, Y. Natural zeolites as effective adsorbents in water and wastewater treatment. *Chem. Eng. J.* **2010**, *156*, 11–24. [CrossRef]

138. Abatal, M.; Quiroz, A.T.V.; Olguín, M.T.; Vázquez-Olmos, A.R.; Vargas, J.; Anguebes-Franseschi, F.; Giácoman-Vallejos, G. Sorption of Pb(II) from Aqueous Solutions by Acid-Modified Clinoptilolite-Rich Tuffs with Different Si/Al Ratios. *Appl. Sci.* **2019**, *9*, 2415. [CrossRef]

139. Zendelska, A.; Golomeova, M.; Blazev, K.; Krstev, B.; Golomeov, B.; Krstev, A. Adsorption of copper ions from aqueous solutions on natural zeolite. *Environ. Prot. Eng.* **2015**, *41*, 17–36.

140. Armagan, B.; Ozdemir, O.; Turan, M.; Celik, M.S. Equilibrium studies on the adsorption of reactive azo dyes into zeolite. *Desalination* **2004**, *170*, 33–39. [CrossRef]

141. Wang, S.; Terdkiatburana, T.; Tadé, M.O. Adsorption of Cu(II), Pb(II) and humic acid on natural zeolite tuff in single and binary systems. *Sep. Purif. Technol.* **2008**, *62*, 64–70. [CrossRef]

142. Yousef, R.I.; El-Eswed, B.; Al-Muhtaseb, A.H. Adsorption characteristics of natural zeolites as solid adsorbents for phenol removal from aqueous solutions: Kinetics, mechanism, and thermodynamics studies. *Chem. Eng. J.* **2011**, *171*, 1143–1149. [CrossRef]

143. Li, L.Y.; Tazaki, K.; Lai, R.; Shiraki, K.; Asada, R.; Watanabe, H.; Loretta, M.C.; Chen, M. Treatment of acid rock drainage by clinoptilolite-Adsorptivity and structural stability for different pH environments. *Appl. Clay Sci.* **2008**, *39*, 1–9. [CrossRef]

144. Babel, S.; Kurniawan, T.A. Low-cost adsorbents for heavy metals uptake from contaminated water: A review. *J. Hazard. Mater.* **2003**, *B97*, 219–243. [CrossRef]

145. Stylianou, M.A.; Inglezakis, V.J.; Moustakas, K.G.; Malamis, S.P.; Loizidou, M.D. Removal of Cu(II) in fixed bed and batch reactors using natural zeolite and exfoliated vermiculite as adsorbents. *Desalination* **2007**, *215*, 133–142. [CrossRef]

146. Markou, G.; Vandamme, D.; Muylaert, K. Using natural zeolite for ammonia sorption from wastewater and as nitrogen releaser for the cultivation of Arthrospira platensis. *Bioresour. Technol.* **2014**, *155*, 373–378. [CrossRef]

147. Andronikashvilf, T.G.; Urushadze, T.F.; Eprikashvili, L.G.; Gamisonia, M.K. Use of natural zeolites in plant growing-transition to biological agriculture. *Bull. Georgian Natl. Acad. Sci.* **2007**, *175*, 112–117.

148. Ahmed, O.H.; Yap, C.H.B.; Muhamad, A.M.N. Minimizing ammonia loss from urea through mixing with zeolite and acid sulphate soil. *Int. J. Phys. Sci.* **2010**, *5*, 2198–2202.

149. Wu, Q.; Xia, G.; Chen, T.; Wang, X.; Chi, D.; Sun, D. Nitrogen use and rice yield formation response to zeolite and nitrogen coupling effects: Enhancement in nitrogen use efficiency. *J. Soil Sci. Plant. Nutr.* **2016**, *16*, 999–1009. [CrossRef]

150. Kavoosi, M. Effects of zeolite application on rice yield, nitrogen recovery and nitrogen use efficiency. *Commun. Soil Sci. Plant. Anal.* **2007**, *38*, 69–76. [CrossRef]

151. Wu, Q.; Chen, T.; Chi, D.; Xia, G.; Sun, Y.; Song, Y. Increasing nitrogen use efficiency with lower nitrogen application frequencies using zeolite in rice paddy fields. *Int. Agrophys.* **2019**, *33*, 263–269. [CrossRef]

152. Peng, Y.; Ma, J.; Jiang, M.; Yan, F.; Sun, Y.; Yang, Z. Effects of slow/controlled release fertilizers on root morphological and physiological characteristics of rice. *J. Plant. Nut. Fert.* **2013**, *19*, 1048–1057. (In Chinese)
153. Sepaskhah, A.R.; Barzegar, M. Yield, water and nitrogen-use response of rice to zeolite and nitrogen fertilization in a semi-arid environment. *Agric. Water Manag.* **2010**, *98*, 38–44. [CrossRef]
154. Ozbahce, A.; Tari, A.F.; Gonulal, E.; Simsekli, N.; Padem, H. The effect of zeolite applications on yield components and nutrient uptake of common bean under water stress. *Arch. Agron. Soil Sci.* **2014**, *61*, 615–626. [CrossRef]
155. Colombani, N.; Giuseppe, D.D.; Faccini, B.; Ferretti, G.; Mastrocicco, M.; Coltorti, M. Estimated water savings in an agricultural field amended with natural zeolites. *Environ. Process.* **2016**, *3*, 617–628. [CrossRef]
156. Malekian, R.; Abedi-Koupai, J.; Eslamian, S.S. Influences of clinoptilolite and surfactantmodified clinoptilolite zeolite on nitrate leaching and plant growth. *J. Hazard. Mater.* **2011**, *185*, 970–976. [CrossRef] [PubMed]
157. Latifah, O.; Ahmed, O.H.; Majid, N.M.A. Effect of mixing urea with zeolite and sago waste water on nutrient use efficiency of maize (*Zea mays* L.). *Afr. J. Microbiol. Res.* **2011**, *5*, 3462–3467. [CrossRef]
158. Rahmani, A.R.; Mahvi, A.H.; Mesdaghinia, A.R.; Nasseri, S. Investigation of ammonia removal from polluted waters by clinoptilolite zeolite. *Int. J. Environ. Sci. Technol.* **2004**, *1*, 125–133. [CrossRef]
159. Rabai, K.A.; Ahmed, O.H.; Kasim, S. Use of formulated Nitrogen, Phosphorus, and Potassium compound fertilizer using clinoptilolite zeolite in Maize (*Zea mays* L.) cultivation. *Emir. J. Food Agric.* **2013**, *25*, 1–5. [CrossRef]
160. Zahedi, H.; Hossein, A.; Rad, S.; Reza, H.; Moghadam, T. Effect of zeolite and selenium foliar application on growth, production and some physiological attributes of three canola (*Brassica napus* L.) cultivars subjected to drought stress. *Rev. Udo Agric.* **2011**, *12*, 136–143. [CrossRef]
161. Abasiyeh, S.K.; Rad, A.H.S.; Delkhoush, B.; Mohammadi, G.N.; Nasrollahi, H. Effect of potassium and zeolite on seed, oil and, biological yield in safflower. *Ann. Biol. Res.* **2013**, *4*, 204–207.
162. Ozbahce, A.; Fuat Tari, A.; Gonulal, E.; Simsekli, N. Zeolite for Enhancing Yield and Quality of Potatoes Cultivated Under Water-Deficit Conditions. *Potato Res.* **2018**, *61*, 247–259. [CrossRef]
163. Mahmoud, S.S.; Croteau, R.B. Menthofuran regulates essential oil biosynthesis in peppermint by controlling a downstream monoterpene reductase. *Proc. Natl. Acad. Sci. USA* **2003**, *100*, 14481–14486. [CrossRef] [PubMed]
164. Tabatabaie, J.; Nazari, J. Compare of essential oil yield and menthol existent in Peppermint (*Mentha piperita* L.) planted in different origin of Iran. *J. Med. Plant. Inst.* **2007**, *3*, 73–78.
165. Flexas, J.; Medrano, H. Drought inhibition of photosynthesis in C3 plants: Stomatal and non-stomatal limitations revisited. *Ann. Bot.* **2009**, *89*, 183–189. [CrossRef] [PubMed]
166. Khalid, K.A. Influence of water stress on growth essential oil and chemical composition of herbs (*Ocimum* sp.). *Int. Agrophys.* **2006**, *20*, 289–296.
167. Sanchez-Mora, F.; Montufar, G.V.; Saltos, F.A.; Chang, J.V.; Reache, R.R.; Coronel, G.D.; Navarrete, E.T.; Lopez, G.J. Zeolites en la fertilizacionquimca del cacao ccn-51 asociado con cuatro species maderables. *Artic. Cient. Cienc.* **2013**, *6*, 21–29.
168. Mehrab, N.; Chorom, M.; Hojati, S. Effect of raw and NH_4^+ enriched zeolite on nitrogen uptake by wheat and nitrogen leaching in soils with different textures. *Commun. Soil Sci. Plant. Anal.* **2016**, *47*, 1306–1316. [CrossRef]
169. Nozari, R.; Tohidi Moghadam, H.R.; Zahedi, H. Effect of cattle manure and zeolite applications on physiological and biochemical changes in soybean (*Glycine max* (L.) Merr.) grown under water deficit stress. *Rev. Cient. Udo Agric.* **2013**, *13*, 76–84.
170. Ghannad, M.; Ashraf, S.; Alipour, Z.T. Enhanching yield and quality of potato (*Solanum tuberosum* L.) tuber using an integrated fertilizer management. *Int. J. Agric. Crop. Sci.* **2014**, *10*, 742–748.
171. Ramesh, K.; Sammi Reddy, K.; Rashmi, I.; Biswas, A.K.; Islam, R. Weather particle size of clinoptolite zeolite has a role in textural properties? An insight through differential pore-volume distribution of Barret, Joyner and Halenda model. *Commun. Soil Sci. Plant. Anal.* **2015**, *46*, 2070–2078. [CrossRef]
172. Ryakhovskaya, N.I.; Gainatulina, V.V. Potato and oat yield in short-cycle crop rotation with zeolite application. *Plant. Ind. Agric. Sci.* **2009**, *35*, 153. [CrossRef]
173. Klaus, A.; Grubisic, M.; Niksic, M. Influence of some zeolits on the growth of mycelia of industrial fungi *Agaricusbisporus, Ganoderma lucidum, Pleurotusostreatus* and *Lentinus edodes*. In Proceedings of the First South East European Congress of Chemical Engineering, Faculty of Technology and Metallurgy, Belgrade, Serbia, 25–28 September 2005.
174. Susana, S. Researches Influence of Zeolite on Productivity Elements and Microbiological Activity on Spring Barley, Soybeans and Maize at Ardsturda. Ph.D. Thesis, University of Agricultural Sciences and Veterinary Medicine Cluj-Napoca, Cluj-Napoca, Romania, 2015.
175. Andronikashvili, T.; Urushadze, T.; Eprikashvili, L.; Gamisonia, M. Towards the biological activity of the natural zeolite-clinoptilolite-containing tuff. *Bull. Georgian Nat. Acad. Sci.* **2008**, *2*, 99–107.
176. Gruener, J.E.; Ming, D.W.; Henderson, K.E.; Galindo, C. Common ion effects in zeoponic substrates: Wheat plant growth experiment. *Microporous Mesoporous Mater.* **2003**, *61*, 223–230. [CrossRef]
177. Rodriguez-Fuentes, G. *The 3rd International Conference on the Occurrence, Properties and Utilization of Natural Zeolites, Havana, Cuba, 9–11 April 1991*; International Conference Center Press: Havana, Cuba, 1992.
178. Henderson, K.E.; Ming, D.W.; Carrier, C.; Gruener, J.E.; Galindo, C., Jr.; Golden, D.C. *Natural Zeolites for the Third Millenium*; De FredeEditore: Napoli, Italy, 2000; pp. 441–447.

179. Barros, M.A.S.D.; Zola, A.S.; Arroyo, P.A.; Sousa-Aguiar, E.F.; Tavares, C.R.G. Binary ion exchange of metal ions in Y and X zeolites. *Braz. J. Chem. Eng.* **2003**, *20*, 413–421. [CrossRef]

180. Rumbos, C.I.; Sakka, M.; Berillis, P.; Athanassiou, C.G. Insecticidal potential of zeolite formulations against three major stored-grain insects: Particle size effect, adherence ratio and effect of test weight on grains. *J. Stored Prod. Res.* **2016**, *68*, 93–101. [CrossRef]

181. Haryadi, Y.; Syarief, R.; Hubeis, M.; Herawati, I. Effect of zeolite on the development of Sitophilus zeamaisMotsch. In *Proceedings of the Sixth International Working Conference on Stored-product Protection, Canberra, Australia, 17–23 April 1994*; CAB International: Wallingford, UK, 1994.

182. Daniel, C.; Dieraurer, H.; Clerc, M. The potential of silicate rock dust to control pollen beetles (*Meligethes* spp.). *Iobc Wprs Bull.* **2013**, *96*, 47–55.

183. Mulla, M.S.; Thavara, U.; Tawatsin, A.; Chompoosri, J. Procedures for evaluation of field efficacy of slow-release formulations of larvicides against Aedes aegypti in water-storage containers. *J. Am. Mosq. Control. Assoc.* **2004**, *20*, 73–92.

184. Colella, C. Recent advances in natural zeolite applications based on external surface interaction with cations and molecules. In Proceedings of the 40th anniversary of international zeolite Conference, Beijing, China, 12–17 August 2007.

185. Petrovic, A.M. Potential for natural zeolite uses on golf courses. *Usga Green Sect.* **1993**, *31*, 11–14.

186. Mumpton, F.A. La rocamajica: Uses of natural zeolites in agriculture and industry. *Proc. Natl. Acad. Sci. USA* **1999**, *96*, 3463–3470. [CrossRef] [PubMed]

187. Bhattacharyya, A.; Bhaumik, A.; Pathipati, U.R.; Mandal, S.; Epidi, T.T. Nanoparticles—A recent approach to insect pest control. *Afr. J. Biotechnol.* **2010**, *9*, 3489–3493.

188. Neethirajan, S.; Jayas, D.S. Nanotechnology for the food and bioprocessing industries. *Food Bioproc. Tech.* **2011**, *4*, 39–47. [CrossRef] [PubMed]

189. Prasad, R.; Kumar, V.; Prasad, K.S. Nanotechnology in sustainable agriculture: Present concerns and future aspects. *Afr. J. Biotechnol.* **2014**, *13*, 705–713.

190. Stadler, T.; Buteler, M.; Weaver, D.K. Novel use of nanostructured alumina as an insecticide. *Pest. Manag. Sci.* **2010**, *66*, 577–579. [CrossRef]

191. Eroglu, N.; Emekci, M.; Athanassiou, C. Applications of natural zeolites on agriculture and food production. *J. Sci. Food Agric.* **2017**, *97*, 3487–3499. [CrossRef]

192. Kljajic, P.; Andric, G.; Adamovic, M.; Bodroza-Solarov, M.; Markovic, M.; Peric, I. Laboratory assessment of insecticidal effectiveness of natural zeolite and diatomaceous earth formulations against three stored-product beetle pests. *J. Stored Prod. Res.* **2010**, *46*, 1–6. [CrossRef]

193. Perez, J.C.; Pino, O.; Ramirez, S.; Suris, M. Evaluation of natural products in the control of *Lasiodermaserricorne* (F.) (Coleopteran: Anobiidae) in chickpea under laboratory conditions. *Rev. Prot. Veg.* **2012**, *27*, 26–32.

194. Kabak, B.; Dobson, A.D.; Var, I. Strategies to prevent mycotoxin contamination of food and animal feed: A review. *Crit. Rev. Food Sci. Nutr.* **2006**, *46*, 593–619. [CrossRef]

195. Phillips, T.D.; Sarr, B.A.; Clement, B.A.; Kubena, L.F.; Harvey, R.B. Prevention of aflatoxicosis infarm animals via selective chemisorptions of aflatoxin. In *Mycotoxins, Cancer and Health*; Bray, G.A., Ryan, D.H., Eds.; Louisiana State University Press: Baton Rouge, LA, USA, 1991; pp. 223–237.

196. Galvano, F.; Piva, A.; Ritieni, A.; Galvano, G. Dietary strategies to counteract the effects of mycotoxins: A review. *J. Food Prot.* **2001**, *64*, 120–131. [CrossRef]

197. Dwyer, M.R.; Kubena, R.B.; Harvey, L.F.; Mayura, K.; Sarr, A.B.; Buckley, S.; Bailey, R.H.; Phillips, T.D. Effects of inorganic adsorbents and cylcopiazonic acid in broiler chicks. *Poult. Sci* **1997**, *76*, 1141–1149. [CrossRef]

198. Phillips, T.D. Dietary clay in the chemoprevention of aflatoxin-induced disease. *Toxicol. Sci.* **1999**, *52*, 118–126. [CrossRef]

199. Parlat, S.S.; Yildiz, A.O.; Oguz, H. Effect of clinoptilolite on performance of Japanese quail Coturnix japonica during experimental aflatoxicosis. *Br. Poult. Sci.* **1999**, *40*, 495–500. [CrossRef] [PubMed]

200. Tomasevic-Canovic, M.; Dacovic, A.; Markovic, V.; Stojcic, D. The effect of exchangeable cations in clinoptililite and montmorillonite on the adsorbtion of aflatoxin B1. *J. Serb. Chem. Soc.* **2001**, *66*, 555–561. [CrossRef]

201. Tomasevic-Canovic, M.; Dakovic, A.; Rottinghaus, G.; Matijasevic, S.; Duricic, M. Surfactant modified zeolite—New efficient adsorbents for mycotoxins. *Micro Meso Mater.* **2003**, *61*, 173–180. [CrossRef]

202. Adamovic, M.; Magdalena, T.C.; Milosevic, S.; Aleksandra, D.; Lemic, J. The contribution of mineral adsorbent in the improvement of animal performance, health and quality of animal products. *Biotechnol. Anim. Husb.* **2003**, *19*, 383–395.

203. Huwig, A.; Freimund, S.; Kappeli, O.; Dutler, H. Mycotoxin detoxication of animal feed by different adsorbents. *Toxicol. Lett.* **2001**, *122*, 179–188. [CrossRef]

204. Król, M. Natural vs. Synthetic Zeolites. *Crystals* **2020**, *10*, 622. [CrossRef]

Article

Water Table Fluctuation and Methane Emission in Pineapples (*Ananas comosus* (L.) Merr.) Cultivated on a Tropical Peatland

Wendy Luta [1], Osumanu Haruna Ahmed [1,2,3], Latifah Omar [1,2,*], Roland Kueh Jui Heng [2,3,4], Liza Nuriati Lim Kim Choo [5], Mohamadu Boyie Jalloh [6], Adiza Alhassan Musah [7] and Arifin Abdu [8]

[1] Department of Crop Science, Faculty of Agricultural and Forestry Science, Universiti Putra Malaysia Bintulu Sarawak Campus, Bintulu 97008, Sarawak, Malaysia; wendyluta90@gmail.com (W.L.); osumanu@upm.edu.my (O.H.A.)

[2] Institut Ekosains Borneo, Universiti Putra Malaysia Bintulu Sarawak Campus, Bintulu 97008, Sarawak, Malaysia; roland@upm.edu.my

[3] Institute of Tropical Agriculture and Food Security (ITAFoS), Universiti Putra Malaysia, Serdang 43400, Selangor Darul Ehsan, Malaysia

[4] Department of Forestry Science, Faculty of Agricultural and Forestry Science, Universiti Putra Malaysia Bintulu Sarawak Campus, Bintulu 97008, Sarawak, Malaysia

[5] Soil Science, Water and Fertilizer Research Centre, Malaysian Agricultural Research and Development Institute (MARDI), Roban, Saratok 95300, Sarawak, Malaysia; lizalim@mardi.gov.my

[6] Faculty of Sustainable Agriculture, Universiti Malaysia Sabah, Sandakan Branch, Locked Bag No. 3, Sandakan 90509, Sabah, Malaysia; mbjalloh@ums.edu.my

[7] Faculty of Business Management and Professional Studies, Management and Science University, University Drive, Off Persiaran Olahraga Section 13, Shah Alam 40100, Selangor, Malaysia; adiza_alhassan@msu.edu.my

[8] Department of Forest Management, Faculty of Forestry and Environment, Universiti Putra Malaysia, Selangor 43400, Malaysia; arifinabdu@upm.edu.my

* Correspondence: latifahomar@upm.edu.my; Tel.: +60-109599147

Citation: Luta, W.; Ahmed, O.H.; Omar, L.; Heng, R.K.J.; Choo, L.N.L.K.; Jalloh, M.B.; Musah, A.A.; Abdu, A. Water Table Fluctuation and Methane Emission in Pineapples (*Ananas comosus* (L.) Merr.) Cultivated on a Tropical Peatland. *Agronomy* **2021**, *11*, 1448. https://doi.org/10.3390/agronomy11081448

Academic Editors: Nikolaos Monokrousos and Efimia M. Papatheodorou

Received: 22 April 2021
Accepted: 10 June 2021
Published: 21 July 2021

Publisher's Note: MDPI stays neutral with regard to jurisdictional claims in published maps and institutional affiliations.

Abstract: Inappropriate drainage and agricultural development on tropical peatland may lead to an increase in methane (CH_4) emission, thus expediting the rate of global warming and climate change. It was hypothesized that water table fluctuation affects CH_4 emission in pineapple cultivation on tropical peat soils. The objectives of this study were to: (i) quantify CH_4 emission from a tropical peat soil cultivated with pineapple and (ii) determine the effects of water table depth on CH_4 emission from a peat soil under simulated water table fluctuation. Soil CH_4 emissions from an open field pineapple cultivation system and field lysimeters were determined using the closed chamber method. High-density polyethylene field lysimeters were set up to simulate the natural condition of cultivated drained peat soils under different water table fluctuations. The soil CH_4 flux was measured at five time intervals to obtain a 24 h CH_4 emission in the dry and wet seasons during low- and high-water tables. Soil CH_4 emissions from open field pineapple cultivation were significantly lower compared with field lysimeters under simulated water table fluctuation. Soil CH_4 emissions throughout the dry and wet seasons irrespective of water table fluctuation were not affected by soil temperature but emissions were influenced by the balance between methanogenic and methanotrophic microorganisms controlling CH_4 production and consumption, CH_4 transportation through molecular diffusion via peat pore spaces, and non-microbial CH_4 production in peat soils. Findings from the study suggest that water table fluctuation at the soil–water interface relatively controls the soil CH_4 emission from lysimeters under simulated low- and high-water table fluctuation. The findings of this study provide an understanding of the effects of water table fluctuation on CH_4 emission in a tropical peatland cultivated with pineapple.

Keywords: drained peat; greenhouse gas; global warming; organic soil; pineapple; water table

1. Introduction

Drained peatlands worldwide emit approximately 2 Gt of carbon dioxide (CO_2) through microbial peat oxidation or peat fires representing 5% of all anthropogenic greenhouse gas (GHG) emissions [1]. Carbon dioxide emitted from peatlands has been implicated in the ongoing global warming debate [2]. Unlike CO_2, which is cycled and released into the atmosphere, methane (CH_4) is emitted mostly from agricultural activities [3]. Methane contributes to significant anthropogenic GHG and the concentration of CH_4 is on the increase [4,5]. The pathways of CH_4 emissions are through aerobic and anaerobic microbial respiration, root respiration, peat oxidation, nitrification, and denitrification although the determinant factors which affect CH_4 emissions are land-use type [6], peat type [7], photosynthetic activities [8], and water table fluctuation [9]. Carbon (C) is transformed and stored in different pools within the C cycle through, for example, burning of fossil fuels or decomposition of soil organic matter in the form of C gases into the atmosphere, whereas photosynthesis locks atmospheric C in plant tissues and deposition of organic-rich sediments on the ocean floor locks C in geologic rocks and sediments.

The increasing interest in reducing CH_4 emissions to meet global temperature targets is because of the short atmospheric life span of CH_4 which is approximately 10 years, but CH_4 has relatively high global warming potential [10]. Current and future regional and global CH_4 budgets and mitigation strategies require better quantitative and process-based understanding of CH_4 sources, pathways, and removals under climate and land-use change [11]. According to Leifeld [12], peatland rewetting is a cost-effective measure to curb GHG emissions, however, increasing water table depth increases CH_4 emissions [13]. Lowering peatland water tables increases peat decomposition rates because of enhanced microbial degradation of organic matter [14]. It must be stressed that the understanding of soil C flux based on studies conducted in boreal and temperate peats is not fully applicable to tropical peatland because of differences in environmental factors, peat soil properties, peat temperature, peatland-use practices, vegetation composition and structure, and microbial diversity and population.

Tropical peatlands are commonly developed for agriculture. Huang et al. [15] reported that agricultural productivity in low latitudes (tropical and semi-tropical) are likely to decline due to climate change which affects world food security and farm incomes because most developing countries, including Malaysia, are located in lower latitude regions. Falling farm incomes will increase poverty and reduce households' ability to invest for a better future [16]. According to Melling et al. [17], peat soil reclamation for agriculture involves drainage which is characterized by lowering water table and soil compaction to aerate crop root zones. Drainage of tropical peatland may cause loss of soil C reserve. Drainage via lowering of water table could change peatlands from being a C sink to a C source because drainage reverses the C flux into net CO_2 emissions [18]. The decomposition of organic materials and microbial activities releases CO_2, CH_4, organic acids, and organic particulates. The rate of C loss is related to the increased intensity of dry and wet periods. The resultant extreme water table fluctuation could affect the amount and nature of aerobic and anaerobic peat material, which subsequently affect the decomposition of peat material, microbial activity, and the crop growth. A study had revealed that the CH_4 emissions from drained tropical peatland for pineapple (*Ananas comosus* (L.) Merr.) cultivation was lower than those emitted from bare peatland and bare peatland fumigated with chloroform [19]. In this study, our approach is to estimate CH_4 emissions from tropical peatland cultivated with pineapple under fluctuating water table. Pineapple could absorb CO_2 for photosynthesis to produce carbohydrates in plant tissues. The emissions of CH_4 could be reduced through maintaining ground water level because the ground water level below the surface alters the CH_4 dynamic by weakening the potential for CH_4 production and increasing the potential for CH_4 oxidation in the upper peat layers [17,20]. Considering the potential importance of tropical peatlands in the global CH_4 budgets [21], it is essential to understand the effects of water table fluctuation on CH_4 emissions from tropical peatlands cultivated with, for example, pineapples.

There is lack of standard procedures to measure CH_4 emissions in tropical peatlands as reported by Ahmed and Liza [19]. Couwenberg [18] and Burrows et al. [22] suggested that GHG emissions should be measured on the soil surface using a closed chamber method [23,24]. Greenhouse gas monitoring which is conducted using the closed chamber method is limited in space (few cm^2) and time (few minutes). To date, there is limited information on CH_4 emissions from peatlands which are cultivated with pineapples that are relatively tolerant to peatlands' acidity. According to a study conducted by Raziah and Alam [25], the contribution of pineapples cultivated on tropical peatlands to CH_4 emissions is important because 90% of pineapples are grown on peatlands of Malaysia. In this study, it was hypothesized that peatland water table fluctuation will affect the emission of CH_4 in pineapple cultivation on tropical peatlands. This assumption is premised on the fact that peatland water table fluctuation could minimize the CH_4 emissions through suppression of the anaerobic decomposition of organic matter (reduction of CO_2 by H_2) after which it is affected by the balance of CH_4 production and oxidation. Also, regulating water table level could control the peatland water temperature because increasing soil temperature leads to an increase in CH_4 emission.

The research questions that were addressed in this study were: (i) does water table depth affect CH_4 emission from tropical peatlands cultivated with pineapples? and (ii) what is the amount of CH_4 emitted from tropical peatland which are cultivated with pineapple in relation to simulated water table fluctuation? The quantification of CH_4 emission was carried out in the dry season (July and August 2015) and wet season (September and December 2015) to take into account the effects of temperature. Warm peatlands transformed soil organic C from a C sink to a C source [26] with well-drained soils releasing CO_2 to the atmosphere [27]. However, the decomposition of organic matter and peatland temperature in relation to water table fluctuation and CH_4 emission from a tropical peatland cultivated with pineapples have scarcely been explored. Thus, this study was carried out to: (i) determine the effects of water table depth and CH_4 loss from a tropical peatland cultivated with pineapples; (ii) quantify CH_4 loss in a tropical peatland under simulated water table fluctuation; and (iii) determine the effects of water table fluctuation on soil temperature during CH_4 emission.

The implication of regulating water table level as an approach to minimizing CH_4 emission from a tropical peat soil cultivated with pineapples is an attempt to hinder CH_4 emission or consumption. It is well known that the pathways of CH_4 emission are diffusion, ebullition, and plant-mediated transport. This study focuses on the loss of CH_4 from a tropical peatland cultivated with pineapples because different vegetation growing on the same peatland results in differences in CH_4 emission or consumption. According to Hu et al. [28], under forest vegetation, soil served as a net sink of CH_4, whereas maize field (*Zea mays* L.) was essentially CH_4 neutral, and a paddy field was a net source of CH_4 diffused to the atmosphere. The findings suggested that the water table fluctuation has significant effects on CH_4 emission apart from the different crop-mediated transport. Most of the crop mediated transport are focused on the aerenchyma and not on the crassulacean acid metabolism plants such as pineapple. Aerenchyma is a type of plant that has porous root tissue, particularly well developed in wetland plants, which enables diffusive flux of gases from above-ground tissues to root tips [29]. Owing to this, most of the CH_4 emission studies on drained peatlands are limited to rice, soybean, and sago [23,30] with little exploration focus on the pineapple [19]. Our approach was not only limited to determining the effects of water table fluctuation on CH_4 emitted from a tropical peatland cultivated with pineapples, but it was also focused on the measurements of soil CH_4 emission. This study also provides information on the mechanism of CH_4 emission from different water table depth and the amount of CH_4 emission from dry and wet seasons. This study partly shows that appropriately reclaimed land use on tropical peatlands favours low CH_4 emission, and benefits pineapple planters, economy, society, and the environment.

2. Materials and Methods

2.1. Site Description for Soil Methane Emission from Field Cultivated with Pineapples

The study was carried out to quantify C losses in the form of CO_2 (data on CO_2 emissions have been published in 2017) and CH_4 in a tropical peatland subjected to water table fluctuation. The study was carried out under field conditions and simulated lysimeter at the Malaysian Agricultural Research and Development Institute (MARDI), Sesang, Saratok, Sarawak, Malaysia (Figure 1). The study area of 387 hectares (ha) was located on the logged-over forest with a flat topography of 5 to 6 m above the mean sea level. Based on the Von Post scale of H7 to H9 [31], the peatland is classified as well decomposed dark brown to dark coloured sapric peat with a strong smell and the thickness of 0.5 to 3.0 m. The average temperature of the area ranges from 22.1 to 31.7 °C while the relative humidity of the area ranges from 61% to 98% humidity with annual mean rainfall of 3749 mm. From November to January (wet season), the monthly rainfall is greater than 400 mm, whereas the mean rainfall during the dry season particularly in July is 189 mm [19]. The area of the peatland cultivated with *Moris* pineapple was 0.21 hectares (Figure 2). The Moris pineapple were planted in two rows with a planting distance of 30 cm × 60 cm × 90 cm, and the pineapples were managed according to the standard agronomic practices for pineapple management on tropical peat soils. Soil CH_4 flux measurements were carried out using the closed chamber method [32] on a 10 m × 10 m plot with five replications. The study was carried in the dry (July and August 2015) and wet (September and December 2015) seasons.

Figure 1. Location of study site in Sesang, Saratok, Sarawak, Malaysia.

Figure 2. Study site of peatland cultivated with *Ananas comosus*.

2.2. Establishment of Lysimeters for Methane Estimation under Simulated Water Table Fluctuation

Ten cylindrical field lysimeters made from high-density polyethylene (HDPE), 0.56 m in diameter and 0.97 m in height, were set up to simulate the natural condition of drained tropical peats (Figure 3). The size of the lysimeters used in this study was designed to ensure satisfactory growth and development of the pineapple. The lysimeters were equipped with a water spillage opening which was attached to clear tubes mounted on the outside of the vessel to regulate and monitor water level. Each lysimeter was filled with peat soil up to 0.90 m depth (Figure 3). Water loss from the soil was replenished by showering each lysimeter with rainwater. The amounts of the rainwater added were based on the volume of the fabricated lysimeter and the mean annual rainfall at Saratok, Sarawak, Malaysia. The lysimeters with the peat soil were left in the open field for five months (January to June 2015) to equilibrate. During the modification of the lysimeter, clear tubing and water spillage openings were attached to one side of the lysimeter to regulate and monitor the water level. Before the lysimeter was filled with peat soil, a polyvinyl chloride (PVC) pipe was installed vertically onto the soil to enable the bailer to reach the bottom of the lysimeter (Figure 3). The water table in the lysimeter was controlled by draining excess water through the water spillage opening or watering the peat soil with rainwater to the desired water table depth to simulate the effects of drainage and rainfall. During rainy days, the lysimeter was covered with a plastic cover to maintain a consistent water level. The depth of the water table in the lysimeter was controlled at 0 m and 0.9 m from the soil surface to represent the driest (low water table) and wettest (high water table) months, respectively.

Figure 3. Fabricated field lysimeter made from high density polyethylene.

2.3. Estimation of Soil Methane Emission during Water Table Fluctuation

Soil CH_4 emissions from the field and lysimeters were measured using the closed chamber method [32]. The CH_4 emissions from peatland were quantified using gas chromatography (Agilent 7890A) equipped with a thermal conductivity detector (TCD). The chambers were placed vertically on the soil surface between pineapple plants. The CH_4 emissions measurements were carried out on the daily basis of dry and wet months, before total draining of the plot, at 2–4 h intervals over 2–3 days duration to reflect the total of CH_4 losses through the soil surface. The size of the closed chamber was 20 cm × 20 cm × 20 cm and made up of acrylic (Figure 4). The top of the chamber was fitted with two sampling ports plugged with a rubber septum for gas sampling and thermometer installation, respectively (Figure 4). A battery-operated fan was also attached to the chamber to allow equilibrium gas pressure in and outside the closed chamber (Figure 4). The chamber was covered with a reflective aluminium foil to minimize the effect of temperature differences within and outside the chamber. The headspace samples of 20 mL were extracted from the chamber at 1, 2, 3, 4, 5 and 6 min using a polypropylene syringe equipped with a three-way stopcock. The extracted CH_4 gas was transferred to a 10 mL vacuum vial bottle by a double-ended hypodermic needle to be quantified using gas chromatography (Agilent 7890A, Agilent Technologies Inc., Wilmington, DE, USA) equipped with a flame ionization detector (FID). The values obtained were averaged and converted into units of t ha^{-1}yr^{-1}.

The CH_4 flux was calculated from the increase in the chamber concentration over time using the chamber volume and soil area covered, using the following equation:

$$\text{Flux} = [d(CH_4)/dt] \times PV/ART$$

where $d(CH_4)/(dt)$ is the evolution rate of CH_4 within the chamber headspace at a given time after putting the chamber into the soil, P is the atmospheric pressure, V is the volume headspace gas within the chamber, A is the area of the soil closed by the chamber, R is the gas constant, and T is the air temperature [24,32].

The CH_4 flux was measured in the early morning I (06:00 to 06:35), afternoon (12:00 to 12:35), evening (18:00 to 18:35), midnight (00:00 to 00:35) and early morning II (06:00 a.m. to 06:35 a.m.) to obtain a 24 h of CH_4 emissions. The 24 h measurement was carried out to meet the gas flux measurement requirement based on the procedure described by Ahmed and Liza [19]. The flux measurements were carried out in July and August 2015 for the dry season and in September and December 2015 to represent the concentrations of CH_4 emitted in the wet season. Soil temperature at 6 cm depth were measured at the same time of the CH_4 flux measurement using a digital thermometer. Rainfall distribution data was collected from a portable weather station (WatchDog 2900ET, Spectrum Technologies Inc., Plainfield, IL, USA) installed at the experimental site. Although CH_4 fluxes were only monitored for two cycles for each weather season and results obtained might not be conclusive enough to confirm the findings on the effect of water table fluctuation on CH_4 emission, it must be emphasized that time allocated for soil CH_4 emission determination per sample was the limitation of this present study. This is because increasing the number of gas flux monitoring cycles are costly and time consuming. For example, a minimum retention time of 6 min is required for a gas sample analysed using gas chromatography, and the total samples for each CH_4 flux monitoring cycle were 450 per month.

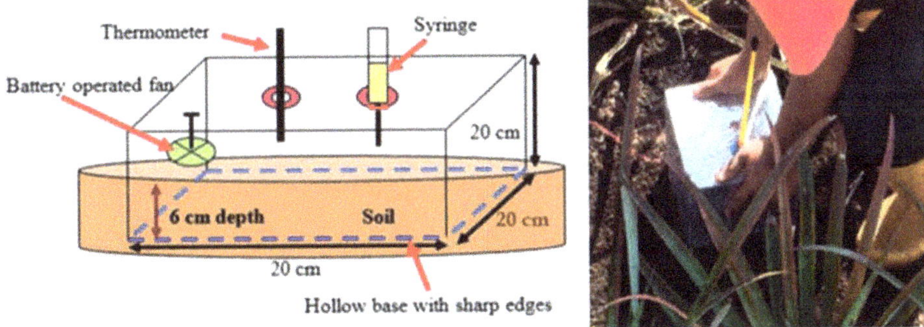

Figure 4. A closed chamber system to estimate soil methane emission from tropical peatland.

2.4. Statistical Analysis

Fluctuation of water table in relation to CH_4 emission was tested using analysis of variance (ANOVA) and means of the water table fluctuations in triplicates were compared using Duncan's new multiple range test (DNMRT) at $p \leq 0.05$. The relationships between CH_4 flux and soil temperature were analyzed using Pearson correlation analysis. The statistical software used for this analysis was the Statistical Analysis System (SAS) Version 9.3.

3. Results

3.1. Soil Methane Emissions from Peat Soils Grown with Pineapple under Open Field Cultivation System in the Dry and Wet Seasons

Soil CH_4 emissions from tropical peat soils cultivated with pineapples in the dry and wet seasons are presented in Figures 5 and 6, respectively. During the dry season (July and August 2015), the soil CH_4 emission showed no specific trend with the time of sampling but CH_4 emissions were higher at midnight (Figure 5). In July 2015, the CH_4 emissions was generally similar, whereas soil CH_4 emissions were lower in the early morning I, afternoon, evening, and early morning II than at midnight in August 2015. Compared with the wet season, soil CH_4 emissions were lower in the afternoon and at midnight during the gas flux monitoring in September and December 2015, respectively (Figure 6).

Averaged soil CH_4 emissions over 24 h from a drained peat soil cultivated with pineapples throughout the dry (July and August 2015) and wet (September and December 2015) seasons are presented in Figure 7. Soil CH_4 emissions were higher in September 2015 but emissions were lower in July, August and December 2015. However, soil CH_4 emissions in July, August and December 2015 were similar.

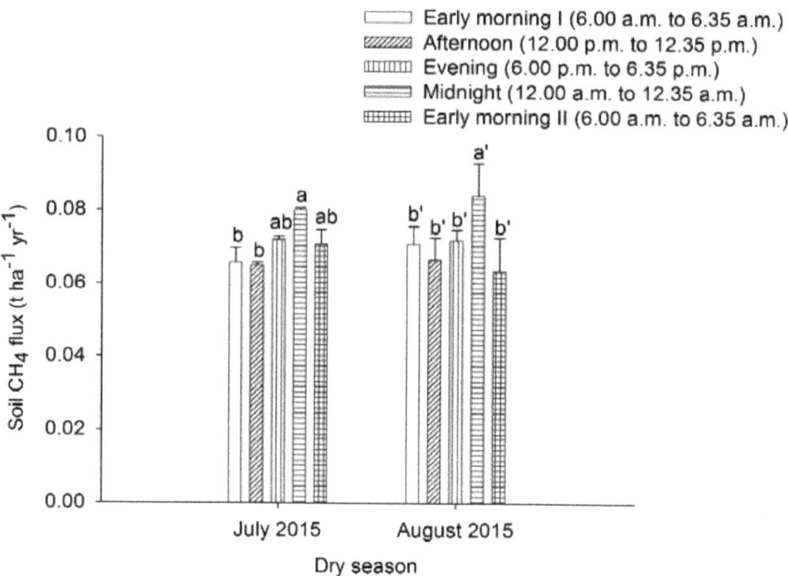

Figure 5. Soil CH_4 emissions (at different times of the day) from a tropical peatland cultivated with pineapples in the dry season (July and August 2015). Error bars represent standard error and soil mean fluxes with different letters and noted by prime are significantly different using Duncan's new multiple range test (DNMRT) at $p \leq 0.05$.

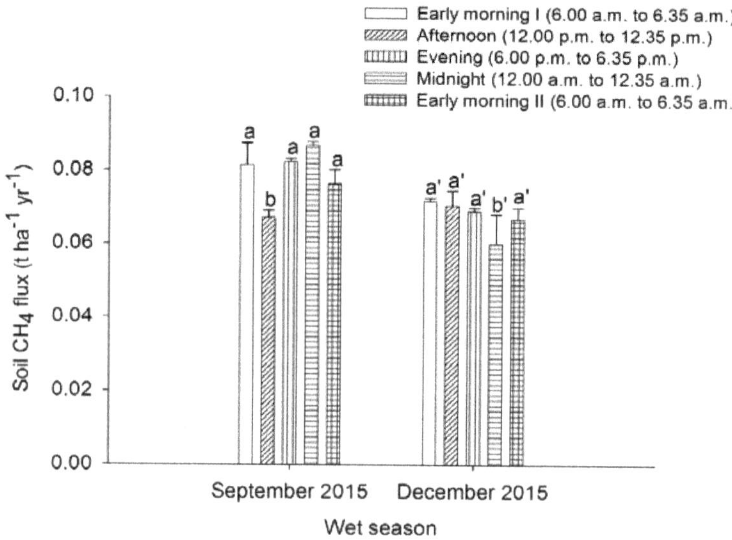

Figure 6. Soil CH_4 emissions (at different times of the day) from a tropical peatland cultivated with pineapples in the wet season (September and December 2015). Error bars represent standard error and soil mean fluxes with different letters and noted by prime are significantly different using DNMRT at $p \leq 0.05$.

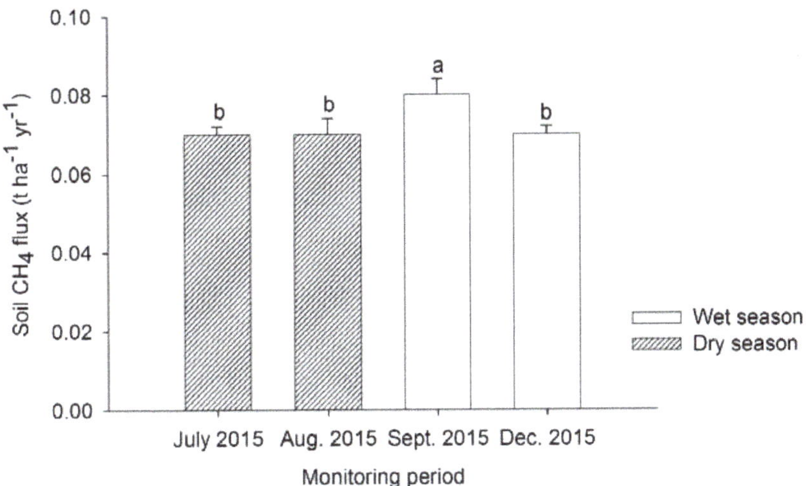

Figure 7. Averaged soil CH_4 emissions over 24 h from a tropical peat soils cultivated with pineapple throughout the dry (July and August 2015) and wet (September and December 2015) seasons. Error bars represent standard error and soil mean fluxes with different letters are significantly different using DNMRT at $p \leq 0.05$.

Throughout the CH_4 flux monitoring, soil temperature was statistically similar during the dry and wet seasons irrespective of sampling time (Table 1). Also, there was no significant correlation between CH_4 emission and soil temperature (Table 1).

Table 1. Relationship between soil CH_4 emission and soil temperature from a peat soil cultivated with pineapples throughout the dry and wet seasons in 2015.

	Soil Temperature (°C)			
Variable		Dry Season	Wet Season	
	July 2015	**August 2015**	**September 2015**	**December 2015**
Early morning I (6:00 a.m. to 6:35 a.m.)	30.7 [a]	28.3 [a]	29.0 [a]	31.0 [a]
Afternoon (12:00 p.m. to 12:35 p.m.)	25.3 [a]	27.0 [a]	25.7 [a]	25.0 [a]
Evening (6:00 p.m. to 6:35 p.m.)	28.3 [a]	29.3 [a]	29.3 [a]	29.3 [a]
Midnight (12:00 a.m. to 12:35 a.m.)	25.3 [a]	28.3 [a]	28.3 [a]	27.0 [a]
Early morning II (6:00 a.m. to 6:35 a.m.)	26.7 [a]	28.5 [a]	28.5 [a]	26.7 [a]
Soil CH_4 emission	$r = -0.1191$ $p = 0.6725$	$r = 0.2209$ $p = 0.4286$	$r = -0.1386$ $p = 0.6224$	$r = -0.0529$ $p = 0.8513$

Mean values with same letters within the same column are not significantly difference between means using DNMRT at $p \leq 0.05$. Top value indicates Pearson's correlation coefficient (r), whereas the bottom values indicate probability level at 0.05 ($n = 600$).

3.2. Soil Methane Emissions from Peat Soils Cultivated with Pineapples in Lysimeters under Simulated Water Table Fluctuation in the Dry and Wet Seasons

Soil CH_4 emission varied with time of sampling throughout the wet and dry seasons under low and high water table fluctuations (Figure 8a,b). At low water table during the dry season (Figure 8a), soil CH_4 emissions decreased from early morning I to early morning II in July 2015, whereas CH_4 emissions were higher in the afternoon but lower at midnight in August 2015. However, at low water table during the wet season (Figure 8a), soil CH_4 emissions were higher in the evening but lower in the early morning II in September 2015, whereas in December 2015, CH_4 emission decreased from early morning I to evening, followed by an increase at midnight, after which the CH_4 emission decreased until early morning II. Compared with lysimeters subjected to a high water table (Figure 8b), soil CH_4 emissions in the dry season were higher at midnight and early morning II in July and

August, respectively. However, at high water table during the wet season (Figure 8b), soil CH_4 emission was higher at noon in September 2015, whereas CH_4 emissions decreased from early morning I to evening, after which the CH_4 emission increased until early morning II.

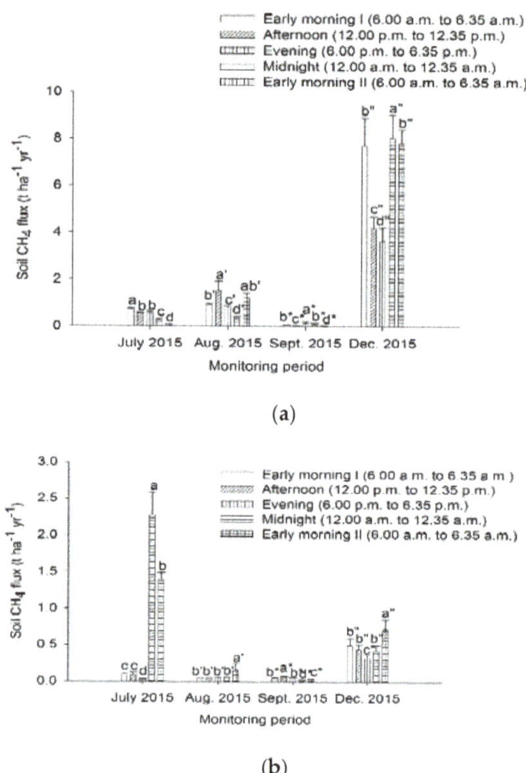

(a)

(b)

Figure 8. Soil CH_4 emissions (at different times of the day) from a peat soil grown with pineapples in lysimeters under simulated water table fluctuation (**a**) low water table and (**b**) high water table throughout the dry (July and August 2015) and wet (September and December 2015) seasons. Error bars represent standard error and soil mean fluxes with different letters and noted by prime, asterisk, and double prime are significantly different using DNMRT at $p \leq 0.05$.

Averaged soil CH_4 emissions over 24 h under different water table depth varied in the dry and wet seasons (Figure 9). At low water table (0.9 m from the soil surface), averaged soil CH_4 emissions were higher in the wet season (December 2015) but lower throughout the monitoring period in July, August and September 2015. Conversely, at high water table (0 m from the soil surface), averaged soil CH_4 emissions were higher in the dry season (July 2015) but emissions were lower in August and September 2015. However, throughout the dry and wet seasons, averaged soil CH_4 emissions were significantly higher under the low water table compared with that of the high water table (Figure 10).

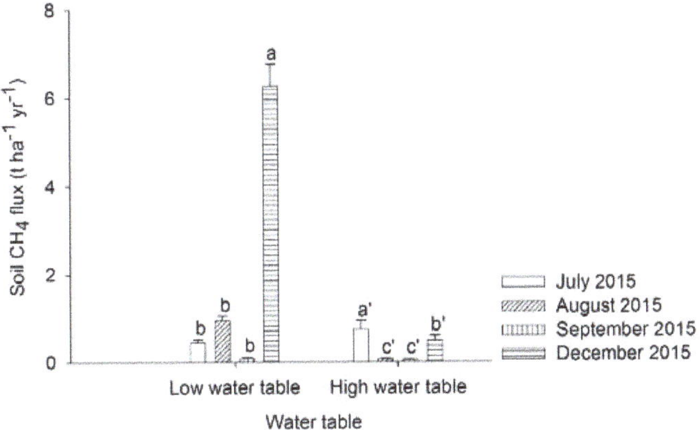

Figure 9. Averaged soil CH_4 emissions over 24 h from a peat soil cultivated with pineapples in lysimeters under low and high water tables throughout the dry (July and August 2015) and wet (September and December 2015) seasons. Error bars represent standard error and soil mean fluxes with different letters and noted by prime are significantly different using DNMRT at $p \leq 0.05$.

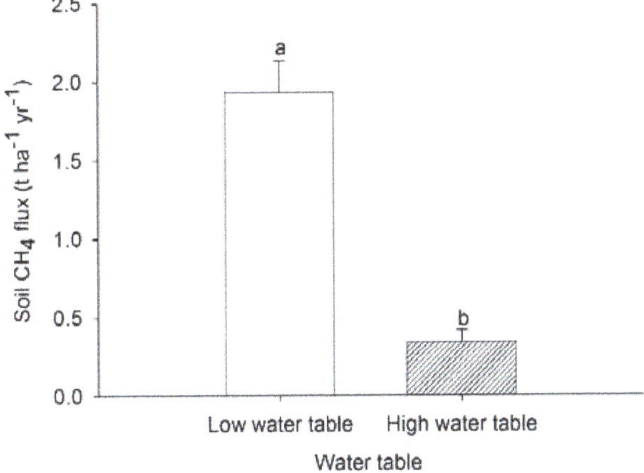

Figure 10. Averaged soil CH_4 emissions over 24 h from a peat soil grown with pineapple in lysimeters under low and high water table. Error bars represent standard error and soil mean fluxes with different letters are significantly different using DNMRT at $p \leq 0.05$.

Throughout the dry and wet seasons, soil temperature was statistically similar irrespective of water table and sampling time (Table 2). Moreover, there was no significant correlation between soil temperature and CH_4 emission (Table 2). This observation is consistent with the results obtained from the soil CH_4 measurement from open field pineapple cultivation (Table 1).

Compared with the peat soils grown with pineapple under open field cultivation system, averaged soil CH_4 emissions from pineapples in lysimeters subjected to water table fluctuation were significantly higher throughout the dry and wet seasons in 2015 (Figure 11).

Table 2. Relationship between soil CH_4 emission and soil temperature from lysimeters cultivated with pineapples under simulated water table fluctuation throughout the dry and wet seasons in 2015.

Variable	Soil Temperature (°C)							
	Dry Season				Wet Season			
	July 2015		August 2015		September 2015		December 2015	
	Low Water Table	High Water Table	Low Water Table	High Water Table	Low Water Table	High Water Table	Low Water Table	High Water Table
Early morning I (6:00 a.m. to 6:35 a.m.)	32.7 [a]	32.3 [a,b]	27.5 [a,b]	28.3 [a]	28.3 [a]	27.7 [a]	29.2 [a]	30.3 [a]
Afternoon (12:00 p.m. to 12:35 p.m.)	24.3 [a]	25.0 [a]	27.2 [b]	29.0 [a]	29.0 [a]	28.7 [a]	27.8 [a]	28.0 [a]
Evening (6:00 p.m. to 6:35 p.m.)	32.7 [a]	36.0 [a]	29.3 [a,b]	29.3 [a]	28.3 [a]	30.0 [a]	26.3 [a]	30.7 [a]
Midnight (12:00 a.m. to 12:35 a.m.)	28.0 [a]	24.7 [b]	28.3 [a,b]	28.2 [a]	29.3 [a]	28.7 [a]	29.7 [a]	28.7 [a]
Early morning II (6:00 a.m. to 6:35 a.m.)	32.0 [a]	28.3 [a,b]	29.7 [a]	27.7 [a]	28.2 [a]	28.0 [a]	27.0 [a]	30.0 [a]
Soil CH_4 emission	r = 0.2904 p = 0.2938	r = −0.0609 p = 0.8292	r = 0.4051 p = 0.1342	r = −0.0299 p = 0.9156	r = −0.0183 p = 0.9484	r = −0.0014 p = 0.9959	r = −0.1583 p = 0.5731	r = −0.4843 p = 0.0674

Mean values with same letters within the same column are not significantly difference between means using DNMRT at $p \leq 0.05$. Top value indicates Pearson's correlation coefficient (r), whereas the bottom values indicate probability level at 0.05 ($n = 1200$).

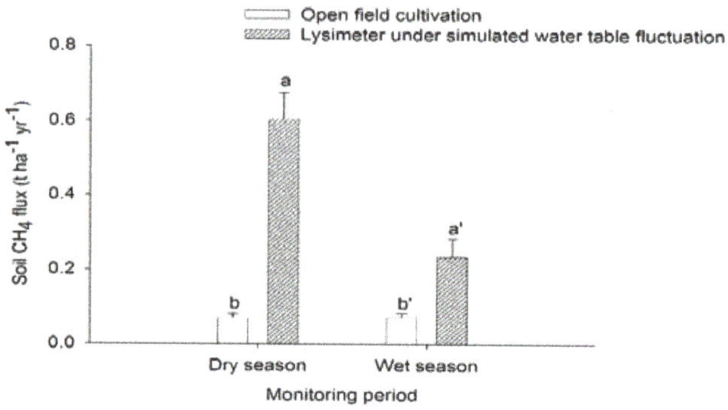

Figure 11. Averaged soil CH_4 emissions from peat soils grown with pineapples under open field cultivation system and lysimeters subjected to water table fluctuation in the dry and wet seasons. Error bars represent standard error and soil mean fluxes with different letters and noted by prime are significantly different using DNMRT at $p \leq 0.05$.

4. Discussion

4.1. Soil Methane Emissions from Peat Soils Grown with Pineapple under Open Field Cultivation System in the Dry and Wet Seasons

Differences in soil CH_4 emission across time (early morning, afternoon, evening and midnight) from pineapple cultivated peat soils could be attributed to the microbial structure in the peat soil that controls the balance between CH_4 production and CO_2 conversion and vice versa by methanogenic bacteria and methanotrophs under anaerobic and aerobic conditions [33], respectively, throughout the dry and wet seasons. Peat soils become net source of CH_4 when CH_4 production by methanogenic bacteria surpasses consumption by methanotrophic bacteria [34]. Moreover, soil CH_4 fluxes are regulated by oxygen supply and availability of labile carbon, where methanogenesis is predominant under anaerobic conditions. Also, soil CH_4 emissions might have been affected by the transportation of CH_4 by molecular diffusion through the aerobic layer of the peat soils, and through ebullition in the form of bubbles at the peat water table interface [35–37].

Although the averaged soil CH_4 emissions were not affected by the flux monitoring period throughout the dry (July and August 2015) and wet (December 2015) seasons, the higher CH_4 emission particularly in September 2015 during the wet season was because of the higher rainfall received at the experimental site amounting to 72 mm compared with the lower rainfall received in July (29 mm), August (52 mm) and September (69 mm) 2015 [38].

This result suggests that higher CH_4 is emitted under anaerobic and waterlogged conditions. The waterlogged condition of the peat soil in September 2015 might have favoured the thriving of methanogenesis bacteria under anoxic conditions, thus causing higher soil CH_4 emission. This result also corroborates previous work by Furukawa et al. [20] and Inubushi et al. [30], who reported that the increase in soil CH_4 emission is due to high rainfall.

Although soil CH_4 emission was affected by the time of sampling, the insignificant correlation between soil CH_4 emission and soil temperature regardless of seasons (dry and wet period) suggest that CH_4 emission was not affected by soil temperature due to the moderate soil temperature fluctuation during CH_4 flux measurement. The peat soil temperature ranged between 25 to 31 °C during the CH_4 sampling (Table 1).

4.2. Soil Methane Emissions from Peat Soils Cultivated with Pineapples in Lysimeters under Simulated Water Table Fluctuation in the Dry and Wet Seasons

Similar to the pineapple cultivated under an open field system, differences in soil CH_4 emission across time from field lysimeters subjected to water table fluctuation (low and high water table) relates to the microbial structure in the peat soils, particularly the methanogenic and methanotrophic microorganisms because these organisms control CH_4 production and consumption. Also, soil CH_4 release might have been influenced by the collapse of peat pores (during the soil excavation and setting up of the lysimeters) that affected CH_4 transportation through molecular diffusion, and subsequent soil subsidence in the lysimeters due to water table fluctuation.

In this present study, there was a discrepancy on the averaged soil CH_4 emissions from lysimeters under low and high water tables in the dry and wet seasons (Figure 9). The findings reported higher soil CH_4 emissions both under low and high water table in the wet (December 2015) and dry (July 2015) seasons, respectively. Moreover, averaged soil CH_4 emission under a low water table were higher compared with that of high water table. These observations were not in agreement with the general belief that soil CH_4 emission increases with a higher water table. There are no specific reasons that explain the anomaly from the findings obtained. However, the inconsistency of soil CH_4 emissions from peat the soils suggest that the factor controlling CH_4 emission from the lysimeters could be attributed to the fluctuation of the water table at the soil–water interface. The water table level and its fluctuation at the soil–water interface may have altered the intensity and duration of CH_4 production and oxidation processes throughout the dry and wet seasons.

The results on the insignificant correlation between soil temperature and CH_4 emission from lysimeters under simulated water table fluctuation irrespective of seasons was consistent with that reported for CH_4 measurement under an open field pineapple cultivation system. These observations are further supported by the fact that CH_4 emission was not affected by soil temperature because of the moderate soil temperature fluctuation (24.7 to 32.7 °C) during CH_4 flux measurement. This finding was in agreement with the study by [39,40] who reported that temperature changes had minimal effects on CH_4 emission from cultivated peatlands.

It is also possible that soil CH_4 from lysimeters and under open field pineapple cultivation was released from non-microbial production of CH_4 sources particularly humic acids and lignin [41]. The emission of non-microbial CH_4 may have occurred under moderate temperature fluctuations of the tropics because of the high amount of organic matter, humic acids, fulvic acids, lignin, humin and carbohydrate in peat soils [42–44]. In this study, the lower soil CH_4 emission from peat soils grown with pineapple under an open field cultivation system compared with that of CH_4 emission from the lysimeters throughout the dry and wet seasons (Figure 11) could be attributed to pineapple fertilization. Compound NPK fertilizers containing ammonium were applied to pineapple at 3, 6 and 9 months after planting in June, September and December 2015, respectively. The compound fertilizers might have increased nitrate content in the peat soils because nitrification increases with peat oxidation. The lower CH_4 emission due to pineapple fertilization relates to the availability of electron acceptors namely nitrate which inhibits CH_4 production [45].

Nitrate is water soluble and leaches to the water table interface leading to decreased CH_4 production in anaerobic condition. Also, water table fluctuation (50 to 70 cm from the soil surface) and lateral water movement in the peat soil (open field cultivation system) might have affected the balance between CH_4 production and consumption by methanogenic bacteria and methanotrophs. Water table depth affects the soil CH_4 emissions because it determines the depth of the oxic or anoxic boundary and redox level within the soil. By contrast, water table fluctuation at the soil–water interface, transportation of CH_4 through molecular diffusion through the aerobic peat layer and ebullition at the peat water table interface relatively explains the higher CH_4 emission from lysimeters under simulated water table fluctuation.

5. Conclusions

Soil CH_4 emission throughout the dry and wet seasons under open field pineapple cultivation and from lysimeters subjected to water table fluctuation were not affected by soil temperature but emissions were influenced by the balance between methanogenic and methanotrophic microorganisms controlling CH_4 production and consumption, CH_4 transportation via molecular diffusion through peat pore spaces, and non-microbial CH_4 production sources in peat soils namely humic acids and lignin. Although it is generally believed that a high water table increases soil CH_4 emission, findings from the study suggest that water table fluctuation at the soil-water interface relatively controls the soil CH_4 emission from lysimeters under simulated low and high water table fluctuations. The outcome of this present study demonstrated that soil CH_4 emission throughout the dry and wet seasons under an open field cultivation system was affected by the availability of nitrate electron acceptors from pineapple fertilization, which restrict CH_4 production, thus leading to lower soil CH_4 emission compared with that of CH_4 emissions from lysimeters. However, the limited number of CH_4 flux monitoring throughout the dry and wet seasons (July, August, September and December 2015) may not be conclusive enough to confirm the findings from the study. Thus, a long-term CH_4 flux monitoring period is required to confirm the findings obtained because rainfall distribution, microbial population, chamber humidity and headspace temperature may influence CH_4 emission and the outcome of the study. The findings of this study provide an understanding on the effects of water table fluctuation on CH_4 emissions in a tropical peatland under pineapple cultivation.

Author Contributions: W.L. was responsible for investigation, writing, and original draft preparation. L.N.L.K.C., R.K.J.H. and M.B.J. were responsible for data analysis and visualization. O.H.A. was responsible for supervising, funding acquisition, project administration, experimental methodology, editing, and reviewing. L.O. was responsible for data arrangement, conceptualization, reviewing, and editing the second draft. A.A., A.A.M. were also involved in funding acquisition and project administration. All authors have read and agreed to the published version of the manuscript.

Funding: This research was funded by Long-term Research Grant Scheme (LRGS) and Translational Research Grant Scheme (TRGS) vote number 6300914 from the Ministry of Higher Education Malaysia.

Institutional Review Board Statement: Not applicable.

Informed Consent Statement: Not applicable.

Data Availability Statement: Not applicable.

Acknowledgments: The authors gratefully acknowledge the financial support provided through Putra Grant UPM. Appreciation also goes to Malaysia Agricultural Research and Development Institute (MARDI) Saratok, Sarawak, Malaysia, for the collaborative research.

Conflicts of Interest: The authors declare no conflict of interest.

References

1. Joosten, H.; Sirin, A.; Couwenberg, J.; Laine, J.; Smith, P. The role of peatlands in climate regulation. In *Peatland Restoration and Ecosystem Services: Science, Policy and Practice*; Cambridge University Press: Cambridge, UK, 2016; pp. 63–76.
2. Chen, X.; Tung, K.K. Varying planetary heat sink led to global-warming slowdown and acceleration. *Science* **2014**, *6199*, 897–903. [CrossRef]
3. Snyder, C.S.; Tom, B.; Jensen, T.L.; Paul, E.F. Review of greenhouse gas emissions from crop production systems and fertilizer management effects. *Agric. Ecosys. Environ.* **2009**, *133*, 247–266. [CrossRef]
4. Nisbet, E.; Manning, M.R.; Dlugockenky, E.J.; Fisher, R.E.; Lowry, D.; Michel, S.E. Very strong atmospheric methane growth in the 4 years 2014–2017: Implications for the Paris Agreement. *Glob. Biogeochem. Cycles* **2019**, *33*, 318–342. [CrossRef]
5. Dlugokencky, E.J.; Lang, P.M.; Crotwell, A.M.; Mund, J.W.; Crotwell, M.J.; Thoning, K.W. Atmospheric Methane Dry Air Mole Fractions from the NOAA ESRL Carbon Cycle Cooperative Global Air Sampling Network, 1983–2016. In Division N-EGM, Editor. Version: 2017-07-28. 2017. Available online: Ftp://aftp.cmdl.noaa.gov/data/trace_gases/ch4/flask/surface/ (accessed on 29 May 2021).
6. Ismail, A.B.; Zulkefli, M.; Salma, I.; Jamaludin, J.; Mohamad, H.M.J. Selection of land clearing technique and crop type as preliminary steps in restoring carbon reserve in tropical peatland under agriculture. In Proceedings of the 12th International Peat Congress, Tullamore, Ireland, 8–13 June 2008; Int Peat Society.
7. Kechavarzi, C.; Dawson, Q.; Bartlett, M.; Leeds-Harrison, B.P. The role of soil moisture, temperature, and nutrient amendment on CO_2 efflux from agricultural peat soil microcosms. *Geoderma* **2010**, *154*, 203–210. [CrossRef]
8. Mäkiranta, P.; Minkkinen, K.; Hytönen, J.; Laine, J. Factors causing temporal and spatial variation in heterotrophic and rhizospheric components of soil respiration in afforested organic soil croplands in Finland. *Soil Biol. Biochem.* **2008**, *40*, 1592–1600. [CrossRef]
9. Fenner, N.; Ostle, J.N.; McNamara, N.; Sparks, T.; Harmens, H.; Reynolds, B.; Freeman, C. Elevated CO_2 effects on peatland plant community carbon dynamics and DOC production. *Ecosystems* **2007**, *10*, 635–647. [CrossRef]
10. Collins, W.J.; Fry, M.M.; Yu, H.; Fuglestvedt, J.S.; Shindell, D.T.; West, J.J. Global and regional temperature-change potentials for near-term climate forcers. *Atmos. Chem. Phys.* **2013**, *13*, 2471–2485. [CrossRef]
11. Saunois, M.; Jackson, R.B.; Bousquet, P.; Poulter, B.; Canadell, J.G. The growing role of methane in anthropogenic climate change. *Environ. Res. Lett.* **2017**, *11*, 120207. [CrossRef]
12. Leifeld, J. Prologue paper: Soil carbon losses from land-use change and the global agricultural greenhouse gas budget. *Sci. Total Environ.* **2013**, *465*, 3–6. [CrossRef]
13. Berglund, Ö.; Berglund, K. Influence of water table level and soil properties on emissions of greenhouse gases from cultivated peat soil. *Soil Biol. Biochem.* **2011**, *43*, 923–931. [CrossRef]
14. Van Huissteden, J.; Petrescu, A.M.R.; Hendriks, D.M.D.; Rebel, K.T. Modelling the effect of water-table management on CO_2 and CH_4 fluxes from peat soils. *Netherland J. Geosci.* **2006**, *85*, 3–18. [CrossRef]
15. Huang, H.; von Lampe, M.; van Tongeren, F. Climate change and trade in agriculture. *Food Policy* **2010**, *36*, S9–S13. [CrossRef]
16. Vaghefi, N.; Mad Nasir, S.; Alias, R.; Khalid, A.R. Impact of climate change on food security in Malaysia: Economic and policy adjustments for rice industry. *J. Integr. Environ. Sci.* **2016**, *13*, 19–35. [CrossRef]
17. Melling, L.; Hatano, R.; Goh, K.J. Methane fluxes from three ecosystems in tropical peatland of Sarawak, Malaysia. *Soil Biol. Biochem.* **2005**, *37*, 1445–1453. [CrossRef]
18. Couwenberg, J.; Dommain, R.; Joosten, H. Greenhouse gas fluxes from tropical peatlands in south-east Asia. *Glob. Chang. Biol.* **2010**, *16*, 1715–1732. [CrossRef]
19. Ahmed, O.H.; Liza Nuriati, L.K.C. *Greenhouse Gas Emission and Carbon Leaching in Pineapple Cultivation on Tropical Peat Soil*; Universiti Putra Malaysia Press: Serdang, Malaysia, 2019; pp. 1–145.
20. Furukawa, Y.; Inubushi, K.; Ali, M.; Itang, A.M.; Tsuruta, H. Effect of changing groundwater levels caused by land-use changes on greenhouse gas fluxes from tropical peat lands. *Nutr. Cycl. Agroecosyst.* **2005**, *71*, 81–91. [CrossRef]
21. Deshmukh, C.; Serça, D.; Delon, C.; Tardif, R.; Demarty, M.; Jarnot, C.; Guérin, F. Physical controls on CH_4 emissions from a newly flooded subtropical freshwater hydroelectric reservoir: Nam Theun 2. *Biogeosciences* **2014**, *11*, 4251–4269. [CrossRef]
22. Burrows, H.E.; Bubier, L.J.; Mosedale, A.; Cobb, W.G.; Crill, M.P. Net ecosystem exchange of carbon dioxide in a temperate poor fen: A comparison of automated and manual chamber techniques. *Biogeochemistry* **2005**, *76*, 21–45. [CrossRef]
23. Abdul, H.; Kazuyuki, I.; Yuichiro, F.; Erry, P.; Muhammad, R.; Haruo, T. Greenhouse gas emissions from tropical peatlands of Kalimantan, Indonesia. *Nutr. Cyc. Agroecosyst.* **2005**, *71*, 73–80.
24. Zulkefli, M.; Nuriati, L.L.; Ismail, A.B. Soil CO_2 flux from tropical peatland under different land clearing techniques. *J. Trop. Agric. Food Sci.* **2010**, *38*, 131–137.
25. Raziah, M.L.; Alam, A.R. Status and impact of pineapple technology on mineral soil. *Econ. Tech. Manag. Rev.* **2010**, *5*, 11–19.
26. Barba, J.; Bradford, M.A.; Brewer, P.E.; Bruhn, D.; Covey, K.; Haren, J.; Vargas, R. Methane emissions from tree stems: A new frontier in the global carbon cycle. *New Phytol.* **2018**, *222*, 18–28. [CrossRef]
27. Sjögersten, S.; Aplin, P.; Gauci, V.; Peacock, M.; Siegenthaler, A.; Turner, B.L. Temperature response of ex-situ greenhouse gas emissions from tropical peatlands: Interactions between forest type and peat moisture conditions. *Geoderma* **2018**, *324*, 47–55. [CrossRef]

28. Hu, Y.G.; Feng, Y.L.; Zhang, Z.S.; Huang, L.; Zhang, P.; Xu, B.X. Greenhouse gases fluxes of biological soil crusts and soil ecosystem in the artificial sand-fixing vegetation region in Shapotou area. *Chin. J. Appl. Ecol.* **2014**, *25*, 61–68.
29. Ehrenfeld, J.G. *Encyclopedia of Biodiversity*, 2nd ed.; Levin, S., Ed.; Academic Press: Cambridge, MA, USA; Princeton University: Princeton, NJ, USA, 2013; pp. 109–128.
30. Inubushi, K.; Otake, S.; Furukawa, Y.; Shibasaki, N.; Ali, M.; Itang, A.M.; Tsuruta, H. Factors influencing methane emission from peat soil: Comparison of tropical and temperate wetlands. *Nutr. Cycl. Agroecosyst.* **2005**, *71*, 93–99. [CrossRef]
31. Kazemian, S.; Huat, B.B.K.; Prasad, A.; Barghchi, M. State of an art review of peat: General perspective. *Int. J. Phys. Sci.* **2011**, *6*, 1974–1981.
32. International Atomic Energy Agency (IAEA). Manual on measurement of methane and nitrous oxide emissions from agriculture. In *Sampling Techniques and Sample Handling*; IAEA-TECDOC-674; IAEA: Vienna, Austria, 1992; pp. 45–67.
33. Rachwal, M.; Fernando, G.; Zanatta, J.A.; Dieckow, J.; Denega, G.L.; Curcio, G.R.; Bayer, C. Methane fluxes from waterlogged and drained Histosols of highland areas. *Rev. Bras. Cienc. Solo* **2014**, *38*, 486–494. [CrossRef]
34. Le Mer, J.; Roger, P. Production, oxidation, emission, and consumption of methane by soils: A review. *Eur. J. Soil Biol.* **2001**, *37*, 25–50. [CrossRef]
35. Farmer, J.; Matthews, R.; Smith, J.U.; Smith, P.; Singh, B.K. Assessing existing peatland models for their applicability for modelling greenhouse gas emissions from tropical peat soils. *Curr. Opin. Environ. Sustain.* **2011**, *3*, 339–349. [CrossRef]
36. Pangala, S.R.; Enrich-Prast, A.; Basso, L.S.; Peixoto, R.B.; Bastviken, D.; Hornibrook, E.R.C.; Gauci, V. Large emissions from floodplain trees close the Amazon methane budget. *Nature* **2017**, *552*, 230–234. [CrossRef]
37. Jauhiainen, J.; Silvennoinen, H. Diffusion GHG fluxes at tropical peatland drainage canal water surfaces. *Suo* **2012**, *63*, 93–105.
38. Luta, W.; Ahmed, O.H.; Heng, R.K.J.; Choo, L.N.K. Water table fluctuation and carbon dioxide emission from a tropical peat soil cultivated with pineapples (*Ananas comosus* (L.) Merr.). *Int. J. Biosci.* **2017**, *10*, 172–178.
39. Kløve, B.; Sveistrup, T.E.; Hauge, A. Leaching of nutrients and emission of greenhouse gases from peatland cultivation at Bodin, Northern Norway. *Geoderma* **2010**, *154*, 219–232. [CrossRef]
40. Kløve, B.; Berglund, K.; Berglund, Ö.; Weldon, S.; Maljanen, M. Future options for cultivated Nordic peat soils: Can land management and rewetting control greenhouse gas emissions? *Environ. Sci. Pol.* **2017**, *69*, 85–93. [CrossRef]
41. Wang, B.; Hou, L.; Liu, W.; Wang, Z. Non-microbial methane emissions from soils. *Atmos. Environ.* **2013**, *80*, 290–298. [CrossRef]
42. Allen, S.J.; McKay, G.; Porter, J.F. Adsorption isotherm models for basic dye adsorption by peat in single binary component systems. *J. Colloid Interf. Sci.* **2004**, *280*, 322–333. [CrossRef]
43. Udin, A.B.M.H.; Sujari, A.M.A.; Nawi, M.A.M. Effectiveness of peat coagulant for the removal of textile dyes from aqueous solution and textile wastewater. *Malay J. Chem.* **2003**, *5*, 034–043.
44. Zulfikar, M.A.; Novita, E.; Hertadi, R.; Djajanti, S.D. Removal of humic acid from peat water using untreated powdered eggshell as a low cost adsorbent. *Int. J. Environ. Sci. Technol.* **2013**, *10*, 1357–1366. [CrossRef]
45. Jassal, R.S.; Black, T.A.; Roy, R.; Ethier, G. Effect of nitrogen fertilization on soil CH_4 and N_2O fluxes and soil and bole respiration. *Geoderma* **2011**, *162*, 182–186. [CrossRef]

Article

On the Use of Multivariate Analysis and Land Evaluation for Potential Agricultural Development of the Northwestern Coast of Egypt

Mohamed El Sayed Said [1,*], Abdelraouf. M. Ali [1,2], Maurizio Borin [3,*], Sameh Kotb Abd-Elmabod [4], Ali A. Aldosari [5], Mohamed M. N. Khalil [6] and Mohamed K. Abdel-Fattah [6]

[1] National Authority for Remote Sensing and Space Sciences (NARSS), Cairo 11843, Egypt; raouf.shoker@narss.sci.eg
[2] Agrarian-Technological Institute of the Peoples Friendship University of Russia, ul. Miklukho-Maklaya 6, 117198 Moscow, Russia
[3] Department of Agronomy, Food, Natural Resources, Animals and Environment—DAFNAE, University of Padua, Agripolis Campus, Viale dell' Università 16, 35020 Legnaro (PD), Italy, Veneto
[4] Soils & Water Use Department, Agricultural and Biological Research Division, National Research Centre, Cairo 12622, Egypt; sk.abd-elmabod@nrc.sci.eg
[5] Geography Department, King Saud University, Riyadh 11451, Saudi Arabia; adosari@ksu.edu.sa
[6] Soil Science Department, Faculty of Agriculture, ZagazigUniversity, Zagazig 44519, Egypt; mmnabil11@gmail.com (M.M.N.K.); mkabdelfattah@zu.edu.eg (M.K.A.-F.)
* Correspondence: elsayed.salama@narss.sci.eg (M.E.S.S.); maurizio.borin@unipd.it (M.B.); Tel.: +20-10-6144-5686 (M.E.S.S.)

Received: 20 July 2020; Accepted: 21 August 2020; Published: 3 September 2020

Abstract: The development of the agricultural sector is considered the backbone of sustainable development in Egypt. While the developing countries of the world face many challenges regarding food security due to rapid population growth and limited agricultural resources, this study aimed to assess the soils of Sidi Barrani and Salloum using multivariate analysis to determine the land capability and crop suitability for potential alternative crop uses, based on using principal component analysis (PCA), agglomerative hierarchical cluster analysis (AHC) and the Almagra model of MicroLEIS. In total, 24 soil profiles were dug, to represent the geomorphic units of the study area, and the soil physicochemical parameters were analyzed in laboratory. The land capability assessment was classified into five significant classes (C1 to C5) based on AHC and PCA analyses. The class C1 represents the highest capable class while C5 is assigned to lowest class. The results indicated that about 7% of the total area was classified as highly capable land (C1), which is area characterized by high concentrations of macronutrients (N, P, K) and low soil salinity value. However, about 52% of the total area was assigned to moderately high class (C2), and 29% was allocated in moderate class (C3), whilst the remaining area (12%) was classified as the low (C4) and not capable (C5) classes, due to soil limitations such as shallow soil depth, high salinity, and increased erosion susceptibility. Moreover, the results of the Almagra soil suitability model for ten crops were described into four suitability classes, while about 37% of the study area was allocated in the highly suitable class (S2) for wheat, olive, alfalfa, sugar beet and fig. Furthermore, 13% of the area was categorized as highly suitable soil (S2) for citrus and peach. On the other hand, about 50% of the total area was assigned to the marginal class (S4) for most of the selected crops. Hence, the use of multivariate analysis, mapping land capability and modeling the soil suitability for diverse crops help the decision makers with regard to potential agricultural development.

Keywords: PCA; land capability; crop suitability; GIS; NWCE; Egypt

1. Introduction

The world's population will increase to reach approximately 9.7 billion by 2050 [1,2]. The huge population increase will impact agricultural resources as it causes global food security pressure on the lack of agricultural lands [3,4]. Two types of challenges put pressure on governments, namely the increasing population on the one hand, and the decrease in productive land on the other hand [5]. There are two ways that governments can counteract overpopulation: the first is to encourage farmers to increase crop yields by using land fertilizers, pesticides, etc., which affect environmental quality, and the second is to rely on imports to fill the food gap [3,5]. Therefore, it is required to increase the efforts to improve living standards to provide safe food to feed citizens [6,7]. The expansion of new agricultural lands is the goal of developing countries such as Egypt where, the annual growing population rate is 84%, while the strategic crops production such as wheat is insufficient, therefore, the government relies on importation from abroad [8–13]. Moreover, agricultural expansion in arable lands aims to achieve sustainable agricultural development, which depends on the integration of land and water resources and the surrounding environmental factors [14,15]. Agriculture lands in Egypt are confined to the Nile Valley and the Delta that represents approximately 4% of the total area of Egypt [16]. The agricultural sector contributes to 14.5% of the gross domestic product (GDP) of Egypt, whilst representing about 30% of the provision of foreign currency as a result of the profits of exporting agricultural products abroad, and it also contributes to reducing unemployment by 41% [17]. The concept of land assessment belongs to the rate of land performance and its capacity for crop production where the capacity of land depends on the climate and location/geography, the inherent soil characteristics (physical, chemical) and also includes the soil potential for agricultural production [18,19]. The importance of land evaluation helps in selecting the suitable crops based on the soil characteristics and assists the decision makers [15,20]. The excess of salt concentration in soil may damage soil structure by decreasing soil aeration and its permeability, and consequently adversely affects the agricultural production. However, appropriate soil management leads to decreased salt concentration, a decline in sodium percentage, and improved drainage conditions [21,22]. There are several agriculture management practices for saline–sodic and calcareous soils in arid and semi-arid regions, such as improving the soil's physiochemical characteristics by adding organic matter, reducing soil salinity by fresh water leaching, and reducing sodium saturation and the alkaline pH using gypsum and sulfur applications [14]. The soil limitation factors for crop suitability differ from one place to another in Egypt, while the dominant limiting factors in the north of the Nile Delta are the soil salinity, poor drainage and compaction [23–28]. Meanwhile, the hardpan layers, shallow depth of the soil profile, and the rock outcrops as well as the steep slope are the most common limiting factors in the desert lands [15,29,30].

The northwestern coast region is vulnerable to land degradation and desertification processes which lead to reduced soil fertility and cause environmental impacts, as the lower areas in the north suffer from rising salinization and alkalinization, meanwhile the valleys are susceptible to runoff and soil erosion, where surface runoff reached 200 mm/year, and the soil erosion of soil has reached 60 t h^{-1} y^{-1} [30]. Over the past five decades, several models for soil capability classification have been proposed to classify the soils according to their chemical and physical properties [18]. De la Rosa et al. [31] suggested micro land evaluation system (MicroLEIS) to test the soil suitability: this system integrates soil characteristics, topography, vegetation cover, land use and climate conditions. Other methods were developed to classify the soil capability for crop production according to their soil profile description and soil characteristics such as slope, texture, salinity and other factors such as the drainage conditions, where each class takes an average value between 0 and 100, where 100 reflects the best conditions and vice versa [32]. The assessment of land capability depends on an evaluation of the soil quality and expresses a capacity of the soil to function in an ecosystem in order to sustain the soil productivity of a crop in parallel with reducing the soil degradation processes, in addition to its ability to perform a number of basic functions such as supporting crop production [24,30], whereas the soil is a complex mixture of organic, inorganic materials and it is influenced by the surrounding factors such as climate, topography and human activities. The integration of soil characteristics, climatic data, remotely sensed data, water analysis and socio-economic data with GIS modeling assist to develop spatial decision support system (SDSS) for soil management [12,33,34]. Due to the obscure nature of the soil system, it is not easy to evaluate the soil by integrating their properties together. Therefore, multivariate analysis is an appropriate tool for evaluating soil capability zones due to its efficiency in modeling and systematically using vague and imprecise situations [7,13,35]. Principal component analysis (PCA) is considered the most popular model to analyze the physical, chemical, socio–economic and other factors in a multistage analysis to develop an indicator that represents the evaluation of land capability. The PCA and fuzzy clustering means (FCM) methods were used for land suitability evaluation for oil palm and soil quality based on soil characteristics and climate data [36,37]. FCM is an unsupervised clustering method for the data components that enables investigating the accumulation of multiple elements. Clustering can extract the homogenous regions based on different phenomena. Moreover, given the gradually changing nature of soil behaviors, it seems that the fuzzy clustering method can interpret the spatial variation of studied phenomena better than the other methods [38]. The PCA was used to decrease the variables to increase the accuracy of FCM land quality evaluation. Coinciding with the availability of computers, satellite data and GIS in the past two decades have led to improved methods of land assessment, where satellite images provided important information on the status of plants, topography and climate [39,40]. In addition, agglomerative hierarchical clustering (AHC) was used to define the distances between points where the similar points are forming one cluster, then finally all clusters are presented in a dendrogram form [41]. The Northern coast of Egypt depends on seasonal rains during the winters, and the citizens' lack of awareness of suitable crops and water irrigation quantity suit those conditions [30]. Therefore, the purpose of this study is to use multivariate analysis to assess the soil of the study area to determine the most appropriate land use.

2. Materials and Methods

2.1. Study Area

The investigated area was allocated in Matrouh governorat, northwestern Egypt, close to the Libyan border. The area includes the Sidi Barrani and Salloum districts (Figure 1). It is located between 25°10′ to 26°55′ East and 31°00′ to 31°37′ North. The area covers approximately 918,000 hectares. The study area is characterized by an arid and semi-arid climate, where the average temperature reaches 18 °C in the whole year; except in June, July and August, the temperature ranges between 25 and 30 °C. The annual rainfall fluctuates between 100 and 200 mm/year.

The elevation ranged between 0 and 250 m above sea level in the south, whereas the low elevation values were observed in the parts close to the Mediterranean Sea and the high values were noticed in the plateau that occupies the southern parts. The north of the study area is characterized by a gentle slope, while the south is characterized by a moderate to steep slope (Figure 2). The micro-relief varies from almost flat to undulating with scattered escarpments, and a flat coastal zone about 1–3 km wide. Some principle wadis dissect the escarpment, especially southwest of Sidi Barrani. Wadi is an Arabic term which traditionally refers to a valley that is located in low land and receives a high amount of rainfall compared with the surrounding land forms [42].

The vegetation cover of the study area is changing during the winter and spring seasons as the rainfall is active, whereas the natural vegetation spreads throughout the study area, especially in the wadis and streams. The natural vegetation spreads on the fine sand stacks that keep rain water during the winter season and leads to natural vegetation growth during the autumn and summer seasons [30].

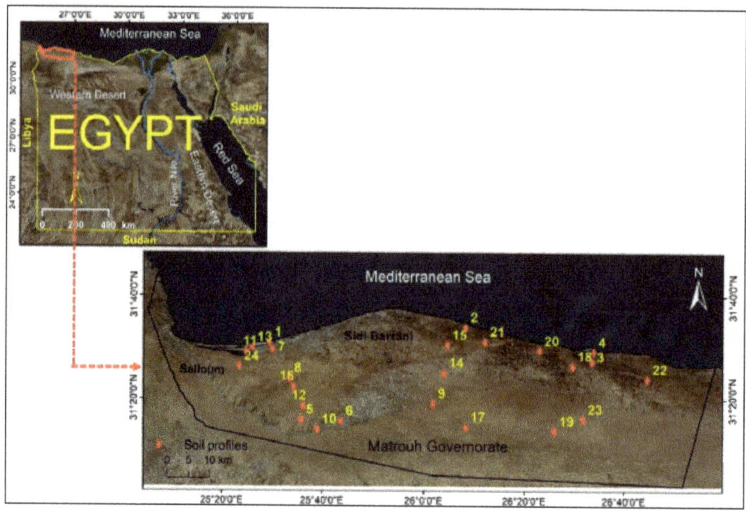

Figure 1. Location of the study area in northwestern Egypt and the soil profiles with a red dot.

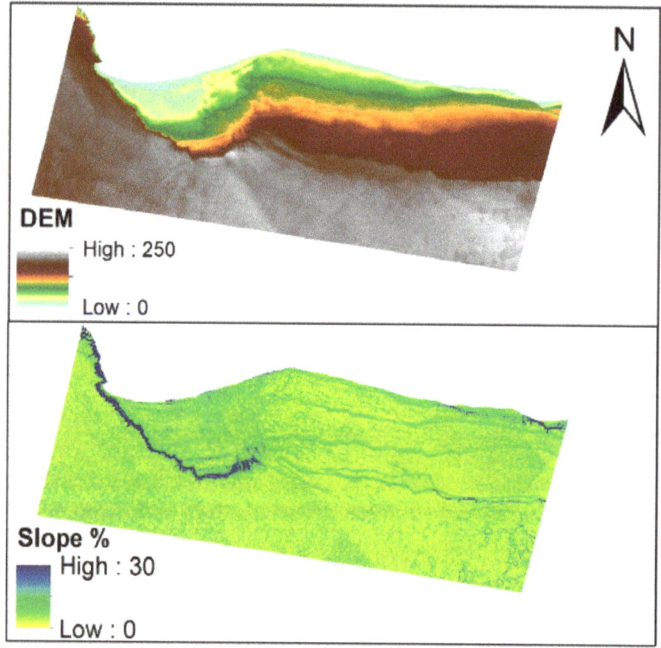

Figure 2. Digital elevation model (DEM) in meters and slope (%) of the study area.

2.2. Remote Sensing and Spatial Analysis

Remote sensing data were used to link the soil capability with land use and geomorphological units of the study areas, whereas the Sentinel 2 image was acquired 1 July 2020 with spatial resolution 10 m for the red, blue and green bands and 20 m for the near infrared band. On the other hand, the digital elevation model (DEM) with 30 m resolution was used to describe the variation in the surface characteristics and its relation with geomorphological units and soil capability.

The slope percentage in the study area reached 30% in the adjacent escarpments of the plateau while low slope values were observed in the wadis (Figure 2). Sentinel 2 image data, DEM and the field survey were integrated to enhance the visibility of the geomorphological map that was produced by [22] using the method described by [43]. This relied on topographic information such as the slope, aspect, and stream networks of the study area that were extracted using DEM. Consequently, a number of 15 geomorphological units were identified to represent the diverse geomorphological features. All geomorphological units were verified based on a field survey using GPS. Then, the produced geomorphological map was used a base map, where each geomorphic unit was homogeneous in its natural characteristics. Generally, the biophysical characteristics were spatially analyzed [44].

2.3. Field Survey and Soil Laboratory Analysis

The field survey relies on the identified geomorphic units of the study area, whereas twenty-four soil profiles were dug to a depth of 150 cm or less according to the presence of hardpan layers, in order to represent geomorphic units (Figure 1). The soil profile samples were determined based on a sample area method that was suggested by [45] where the sample area should be distributed on the lines that pass the geomorphological units without bias. Therefore, some units have more than one profile and another unit has only one profile.

In addition, the units that have large areas have more representative soil profiles compared with the small unit. The descriptions of soil profiles were done in the field using [19]. Soil samples were collected from different layers of soil profiles. Then, the soil physiochemical parameters were analyzed. Table 1 showed the mean soil characteristics of the soil profiles. The following chemical soil properties were determined: salinity (electric conductivity) [46] and soil acidity (pH) in saturated paste [47], cation exchange capacity (CEC, [48]), soil organic matter content by the acid-dichromate potassium and titration method [49]. In addition, the particle size distribution and macronutrients were measured according to [40]. Soil nitrogen (N) was determined using the Kjeldahl method [50]. The soil phosphorus was determined using the spectrophotometer device according to [51] (P) and potassium (K) was determined using a flame photometer. The available soil potassium contents were measured using flame photometry [50]. The exchangeable sodium percentage (ESP) was determined using methods of [52,53]. The soil classification of each soil profile was done according to [54].

2.4. Statistical Analysis

PCA is a statistical procedure that uses orthogonal transformation to convert a set of observations of possibly correlated variables (entities each of which takes on various numerical values) into a set of values of linearly uncorrelated variables called principal components (PCs). PCs having eigenvalues greater than one were retained whereas PCs less than 1 were subtracted away. Soil properties were summarized using PCA.

Before performing the principal component analysis (PCA), a linear relationship between the soil variables were checked using the Pearson correlation coefficient. This analysis requests sampling adequacy, therefore, the Kaiser–Meyer–Olkin (KMO) was done to measure the sampling adequacy for the overall data set. If the KMO value is greater than 0.50, the PCA would be suitable. In addition, the Bartlett's test was performed, and if the p value of Bartlett's test value is less than 0.05, this indicates that the PCA may be suitable for the work [55,56]. Generally, all the statistical analyses were performed using SPSS software version 25.

Based on the PCA results, the soil profiles which were considered as objects for soil capability evaluation were classified into dissimilar clusters using the AHC methods according to each location characterized by a group of soil variables (chemical, physical, biological). Through this analysis, dissimilar groups of soil variables were arranged together graphically in a structure called a dendrogram of dissimilarity.

Table 1. The variation of soil characteristics for each geomorphic unit of the study area.

Profile No.	Depth, cm	pH	ESP, %	OM, %	EC, dS/m	CaCO$_3$, %	CEC cmolc/kg	Texture	Drainage	Rock Fragment, %	AN, mg kg^{-1}	AP, mg kg^{-1}	AK, mg kg^{-1}	Erodibility, ton/ha/Year
1	120	8.5	4.41	0.64	10.03	32.55	23.5	LS	WD	6.4	11.4	10	29	0.18
2	115	8.32	4.42	0.64	2.88	25.86	20.5	SCL	WD	7.3	13	10	29	0.23
3	110	8.6	4.45	0.22	3.01	57.75	14.6	SL	WD	7.4	5.38	3.7	9.7	0.03
4	35	8.59	1.00	0.48	2.57	55.35	5.6	LS	PD	4.6	10.4	7.8	22	0.02
5	150	8.28	2.95	0.14	1.47	40.86	5.6	S	WD	6.5	3.69	2.6	6.3	0.06
6	60	8.2	2.06	0.24	2.09	12.57	9.9	S	ID	3.6	5.82	3.9	11	0.07
7	120	7.8	1.65	0.33	16.67	4.60	9.0	SL	ID	6.8	8.21	5.1	15	0.02
8	55	7.7	8.31	0.54	5.96	5.95	6.3	SL	PD	4.1	11.6	8.6	25	0.04
9	45	8.27	3.52	0.34	3.07	51.73	11.3	S	PD	5.3	8.92	5.4	14	0.10
10	35	7.6	1.25	0.23	2.68	8.03	5.8	S.	PD	6.2	5.2	3.7	11	0.19
11	150	8.24	3.99	0.50	42.99	48.62	13.9	SL	WD	8.3	12.4	7.9	21	0.01
12	85	8.3	2.31	0.37	5.35	51.57	11.3	SL	ID	6.2	8.93	6	17	0.02
13	115	8.2	2.52	0.42	96.56	23.39	10.7	SL	WD	4.2	10.4	6.7	19	0.04
14	95	7.9	3.10	0.50	16.50	7.80	12.1	LS	WD	3.8	8.1	7.2	20	0.11
15	120	8.41	1.18	0.19	3.91	69.50	6.8	S	ID	7.4	5.64	3.1	9.1	0.06
16	90	7.59	1.10	0.30	3.01	20.70	5.4	S	WD	5.5	6.86	4.7	14	0.07
17	20	7.82	1.18	0.38	4.58	10.80	5.6	S	PD	4.8	9.44	6.3	17	0.03
18	30	7.95	2.40	0.38	4.80	9.80	6.2	S	PD	4.7	15	7.2	14	0.10
19	55	8.2	1.20	0.20	9.50	8.40	8.1	SL	ID	3.9	9.2	8.8	16	0.08
20	120	8.13	0.95	0.40	2.54	25.67	9.2	S	WD	5.9	10.3	6.7	18	0.21
21	115	8.13	1.13	0.28	1.91	45.97	6.2	S	WD	6.8	7.62	4.6	13	0.15
22	80	8.08	0.90	0.53	2.16	38.86	4.2	LS	ID	4.3	11.7	8.5	24	0.10
23	150	8.19	0.74	0.18	2.71	57.51	5.0	S	WD	6.7	4.68	3	8.3	0.15
24	120	7.88	0.90	0.20	3.18	25.00	10.1	S	WD	5.3	8.4	3.5	9.2	0.23

ESP, exchangeable sodium percentage; OM, organic matter; EC; electric conductivity; CEC, cation exchange capacity; Drainage: WD, well drained; PD, poorly drained; ID, imperfectly drained. Texture: L, loamy; SCL, sandy clay loam; SL, sandy loam; LS, loamy sand; S, sandy. AN, available nitrogen; AP, available phosphorous; AK, available potassium.

2.5. Land Evaluation

The assessment of soil capability depends on determining the relationship of soil proprieties and agriculture suitability. In this context, the multivariate analysis aims to classify soil capability based on the harmony of the soil properties in each class. Therefore, determining the capability of soil was based on the integration between PCA and AHC methods, as the PCA shows a visual representation of the dominant patterns in order to identify the similarities and differences among soil properties [57]. The assessment output was grouped into five broad classes: C1—a highly capable land, C2—moderately high class, C3—moderate class, C4—low and C5—marginal.

The Almagra model of MicroLEIS [58,59] is a qualitative approach characterized by being built on using soil factors and the favorable conditions for each crop. The soil factors considered in the model are: profile depth (p), texture (t), carbonate (c), salinity (s), drainage (d), and sodium saturation (a). The model was used to assess the soil suitability for ten traditional crops, annuals, e.g., wheat, sunflower, soybean, maize, sugar beet, potato; semiannual, e.g., alfalfa and perennials, e.g., citrus, olive and peach. Almagra defines the soil suitability qualitatively through five classes: S1—optimum, S2—high, S3—moderate, S4—marginal and S5—not suitable. Almagra was calibrated previously in several studies within the Mediterranean, semi-arid, and arid regions [14,60,61]. Figure 3 illustrates the flowchart methodology of soil evaluation based on the integration of soil information remote sensing data and GIS using multivariate analysis as following.

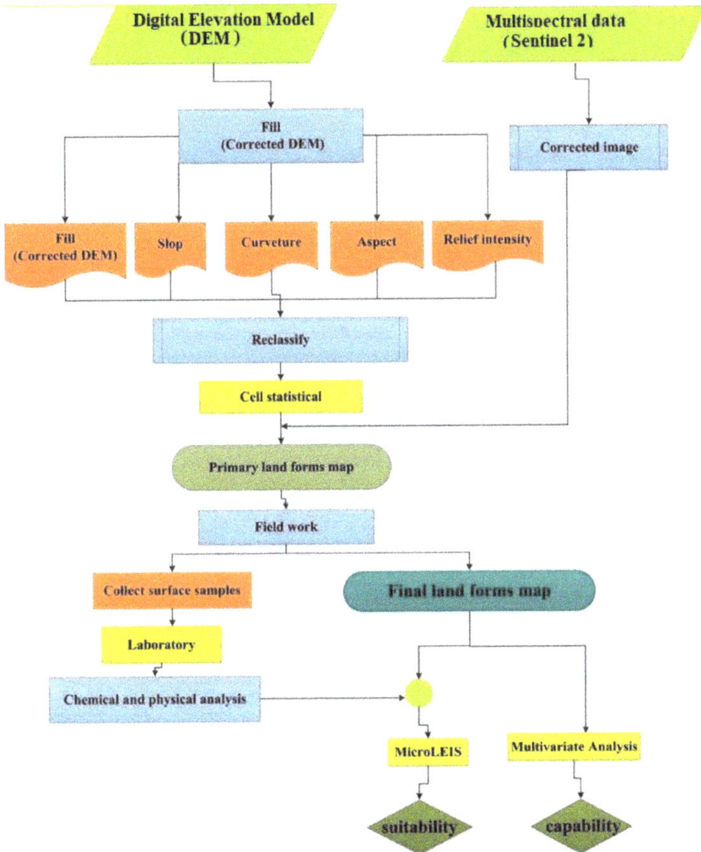

Figure 3. Flowchart of the land evaluation of the study area.

3. Results

3.1. Geomorphological Map and Soil Characteristics

Fifteen geomorphological units were recognized and modified using the integration of DEM, sentenal2 images and ground truth data. Figure 4 and Table 2 show the fifteen geomorphic units that describe the variations of the study area. The landscape of the study area was characterized by an almost flat to undulating surface, while the ridges that were located at the north were intersected by short and long wadis.

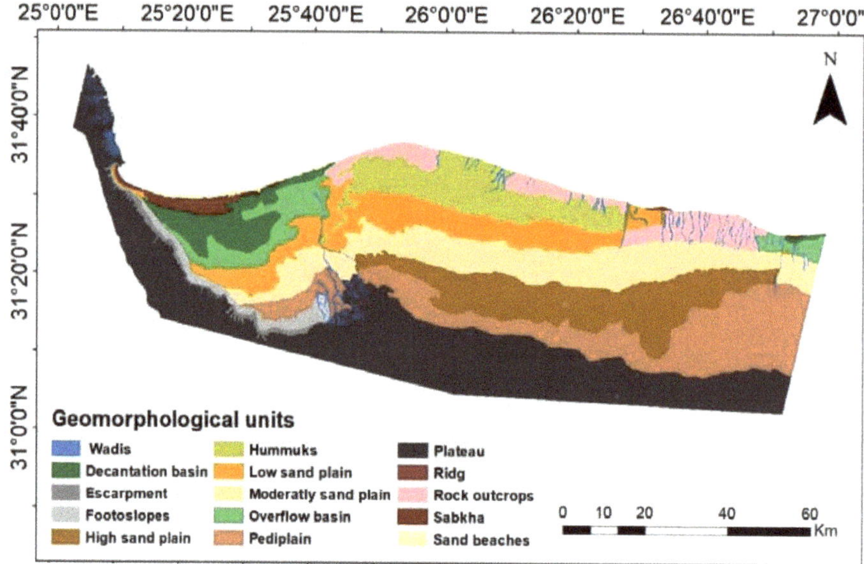

Figure 4. Geomorphological map of the study area. Source: modified after Mohamed et al. [30].

Table 2. The areas of geomorphological units.

Geomorphologic Units	Area, ha	Area, %
Wadis	7201.39	0.93
Decantation basin	28,015.05	3.62
Escarpment	6905.64	0.89
Foot slopes	13,015.67	1.68
High sand plain	86,488.80	11.18
Hummuks	57,821.26	7.47
Low sand plain	71,629.91	9.26
Moderatly sand plain	101,777.62	13.16
Overflow basin	31,048.51	4.015
Pediplain	108,714.35	14.05
Plateau	209,502.84	27.09
Ridg	1091.26	0.14
Rock outcrops	36,830.13	4.76
Sabkha	8095.55	1.04
Sand beaches	5095.41	0.65
Total	**773,233.46**	**100**

The following lines describe the dominant geomorphic units of the study area

Wadis—this unit is widespread in the northwestern coast of Egypt and represents one of the most diagnosed geomorphological units. It is located in gentle slopes and receives high amounts of runoff compared with the surrounding upland. Wadis of the study area include different lithological material, and their sedimentary structures vary widely from gravel to mud. Generally, the soils of wadis are considered the most suitable for agriculture development in the northwestern coast. The dominants soil textures of the Wadis were sand clay loam, sandy loam and sand. The soil depth ranges between 110 and 150 cm based on their position in the valley and slope. The low values of soil salinity were observed where it varied between 1.5 and 3.0 dS/m. The soil pH fluctuated between 8.3 and 8.6. The organic matter (OM) content is very low where it ranges between 0.14 and 0.64%. The carbonate content ($CaCO_3$) is high (25–40%). The CEC depends upon the clay and organic matter contents, while it ranged from 5.6 to 20 cmolc kg^{-1}. Meanwhile, the soil classification of wadis was determined to be *Typic Haplocalcids*.

Basins—basin units are described as low lands where rainfall and drained water accumulate into the outlet. Therefore, basins include the accumulative surface runoff, and nearby streams that run downslope towards the shared outlet. This unit is divided into decantation basins and overflow basins:

I-Decantation basins are characterized by deep and moderate soil depth, which ranges between 100 and 120 cm. The soil texture fluctuates between sandy loam and loamy sand. The soil salinity is slightly moderate to high, varying between 3 and 16 dS/m. The OM is low as it ranges between 0.2 and 0.3%. The soil pH is categorized as moderately alkaline and ranges between 7.8 and 7.9. The carbonate content is slightly moderate to high when it ranges between 4.5 and 25%. The CEC ranged between 9.0 and 10.1 meq/100 g, and the soils are classified as *Typic Torripsamments*.

II-Overflow basin—this unit is the upper areas of basins. The soil texture ranges between sand and loamy sand. Soil salinity fluctuates from low to moderate, where it reached to 3 dS/m. The OM is very low where it is less than 0.3%. The $CaCO_3$ content varies from moderate to high where it reached 20%. CEC was low (5 cmol/kg). Soils are classified as *Typic Torrifluvents*.

Foot slopes are mainly located in the marginal land as plateau areas, were the surface is covered by weathered fragments. The soil depth is shallow as a result of the presence of hardpan layers that are observed at depths ranged between 35 and 55cm. Coarse sand and sandy loam are the dominant soil textures, $CaCO_3$ content is slightly moderate, ranging between 5.95 and 8%. The electric conductivity (EC) is slightly low as it ranges between 2.6 and 5.9 dS/m, and the pH is slightly alkaline (7.6 to 7.7). The OM content is relatively low (0.23 and 0.54). The CEC varied between 5.6 and 6.6 cmol kg^{-1}, while the soil was classified as *Lithic Torriorthents*.

Sand plains occupy an area of about 33.6% of the total area, classified into **I—high, II—moderate and III—low sand plains**. Soil textures were varied between sandy, loamy sand and sandy loam. Soil profile depth ranged between 80 and 150 cm. The EC value fluctuated from low to high (2.1 to 16.5 dS/m). The Soil pH was moderately alkaline except for some parts of the unit, and it ranged between 7.9 and 8.3. The OM content is very low to low where it ranged between 0.18 and 0.53%. The $CaCO_3$ content is varied in a wide range from moderate to high (7.8–57.51%). The CEC was low where it ranged between 4.2 and 12.1 cmol kg^{-1}, unlike the soils classified as *Typic Torripsamments* and *Typic Torrifluvents*.

Hummuks areas are characterized by the undulating surface and it has an accumulation of sand dunes due to the active wind, the units of which occupy an area of about 7% of the total area. The dominated soil textures varied between sandy and loamy sand. The soil depths were moderately deep and deep (115–150 cm). The EC was low as it ranged between 1.9 and 3.9 dS/m. Soil pH is mostly moderate alkaline (8.13–8.4). The OM content is very low (0.19 and 0.40%). The $CaCO_3$ content is high (45–69%). The CEC was low where it ranged between 4.2 and 9 cmol kg^{-1}, while the soil was classified as *Typic Torripsamments*.

Pediplain unit occupies 14% of the total area, gently undulating and almost featureless in their surface, which was formed by the erosion processes over a long time. This unit has a shallow soil depth (20–60 cm). Soil is slightly saline where the EC values ranged between 2.09 and 9.5 dS/m. Soil pH is moderately alkaline (7.8–8.2). The OM content was very low (0.2–0.38%). The CaCO$_3$ content ranged between 8.4 and 10.8%. The CEC was low where it ranged between 5.6 and 9.9 cmol kg^{-1}, whereas the soils were classified as *Lithic Torripsamments*.

Rock outcrops—this unit was a gently undulating surface with rock fragments with a diameter of a few centimeters to a few meters. Soil depths were shallow and ranged between 30 and 45 cm. The EC ranged between 2.57 and 4.8 dS/m, and the soil pH fluctuated between 7.95 and 8.59. The OM content was low where it ranged between 0.34 and 0.48%. The CaCO$_3$ content ranged between 9.80 and 55.35%. The CEC ranged between 5.6 and 11.3 cmol kg^{-1}, while the soils of this unit were classified as *Lithic Torripsamments*.

Sabkha—this unit was located in the low land at the north of the study area, and the soil texture of this unit was loamy sand. The soil salinity was varied between high and very high (42–96 dS/m). The OM was low (0.4 to 0.5%). The CaCO$_3$ content was high and it ranged between 23 and 48%. The CEC was varied from 10.7 to 13.9 cmol kg^{-1}, and the soil was classified as *Typic Haplosalids*.

Sand beaches—this unit is a strip of sand close to the Mediterranean Sea, and the soil texture of this unit was loamy sand. The soil salinity reached 10 dS/m. The OM was low as it reached 0.64%. The CaCO$_3$ content was high and it reached 32.5%. The CEC was 23.5 cmol/kg. The soil was classified as *Typic Torripsamments*.

The study area has others geomorphological units that were not considered for the assessment of the soil capability and crop suitability such as Ridges (narrow strip of the hardpan layer of calcium carbonate that resisted erosion) and the plateau as it has a very shallow soil depth.

3.2. Multivariate Statistical Analysis

The Pearson correlation analysis illustrates the correlations between the soil variables and among them as shown in Table 3. There is a positive significant correlation between the pH and CaCO$_3$ where ($r = 0.72$). Moreover, there was a positive significant correlation between the ESP and CEC ($r = 0.88$) and with available K ($r = 0.46$). While there was a logical positive correlation between the organic matter and CEC ($r = 0.44$), available N ($r = 0.89$), available P ($r = 0.80$) and available K ($r = 0.78$). Moreover, there was a correlation between the CEC and the available P ($r = 0.54$) and the available K ($r = 0.62$).

Table 3. Correlation matrix (Pearson).

Variables	Depth	pH	ESP	OM	EC	CaCO$_3$	CEC	Rock fr.	AN	AP	AK	Er.
pH	0.28											
ESP	0.03	0.08										
OM	−0.23	0.01	**0.41**									
EC	0.25	0.05	0.12	0.15								
CaCO$_3$	0.38	**0.72**	−0.10	−0.25	−0.09							
CEC	0.13	0.15	**0.88**	**0.44**	0.12	−0.13						
Rock fr.%	**0.62**	0.30	0.05	−0.17	−0.08	**0.55**	0.10					
Av. N	−0.25	0.00	0.31	**0.89**	0.23	−0.25	0.39	−0.19				
Av. P	−0.21	0.12	0.40	**0.80**	0.18	−0.27	**0.54**	−0.24	**0.84**			
Av. K	−0.11	0.10	**0.46**	**0.78**	0.22	−0.19	**0.62**	−0.14	**0.76**	**0.94**		
Er.	0.19	−0.11	−0.16	0.07	−0.29	−0.13	0.11	0.10	0.05	0.06	0.07	

ESP, exchangeable sodium percentage; OM, organic matter (%); EC; electric conductivity (dSm^{-1}); CEC, cation exchange capacity (cmolc kg^{-1}); Rock fr., rock fragments; AN, available nitrogen; AP, available phosphorous; AK, available potassium; Er., erodibility (ton/ha/year). Note: values in bold are different from 0 with a significance level alpha = 0.05.

Table 4 shows the factor loadings and component score coefficient outputs that illustrate the higher factor loads using the varimax method. The most representative physical and chemical indicator for PC1, as it is closely correlated with the ESP, OM, CEC, N, P and K, while the second factor (PC2) was correlated with soil depth, pH, CaCO$_3$, and rock fragment, whereas the third (PC3) was correlated with K, the fourth factor (PC4) contributes with the first in ESP and the fifth (PC5) was correlated with EC.

Table 4. Summarization of the principal component analysis (PCA) analysis.

PCs parameters	PC1	PC2	PC3	PC4	PC5
Eigenvalue	4.43	2.45	1.35	1.23	1.06
Variability, %	36.9	20.42	11.27	10.28	8.85
Cumulative, %	36.9	57.32	68.59	78.86	87.72

	Factor loadings					Component Score Coefficient				
	PC1	PC2	PC3	PC4	PC5	PC1	PC2	PC3	PC4	PC5
Depth cm	−0.25	0.70	0.20	0.25	0.48	−0.06	0.29	0.15	−0.19	0.45
pH	−0.01	0.73	−0.28	−0.41	−0.19	0.00	0.30	−0.22	0.33	−0.18
ESP,%	0.61	0.34	0.02	0.55	−0.41	0.14	0.14	0.02	−0.45	−0.38
OM, %	0.88	0.00	0.02	−0.24	0.07	0.20	0.00	0.01	0.19	0.07
EC, dS/m	0.24	0.14	−0.56	0.34	0.65	0.06	0.06	−0.41	−0.28	0.61
$CaCO_3$, %	−0.38	0.74	−0.21	−0.39	−0.16	−0.09	0.3	−0.15	0.31	−0.15
CEC, cmolc kg^{-1}	0.69	0.41	0.22	0.44	−0.25	0.16	0.17	0.17	−0.35	−0.24
Rock fr.,%	−0.28	0.74	0.26	0.02	0.10	−0.06	0.30	0.20	−0.01	0.09
AN, mg kg^{-1}	0.87	−0.03	−0.05	−0.29	0.18	0.20	−0.01	−0.04	0.23	0.16
AP, mg kg^{-1}	0.93	0.04	−0.01	−0.23	0.06	0.21	0.02	−0.01	0.18	0.05
AK, mg kg^{-1}	0.91	0.15	0.02	−0.14	0.07	0.21	0.06	0.02	0.11	0.07
Er., ton/ha/year	0.03	−0.01	0.87	−0.20	0.26	0.01	0.00	0.64	0.18	0.24

ESP, exchangeable sodium percentage; OM, organic matter; EC; electric conductivity; CEC, cation exchange capacity; AN, available nitrogen; AP, available phosphorous; AK, available potassium.

Figure 5 shows the correlations between the different variables based on the angle changes between the vectors, where the degree of the angle expresses the correlations between the variables and each other. The results showed the linkage between some variables by the low angle that means there is a high correlation in the positive direction such as $CaCO_3$ with OM, while the variables were linked together by angle around 90°, and in this case there was no correlation between the variables; on the other hand, there were some variables linked by angle close to 180°, which indicates a correlation in the negative direction. Bartlett's sphericity and KMO test (Table 5) show suitable p values as ($p < 0.05$) of the acceptable level. Furthermore, Bartlett's sphericity test showed that the p value was lower than 0.001. According to the Bartlett's sphericity and KMO test, the PCA was applicable to the current data.

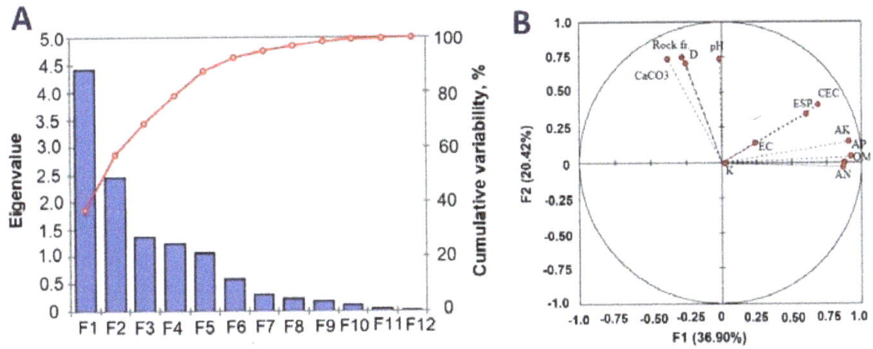

Figure 5. The scree plot of eigenvalue (**A**) and principal component analysis bi-plot (**B**) for soil properties.

Table 5. Bartlett's sphericity and the Kaiser–Meyer–Olkin (KMO) test.

Kaiser–Meyer–Olkin measure of sampling adequacy:	0.52
Bartlett's sphericity test:	
Chi-square (Observed value)	206.90
Chi-square (Critical value)	85.96
Degree of freedom (DF)	66.00
p-value	<0.0001
Alpha	0.05

3.3. Land Capability Based on PCA

The PCA results were used to evaluate the soil capability taking in consideration the variation of topography and climate conditions, therefore the geomorphological units of the study area were used as a base map for the land capability evaluation as each geomorphic unit has specific characteristics such as elevation, slope, aspect and geomorphic features. PCA was performed using the soil physicochemical properties of 24 soil profiles. PCA was applied for the soil characteristics based on the eigenvalues, proportions of variance and cumulative factors by the PCs. Table 4 showed the soil indicator groups. Furthermore, the PCs that had eigenvalues > 1 were retained whereas the PCs < 1 were neglected. Therefore, the first five groups were selected as their eigenvalues were bigger than 1. Table 4 and Figure 5 show these five PCs, which explain the cumulative variance of 87.72% of the studied variables, as the first component explains about 36.9%, the second 20.52%, third 11.27%, fourth 10.28% and the fifth 8.89% of the total variance. Figure 6 illustrates the hierarchical dendrograms for the classification of the soil properties, where each of the five clusters was represented by soil profiles that contain a set of similar soil properties.

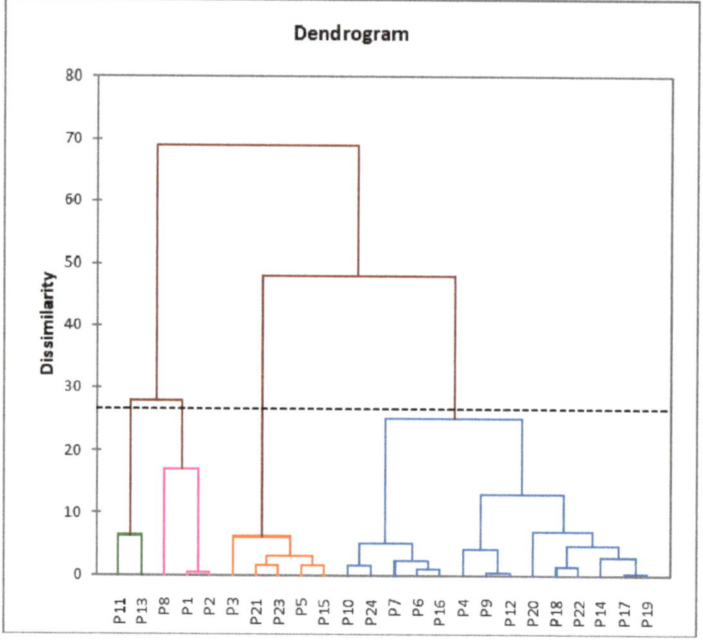

Figure 6. Dendrogram for agglomerative hierarchical clustering.

The land capability map of the study area was produced based on the results of the PCA as the map reflects the five groups previously obtained as shown in Figure 7. Table 6 illustrates the main statistical analysis of the soil properties for soil callability classes (C1 to C5) where;

Highly capable class (C1) represents 7% of the total area, a class characterized by deep soil profiles (<115 cm), a high concentration of macronutrients (N, P, K) and low salinity values. On the other hand, the high contents of calcium carbonates (21.45%) were considered as a limiting factor and the soil of this class is vulnerable to erosion where the erodibility factor reached 0.15 ton/ha/year.

Moderately high class (C2) occupies 52% of the total area, the soils of this class were characterized by a deep soil profile and low soil salinity. The limiting factors were the low contents of macronutrients and high contents of calcium carbonates ($\approx$54%).

Moderate class (C3) occupies 29% of the total area. The soil chemical analysis of this class showed low values of salinity, CEC and OM, while the limiting factors of this unit were high contents of calcium carbonates and the shallow soil depth (56 cm).

Low class (C4) represents 11% of the total study area. This unit has a certain number of limitations, such as the soil depth and very low contents of macronutrients. In addition, the soils of this class are vulnerable to soil erosion where, the erodibility factor reached 0.11 ton/ha/year.

Marginal class (C5) occupies a small area (0.9%) and is characterized by the deep soil and high pH values, while the high content of salts (69 dS/m) represents the major limiting factor for soil capability.

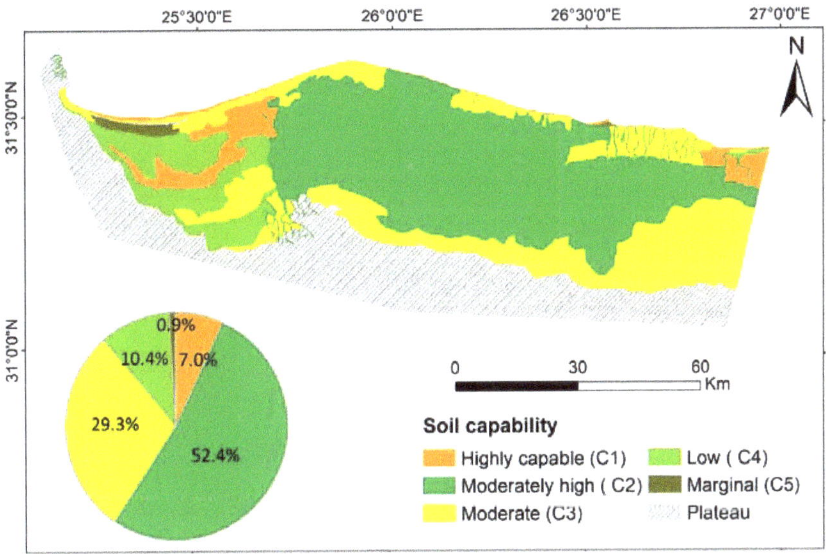

Figure 7. Land capability units and the pie chart represent the percentage area.

3.4. Soil Suitability

Soil profiles were evaluated based on their suitability for crop production taking into consideration the crop requirements, whilst each crop has a specific requirement for soil properties to achieve the maximum yield production. Therefore, the application of the Almagra model is based on the use of soil characteristics such as profile depth (p), texture (t), carbonate (c), salinity (s), drainage (d), and sodium saturation (a).

The assessment results showed that the soil suitability for the studied crops was varied from S2 to S5 with different limiting factors in each class based on the geomorphological unit. The suitability assessment was examined for ten horticultural and field crops (i.e., wheat, maize, olive, alfalfa, potato, sugar beet, peach, citrus, soybean and fig), Figure 8.

Table 6. The main statistical characteristics of the soil properties in five clusters.

Classes	Depth, cm	pH	ESP, %	OM, %	EC, dSm^{-1}	CaCO$_3$, %	CEC, cmolc kg^{-1}	Rock Fragment, %	AN, mg kg^{-1}	AP, mg kg^{-1}	AK, mg kg^{-1}	Erodibility, ton/ha/Year	Limiting Factors
1	118.3 ± 2.	8.2 ± 0.42	5.7 ± 2.25	0.6 ± 0.06	6.3 ± 3.59	21.45 ± 13.84	23.4 ± 2.9	5.9 ± 1.65	11.98 ± 0.85	9.75 ± 0.96	27.44 ± 2.23	0.15 ± 0.1	C, Er
2	129 ± 19.49	8.3 ± 0.19	2.1 ± 1.57	0.2 ± 0.05	2.6 ± 0.96	54.3 ± 11.22	7.6 ± 3.95	6.96 ± 0.42	5.4 ± 1.45	3.4 ± 0.77	9.2 ± 2.33	0.08 ± 0.06	C, Fer
3	56. 7 ± 25.98	8.13 ± 0.24	1.9 ± 1.01	0.44 ± 0.17	5.7 ± 4.64	28.9 ± 20.62	8.2 ± 2.94	4.8 ± 0.83	10.2 ± 2.07	7.1 ± 1.12	17.96 ± 3.36	0.08 ± 0.06	D, Ca
4	90 ± 30	7.8 ± 0.25	1.4 ± 0.46	0.25 ± 0.05	5.5 ± 6.25	12.2 ± 6.32	8. ± 2.28	5.48 ± 1.21	6.9 ± 1.42	4.2 ± 0.7	11.86 ± 2.22	0.11 ± 0.09	D, Er, Fer
5	132.5 ± 24.75	8.2 ± 0.03	3.3 ± 1.04	0.45 ± 0.05	69.8 ± 37.7	36 ± 17.84	12.3 ± 2.20	6.25 ± 2.9	11.4 ± 1.46	7.3 ± 0.78	20.35 ± 1.58	0.02 ± 0.2	S, C

ESP, exchangeable sodium percentage; OM, organic matter; EC, electric conductivity; CEC, cation exchange capacity; AN, available nitrogen; AP, available phosphorous; AK, available potassium. Limiting factors: C, calcium carbonate; Er, erodibility; Fer, fertility; D, depth; S, salinity.

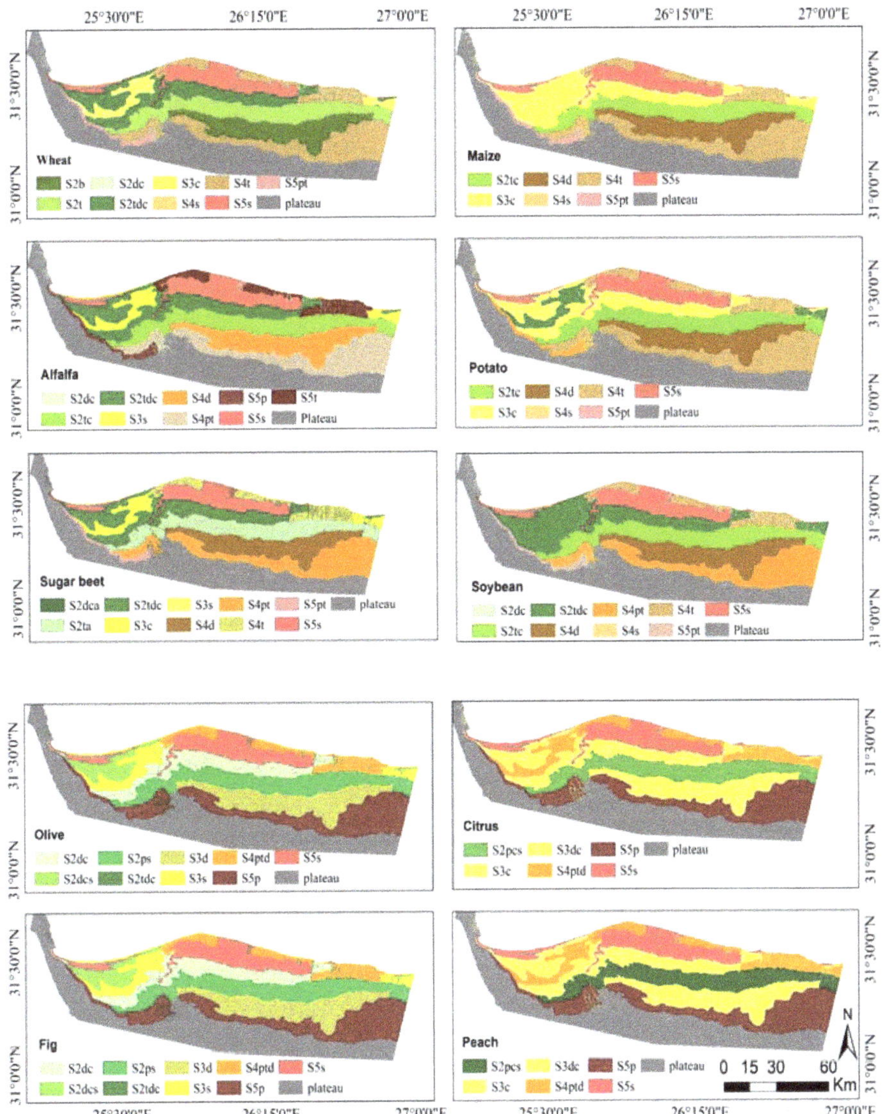

Figure 8. Crop suitability map for some field and horticultural crops. The main limiting factors; s, salinity; t, texture; a, sodium saturation; d, drainage; c, carbonate content; p, profile depth.

The geomorphic units such as the plateau, escarpment and ridge were not considered in the suitability evaluation, which represents 28% of the total study area. The results indicate that 37% of the total area was highly suitable for olive, fig, wheat, alfalfa, sugar beet, soybean crops. Furthermore, 18% of the area was highly suitable (S2) for citrus and peach. An area of about 37% of the arable cultivated area was moderately suitable (S3) for the following crops: wheat, olive, alfalfa, sugar beet and fig, but the limiting factors differed in each crop, while 50% of the total area was classified as S4 for most crops. On the other hand, the results showed that about 56% of the study area has marginal suitability (S4) for both peach and citrus crops.

Generally, the area has good potential for cultivating field crops in order to achieve sustainable agricultural development, but water for irrigation is highly needed.

4. Discussion

4.1. Geomorphology and Soils

Using Sentinel 2 images leads to increased clarification details of landscape where the high and low lands, basin and wadis boundaries were identified [30,62,63]. The soils that are located in southern parts of the study area were classified as *Lithic Torripsamments*, whereas the soil profiles' depth is very shallow and this is considered as a main obstacle to soil capability and crop suitability [64,65]. On the contrary, the soil types *Typic Torripsaments*, *Typic Torrifluvents* and *Typic Haplocalcids* were allocated in the wadi, decantation basins and sand plain geomorphic units with deep soil profile and low soil salinity [66]. Generally, the organic matter content in the study area is very low due to the lack of agricultural activity and monsoon rains, which is consistant with [67].

4.2. Multivariate Statistical for Land Evaluation

There is no doubt that due to the similarity of soil properties, grouping them into similar clusters is not easy, as it depends on understanding whether each soil characteristic is increasing or decreasing. For example, the increase in salt concentration leads to negative impacts on crop productivity, contrary to the increases in soil nutrient concentration and organic matter which aid to improve soil fertility. Hence, it is necessary to classify soil properties and link the classes with soil capability and crop suitability. Consequently, the multivariate statistical analysis is suitable to classify multiple variables of soils [68]. The results reflected that there are positive correlations between soil characteristics such as pH and $CaCO_3$ ($r = 0.72$) or ESP and CEC ($r = 0.88$). These results are logical, as the soil pH is affected by increasing the percentage of calcium carbonate of soils [69]. The obtained factor loading illustrates the acceptable grouping of soil properties and confirm the ability of PCA for group soil properties in different clusters [70]. The first PCs showed an accumulation of 36.9% of the soil characteristics where the ESP, OM, CEC, N, P and K were allocated in these PCs due to the association of the natural conditions and soil formation processes in the study area [71,72]. The second PCs showed a grouping of 20% of the data of the soil depth, pH, $CaCO_3$, and rock fragments; this cluster deals with the natural environment of the areas that have a shallow soil depth have also a high percentage of rock fragment. In addition, this cluster has a high concentration of pH and $CaCO_3$%.

The highly capable class (C1) represents the soils that have good physical and chemical characteristics, in addition to a low erosion vulnerability, Furthermore, the soils of this class were located in the areas with active agricultural management [73]. The soils of the moderately high class (C2) occupy most of the study area and are characterized by low soil fertility due to the lack of agricultural usage. As the study area was located in arid climate conditions with neglected rainfall in the whole year except in the winter season, cultivation is limited by the winter season and the availability of irrigation water. During this cultivation season, farmers may use fertilizers in order to enhance crop growth, also activating microorganisms. Therefore, the lack of farming usage during the whole year may cause low soil fertility [74]. The main limiting factors of the moderate class (C3) and low class (C4) were the soil depth, high percentage of calcium carbonates, and hardpan layers [75,76], yet C4 has more limitations compared to C3. The soils of the marginal class (C5) occupied the low land (sabkhas) whereas the high salt content was due to the sea water percolation and high evaporation [77]. Therefore, the soils of C5 cannot be used for agriculture consistently, as the agriculture management process is difficult.

4.3. Crop Suitability

The assessment of soil suitability for crops was performed using the soil characteristics and crop requirements [78]. About 38% of the study area was classified as S2 for the following crops:

wheat, alfalfa, sugar beet and soybean, which were located in the geomorphic units of wadis, decantation basin, high sand plain, moderate sand plain, and low sand plain. The soil characteristics meet the requirements of those crops, and this is consistent with [79–83]. On the other hand, the rest of the study area are varied in its suitability for wheat crop, between S3 and S5 [63,82]. The highly suitable class (S2) for the maize and potato crops was assigned only to the moderate sand plain unit, while the rest of the study area ranged between S3 and S5 based on the variation of soil limitations and climatic conditions [63,83]. The degree of the soil suitability of the olive and figs crops were classified as S2 in soils of wadis, decantation basin, high sand plain, moderate sand plain, low sand plain, and the overflow basin.

The limiting factors were the soil depth, soil texture, and climate [81]. About 13% of the soils of the study area were classified as S2 for peach and citrus crops, 25% as S3 and the rest of the area was classified as S4 and S5, due to the many limitations that reduce the suitability for both peaches and citrus crops such as soil profile depth and soil texture [84]. Generally, the results indicated that the optimum soil suitability class (S1) was not observed for any crop, as always there is one or more soil characteristics considered as a limiting factor, which is in agreement with [66]. Moreover, the northwest coast is one of the desert areas in Egypt suffering from land degradation which leads to decreased agricultural production. The assessment of soil capability, crop suitability, remote sensing analysis and GIS helps the decision making for sustainable agriculture development [85–90].

5. Conclusions

Multivariate analysis classification techniques can deal with various soil variables. The principal component analysis (PCA) and agglomerative hierarchical cluster analysis aid to classify soil capability based on the correlations and interactions between soil proprieties. The PCA classified the soil capability into five classes, which differed according to the number of soil limitations. The main limiting factors for soil capability in the southern parts of the study area was the shallow soil depth where hardpan layers were observed in the subsurface, while the high salinity and alkalinity represent the major limitations in the low elevated land that located in north of the study area. Whilst the area is vulnerable to the wind and water erosion processes, the crop suitability varied between high and not suitable classes for wheat, maize, fig, potato, and citrus. However, the soils that were located in the wadis, sandy plains and basins geomorphic units were of a highly suitable class for most of the field and horticultural crops.

The use of multivariate analysis, soil capability and crop suitability based on soil physiochemical properties assist in understanding the soil function and to assess the soils under different conditions. Likewise, remote sensing data contribute to map the geomorphic unites that were used as a base map for soil assessment. GIS techniques were considered as a main tool to spatialize the variations of soil capability and crop suitability in order to achieve the optimal land use planning in such new reclaimed areas.

Author Contributions: All the authors substantially contributed to this article M.K.A.-F. and A.M.A. conceptualized the study and developed the methodology. The satellite imagery was analyzed by M.E.S.S., M.K.A.-F. and M.K.A.-F. accomplished the data analysis and wrote a draft of the manuscript. A.A.A., S.K.A.-E., M.M.N.K. and M.B. contributed to reviewing and editing the manuscript. All authors have read and agreed to the published version of the manuscript.

Funding: The Research was supported by University of Padova, DOR project.

Acknowledgments: The authors would like to thank the National Authority for Remote Sensing and Space Sciences for supporting the satellite data and image processing. In addition, we thank, Faculty of Agriculture, Zagazig University for the laboratory analysis. The authors would like to extend the thank to National Research Center, Egypt for its support. The authors would like to extend their sincere appreciation to the Deanship of Scientific Research at King Saud University, through the Research Group Project no. RGP-VPP-275 for support. We would like to extend our gratitude to the Research Group Project no 5-100, RUDN University, Russia, for their support. Finally, we would like to thank the Department of Agronomy, Food, Natural Resources, Animals and Environment, DOR project, University of Padova, for the guidance and the progress of the work.

Conflicts of Interest: The authors would like to hereby certify that there was no conflict of interest in the data collection, processing the data, the writing of the manuscript, and the decision to publish the results.

References

1. Tahmasebinia, F.; Tsumura, Y.; Wang, B.; Wen, Y.; Bao, C.; Sepasgozar, S.; Alonso-Marroquin, F. Floating Cities Bridge in 2050. In *Smart Cities and Construction Technologies*; IntechOpen: London, UK, 2020.
2. Debiagi, F.; Madeira, T.B.; Nixdorf, S.L.; Mali, S. Pretreatment efficiency using autoclave high-pressure steam and ultrasonication in sugar production from liquid hydrolysates and access to the residual solid fractions of wheat bran and oat hulls. *Appl. Biochem. Biotechnol.* **2020**, *190*, 166–181. [CrossRef] [PubMed]
3. Xiang, T.; Malik, T.H.; Nielsen, K. The impact of population pressure on global fertilizer use intensity, 1970–2011: An analysis of policy-induced mediation. *Technol. Forecast. Soc.* **2020**, *152*, 119895. [CrossRef]
4. Gerten, D.; Heck, V.; Jägermeyr, J.; Bodirsky, B.L.; Fetzer, I.; Jalava, M.; Kummu, M.; Lucht, W.; Rockström, J.; Schaphoff, S.; et al. Feeding ten billion people is possible within four terrestrial planetary boundaries. *Nat. Sustain.* **2020**, *3*, 200–208. [CrossRef]
5. Tir, J.; Diehl, P.F. Demographic Pressure and Interstate Conflict. In *Environmental Conflict*; Routledge: London, UK, 2018; pp. 58–83.
6. El Nahry, A.H.; Mohamed, E.S. Potentiality of land and water resources in African Sahara: A case study of south Egypt. *Environ. Earth Sci.* **2011**, *63*, 1263–1275. [CrossRef]
7. Belal, A.A.; Mohamed, E.S.; Abu-hashim, M.S.D. Land evaluation based on GIS-spatial multi-criteria evaluation (SMCE) for agricultural development in dry Wadi, Eastern Desert, Egypt. *Int. J. Soil. Sci.* **2015**, *10*, 100–116. [CrossRef]
8. Abd-Elmabod, S.K.; Mansour, H.; Hussein, A.A.E.F.; Mohamed, E.S.; Zhang, Z.; Anaya-Romero, M.; Jordán, A. Influence of irrigation water quantity on the land capability classification. *Plant. Arch.* **2019**, *19*, 2253–2261.
9. Abd-Elmabod, S.K.; Jordán, A.; Fleskens, L.; Phillips, J.D.; Muñoz-Rojas, M.; van der Ploeg, M.; de la Rosa, D. Modeling agricultural suitability along soil transects under current conditions and improved scenario of soil factors. In *Soil Mapping and Process Modeling for Sustainable Land Use Managemen*; Elsevier: Amsterdam, The Netherlands, 2017; pp. 193–219.
10. Muñoz-Rojas, M.; Abd-Elmabod, S.K.; Zavala, L.M.; De la Rosa, D.; Jordán, A. Climate change impacts on soil organic carbon stocks of Mediterranean agricultural areas: A case study in Northern Egypt. *Agric. Ecosyst. Environ.* **2017**, *238*, 142–152. [CrossRef]
11. Anaya-Romero, M.; Abd-Elmabod, S.K.; Muñoz-Rojas, M.; Castellano, G.; Ceacero, C.J.; Alvarez, S.; De la Rosa, D. Evaluating soil threats under climate change scenarios in the Andalusia Region, Southern Spain. *Land Degrad. Dev.* **2015**, *26*, 441–449. [CrossRef]
12. Mohamed, E.S.; Saleh, A.M.; Belal, A.A. Sustainability indicators for agricultural land use based on GIS spatial modeling in North of Sinai-Egypt. *Egypt. J. Remote Sens. Space Sci.* **2014**, *17*, 1–15. [CrossRef]
13. Mohamed, E.S.; Ali, A.; El-Shirbeny, M.; Abutaleb, K.; Shaddad, S.M. Mapping soil moisture and their correlation with crop pattern using remotely sensed data in arid region. *Egypt. J. Remote Sens. Space Sci.* **2019**. [CrossRef]
14. Abd-Elmabod, S.K.; Alice, C.F.; Zhenhua, Z.; Ali, R.R.; Laurence, J. Rapid urbanisation threatens fertile agricultural land and soil carbon in the Nile delta. *J. Environ. Manag.* **2019**, *252*, 109668. [CrossRef] [PubMed]
15. Saleh, A.M.; Belal, A.B.; Mohamed, E.S. Land resources assessment of El-Galaba basin, South Egypt for the potentiality of agriculture expansion using remote sensing and GIS techniques. *Egypt. J. Remote Sens. Space Sci.* **2015**, *18*, S19–S30. [CrossRef]
16. Bakr, N.; Bahnassy, M.H. Egyptian Natural Resources. In *The Soils of Egypt*; Springer: Cham, Switzerland, 2019; pp. 33–49.
17. Satoh, M.; Aboulroos, S. *Irrigated Agriculture in Egypt: Past, Present and Future*; Satoh, M., Aboulroos, S., Eds.; Springer: Cham, Switzerland, 2017.
18. FAO. A Framework for Land Evaluation System. In *Soil Bull No. 32*; Food and Agriculture Organization: Rome, Italy, 1976; p. 72.
19. FAO. *Guidelines for Soil Profile Description*, 3rd ed.; Soil Resources Management and Conservation Service, Land and Water Development Division, FAO: Rome, Italy, 2006.

20. Mandal, S.; Choudhury, B.U.; Satpati, L. Soil site suitability analysis using geo-statistical and visualization techniques for selected winter crops in Sagar Island, India. *Appl. Geogr.* **2020**, *122*, 102249. [CrossRef]

21. Basak, N.; Barman, A.; Sundha, P.; Rai, A.K. Recent Trends in Soil Salinity Appraisal and Management. In *Soil Analysis: Recent Trends and Applications*; Springer: Singapore, 2020; pp. 143–161.

22. FAO (Food and Agriculture Organization). Salt-Affected Soils and Their Management. In *Soils Bulletin*; FAO: Rome, Italy, 1988; p. 39.

23. Bakr, N.; Elbana, T.A.; Arceneaux, A.E.; Zhu, Y.; Weindorf, D.C.; Selim, H.M. Runoff and water quality from highway hillsides: Influence compost/mulch. *Soil. Till. Res.* **2015**, *150*, 158–170. [CrossRef]

24. AbdelRahman, M.A.; Shalaby, A.; Mohamed, E.S. Comparison of two soil quality indices using two methods based on geographic information system. *Egypt. J. Remote Sens. Space Sci.* **2019**, *22*, 127–136. [CrossRef]

25. Mohamed, E.S.; Belal, A.; Saleh, A. Assessment of land degradation east of the Nile Delta, Egypt using remote sensing and GIS techniques. *Arab. J. Geosci.* **2013**, *6*, 2843–2853. [CrossRef]

26. Hammam, A.A.; Mohamed, E.S. Mapping soil salinity in the East Nile Delta using several methodological approaches of salinity assessment. *Egypt. J. Remote Sens. Space Sci.* **2018**. [CrossRef]

27. Hassan, A.M.; Belal, A.A.; Hassan, M.A.; Farag, F.M.; Mohamed, E.S. Potential of thermal remote sensing techniques in monitoring waterlogged area based on surface soil moisture retrieval. *J. Afr. Earth Sci.* **2019**, *155*, 64–74. [CrossRef]

28. Elbeih, S.F.; El-Zeiny, A.M. Qualitative assessment of groundwater quality based on land use spectral retrieved indices: Case study Sohag Governorate, Egypt. *Remot. Sens. Appl. Soci. Environ.* **2018**, *10*, 82–92. [CrossRef]

29. Bakr, N.; Ali, R.R. Statistical relationship between land surface altitude and soil salinity in the enclosed desert depressions of arid regions. *Arab. J. Geosci.* **2019**, *12*, 715. [CrossRef]

30. Mohamed, E.S.; Schütt, B.; Belal, A. Assessment of environmental hazards in the north western coast-Egypt using RS and GIS. *Egypt. J. Remote Sens. Space Sci.* **2013**, *16*, 219–229. [CrossRef]

31. De la Rosa, D.; Moreno, J.A.; García, L.V.; Almorza, J. MicroLEIS: A microcomputer-based Mediterranean land evaluation information system. *Soil Use Manag.* **1992**, *8*, 89–96. [CrossRef]

32. Storie, R.E. *The Storie Index Soil Rating Revised*; Special Publication Division of Agricultural Science, University of California: Berkeley, CA, USA, 1978.

33. Ghabour, T.K.; Ali, R.R.; Wahba, M.M.; El-Naka, E.A.; Selim, S.A. Spatial decision support system for land use management of newly reclaimed areas in arid regions. *Egypt. J. Remote Sens. Space Sci.* **2019**, *22*, 219–225. [CrossRef]

34. El-Zeiny, A.M.; Effat, H.A. Environmental analysis of soil characteristics in El-Fayoum Governorate using geomatics approach. *Environ. Monit. Assess.* **2019**, *191*, 463. [CrossRef]

35. Mohamed, E.S.; Saleh, A.M.; Belal, A.B.; Gad, A. Application of near-infrared reflectance for quantitative assessment of soil properties. Egypt. *J. Remote Sens. Space Sci.* **2018**, *21*, 1–14.

36. Amini, S.; Rohani, A.; Aghkhani, M.H.; Abbaspour-Fard, M.H.; Asgharipour, M.R. Assessment of land suitability and agricultural production sustainability using a combined approach (Fuzzy-AHP-GIS): A case study of Mazandaran province, Iran. *Info. Process. Agric.* **2019**. [CrossRef]

37. Gusti, S.K.; Abdillah, R. *Oil Palm Plantation Land Suitability Classification using PCA-FCM*; Seminar Nasional Teknologi Informasi, Komunikasi dan Industri: Pekanbaru, Indonesia, 2018.

38. Tan, M.-Z.; Xu, F.-M.; Chen, J.; Zhang, X.L.; Chen, J.-Z. Spatial prediction of heavy metal pollution for soils in Peri-Urban Beijing, China based on Fuzzy set theory. *Pedosphere* **2006**, *16*, 545–554. [CrossRef]

39. Astel, A.M.; Chepanova, L.; Simeonov, V. Soil contamination interpretation by the use of monitoring data analysis. *Water Air Soil. Pollut.* **2011**, *216*, 375–390. [CrossRef]

40. Masto, R.E.; Chhonkar, P.K.; Purakayastha, T.J.; Patra, A.K.; Singh, D. Soil quality indices for evaluation of long-term land use and soil management practices in semi-arid sub-tropical India. *Land Degrad. Dev.* **2008**, *19*, 516–529. [CrossRef]

41. Irpino, A.; Verde, R.A. A New Wasserstein Based Distance for the Hierarchical Clustering of Histogram Symbolic Data. In *Data Science and Classification*; Springer: Berlin/Heidelberg, Germany, 2006; pp. 185–192.

42. Abdel-Fattah, M.; Kantoush, S.A.; Saber, M.; Sumi, T. Rainfall-runoff modeling for extreme flash floods in wadi samail, oman. *J. Jpn. Soc. Civil. Eng.* **2018**, *74*, I691–I696. [CrossRef]

43. Dobos, E.; Norman, B.; Worstell, B. The Use of DEM and Satellite Images for Regional Scale Soil Database. *Agrokem Talajt.* **2002**, *51*, 263–272. [CrossRef]

44. Sys, C.; Van Ranst, E.; Debaveye, J. Land Evaluation, Part II. In *Methods in Land Evaluation*; Agricultural Publication: Brussels, Belgium, 1991; p. 247.

45. Zink, J.A. *Physiography and Soils*; ITC Lecture Note, K6 (SOL41); ITC: Enshede, The Netherlands, 1997.

46. Rhoades, J.D. Salinity: Electrical Conductivity and Total Dissolved Solids. In *Methods of Soil Analysis Part 3, Chemical Methods*; Soil Science Society of America Book Series, No. 5; Sparks, D.L., Ed.; Soil Science Society of America, American Society of Agronomy: Madison, WI, USA, 1996; pp. 417–435.

47. Thomas, G.W. Soil pH and Soil Acidity. In *Methods of Soil Analysis Part 3, Chemical Methods*; Soil Science Society of America Book Series, No. 5; Sparks, D.L., Ed.; Soil Science Society of America, American Society of Agronomy: Madison, WI, USA, 1996; pp. 475–490.

48. Summer, M.E.; Miller, W.P. Cation Exchange Capacity and Exchange Coefficients. In *Methods of Soil Analysis Part 3. Chemical Methods*; Soil Science Society of America Book Series, No. 5; Sparks, D.L., Ed.; Soil Science Society of America, American Society of Agronomy: Madison, WI, USA, 1996; pp. 1201–1229.

49. Walkley, A.; Black, I.A. An examination of the Degtjareff method for determining organic carbon in soils: Effect of variations in digestion conditions and of inorganic soil constituents. *Soil Sci.* **1934**, *63*, 251–263. [CrossRef]

50. Jackson, M.L. *Soil Chemical Analysis*; Prentice Hall Inc.: Englewood Cliffs, NJ, USA, 1967.

51. Van Reeuwijk, L.P. Procedures for soil analysis: Wageningen. *Int. Soil Ref. Inf. Cent. Tech. Pap.* **2002**.

52. Lavkulich, L.M. *Methods Manual: Pedology Laboratory*; Department of Soil Science, University of British Colum-bia: Vacouver, BC, Canada, 1981.

53. Page, A.L.; Miller, R.H.; Keeney, D.R. *Methods of Soil Analysis (Part. 2): Chemical and Microbiological Properties*, 2nd ed.; The American Society of Agronomy: Madison, WI, USA, 1982.

54. Soil Survey Staff. Soil Survey Field and Laboratory Methods Manual. In *Soil Survey Investigations Report Version 2.0*; No. 51; Burt and Soil Survey Staff, Ed.; U.S. Department of Agriculture, Natural Resources Conservation Service: Washington, DC, USA, 2014.

55. Borůvka, L.; Vacek, O.; Jehlička, J. Principal component analysis as a tool to indicate the origin of potentially toxic elements in soils. *Geoderma* **2005**, *128*, 289–300. [CrossRef]

56. Jolliffe, I.T.; Cadima, J. Principal component analysis: A review and recent developments. *Philosoph. Transact. R. Soc. A Mathemat. Phys. Eng. Sci.* **2016**, *374*, 20150202. [CrossRef] [PubMed]

57. Jagadamma, S.; Lal, R.; Hoeft, R.G.; Nafziger, E.D.; Adee, E.A. Nitrogen fertilization and cropping system impacts on soil properties and their relationship to crop yield in the central Corn Belt, USA. *Soil. Till. Res.* **2008**, *98*, 120–129. [CrossRef]

58. De la Rosa, D.; Mayol, F.; Diaz-Pereira, E.; Fernandez, M. Aland evaluation decision support system (MicroLEIS DSS) for agricultural soil protection. Environ. *Model. Softw.* **2004**, *19*, 929–942. [CrossRef]

59. De la Rosa, D. *Soil Survey and Evaluation of Guadalquivir River Terraces, in Sevilla Province*; Cent. Edaf. Cuarto Pub.: Seville, Spain, 1974.

60. Aldabaa, A.A.; HaiLin, Z.; Shata, A.; El-Sawey, S.; Abdel-Hameed, A.; Schroder, J.L. Land suitability classification of a desert area in Egypt for some crops using MicrolEIS program. *Am. Eurasian J. Agric. Environ. Sci.* **2010**, *8*, 80–94.

61. De la Rosa, D.; Cardona, F.; Paneque, G. Evaluación de suelos para diferentes usos agrícolas Un Sistema desarrollado para regiones mediterráneas. *An. Edafol. Agrobiol.* **1977**, *36*, 1100–1112.

62. Abdel-Kader, F.H. Digital soil mapping at pilot sites in the northwest coast of Egypt: A multinomial logistic regression approach. *Egypt. J. Remote Sens. Space Sci.* **2011**, *14*, 29–40. [CrossRef]

63. Shoman, M.M.; Yacoub, R.K.; El Ghonamey, Y.K. Land evaluation of the North western coast of Egypt using microleis and sys models. *Minufiya. J. Agric. Res.* **2013**, *38*, 1779–1800.

64. Abd El-Aziz, S.H. Evaluation of land suitability for main irrigated crops in the North-Western Region of Libya. *Eurasian J. Soil. Sci.* **2018**, *7*, 73–86. [CrossRef]

65. Ali, R.R.; Ageeb, G.W.; Wahab, M.A. Assessment of soil capability for agricultural use in some areas west of the Nile Delta, Egypt: An application study using spatial analyses. *J. Appl. Sci. Res.* **2007**, *3*, 1622–1629.

66. Abdel-Kader, F.H. Digital soil mapping using spectral and terrain parameters and statistical modelling integrated into GIS-northwestern coastal region of Egypt. In *Developments in Soil Classification, Land Use Planning and Policy Implications*; Springer: Dordrecht, The Netherland, 2013; pp. 353–371.

67. Elbasiouny, H.; Abowaly, M.; Abu Alkheir, A.; Gad, A. Spatial variation of soil carbon and nitrogen pools by using ordinary Kriging method in an area of north Nile Delta, Egypt. *Catena* **2014**, *113*, 70–78. [CrossRef]

68. Habeck, C.; Stern, Y. Alzheimer's Disease Neuroimaging Initiative. Multivariate data analysis for neuroimaging data: Overview and application to Alzheimer's disease. *Cell Biochem. Biophys.* **2010**, *58*, 53–67. [CrossRef]

69. Saksono, N.; Yuliusman, Y.; Bismo, S.; Soemantojo, R.; Manaf, A. Effects of pH on calcium carbonate precipitation under magnetic field. *Makara J. Technol.* **2010**, *13*, 79–85. [CrossRef]

70. Nehrani, S.H.; Askari, M.S.; Saadat, S.; Delavar, M.A.; Taheri, M.; Holden, N.M. Quantification of soil quality under semi-arid agriculture in the northwest of Iran. *Ecol. Indic.* **2020**, *108*, 105770. [CrossRef]

71. El Ghani, M.A.; Hamdy, R.; Hamed, A. Aspects of vegetation and soil relationships around athalassohaline lakes of Wadi El-Natrun, Western Desert. *Egypt. J. Biol.* **2014**, *4*, B21–B35.

72. Mohamed, E.S.; Abu-hashim, M.; AbdelRahman, M.A.; Schütt, B.; Lasaponara, R. Evaluating the effects of human activity over the last decades on the soil organic carbon pool using satellite imagery and GIS techniques in the Nile Delta Area, Egypt. *Sustainability* **2019**, *11*, 2644. [CrossRef]

73. Elzahaby, E.M.; Suliman, A.; Bakr, N.; Al-janabi, A. Land Cover Change Detection and Land Evaluation of Burg El Arab Region, North West Coast, Egypt. *Alex. J. Agric. Sci.* **2015**, *3*, 193–204.

74. Abdel-Kader, F.H.; Rafaat, K.Y. Land resources assessment of landmine-affected areas, Northwest of Egypt. In *Suitma Symposium*; The National Information and Documentation Centre (NIDOC): Cairo, Egypt, 2005; Volume 19.

75. Belal, A.A.; Mohamed, E.S.; Saleh, A.; Jalhoum, M. Soil Geography. In *The Soils of Egypt 2019*; Springer: Cham, Switzerland, 2019; pp. 111–136.

76. Mohamed, E.S.; Belal, A.A.; Ali, R.R.; Saleh, A.; Hendawy, E.A. Land degradation. In *The Soils of Egypt 2019*; Springer: Cham, Switzerland, 2019; pp. 159–174.

77. Abduljauwad, S.N.; Al-Amoudi, O.S.B. Geotechnical behavior of saline sabkha soils. *Geotechnique* **1995**, *45*, 425–445. [CrossRef]

78. Sys, C. *Land Evaluation Part I, II, III*; Publication Agricoles State University of Ghent: Ghent, Belgium, 1985.

79. Yousif, I. Land Capability and Suitability Mapping in Some Areas of North-Western Coast. *Egypt. J. Soil. Sci. Agric. Eng.* **2018**, *9*, 111–118. [CrossRef]

80. Mohamed, E.S.; Belal, A.; Shalaby, A. Impacts of soil sealing on potential agriculture in Egypt using remote sensing and GIS techniques. *Eurasian Soil. Sci.* **2015**, *48*, 1159–1169. [CrossRef]

81. Abu-Hashim, M.; Mohamed, E.; Belal, A.E. Identification of potential soil water retention using hydric numerical model at arid regions by land-use changes. *Int. Soil. Water Conse.* **2015**, *3*, 305–315. [CrossRef]

82. Elbasyoni, I.S. Performance and stability of commercial wheat cultivars under terminal heat stress. *Agronomy* **2018**, *8*, 37. [CrossRef]

83. Yialouris, C.P.; Kollias, V.; Lorentzos, N.A.; Kalivas, D.; Sideridis, A.B. An integrated expert geographical information system for soil suitability and soil evaluation. *J. Geo. Inf. Decis. Anal.* **1997**, *1*, 89–99.

84. FAO. Guidelines for Land Use Planning. In *FAO development Series-1*; FAO: Rome, Italy, 1993.

85. Amato, F.; Havel, J.; Gad, A.; El-Zeiny, A. Remotely Sensed Soil Data Analysis Using Artificial Neural Networks: A Case Study of El-Fayoum Depression, Egypt. *ISPRS Int. J. Geoinf.* **2015**, *4*, 677–696. [CrossRef]

86. Abd-Elmabod, S.K.; Muñoz-Rojas, M.; Jordán, A.; Anaya-Romero, M.; Phillips, J.D.; Laurence, J.; Zhang, Z.; Pereira, P.; Fleskens, L.; van der Ploeg, M.; et al. Climate change impacts on agricultural suitability and yield reduction in a Mediterranean region. *Geoderma* **2020**, *374*, 114453. [CrossRef]

87. El-Zeiny, A.; El-Kafrawy, S. Assessment of water pollution induced by human activities in Burullus Lake using Landsat 8 operational land imager and GIS. *Egypt. J. Remote Sens. Space Sci.* **2017**, *20*, S49–S56. [CrossRef]

88. Shaddad, S.M.; Buttafuoco, G.; Castrignanò, A. Assessment and Mapping of Soil Salinization Risk in an Egyptian Field Using a Probabilistic Approach. *Agronomy* **2020**, *10*, 85. [CrossRef]

89. Abd-Elmabod, S.K.; Ali, R.R.; Anaya-Romero, M.; de la Rosa, D. Evaluating soil contamination risks by using MicroLEIS DSS in El-Fayoum Province, Egypt. In Proceedings of the 2010 2nd International Conference on Chemical, Biological and Environmental Engineering, Cairo, Egypt, 2–4 November 2010.

90. Mohamed, E.S. Spatial assessment of desertification in north Sinai using modified MEDLAUS model. *Arab. J. Geosci.* **2013**, *12*, 4647–4659. [CrossRef]

MDPI

St. Alban-Anlage 66

4052 Basel

Switzerland

Tel. +41 61 683 77 34

Fax +41 61 302 89 18

www.mdpi.com

Agronomy Editorial Office

E-mail: agronomy@mdpi.com

www.mdpi.com/journal/agronomy